OXFORD MONOGRAPHS ON GEOLOGY AND GEOPHYSICS NO. 6

Series editors

P. Allen
H. Charnock
E. R. Oxburgh
B. J. Skinner

OXFORD MONOGRAPHS ON GEOLOGY AND GEOPHYSICS

Precambrian Geology of India

S. MAHMOOD NAQVI

National Geophysical Research Institute
Hyderabad

and

JOHN J. W. ROGERS

University of North Carolina
Chapel Hill

New York • Oxford
CLARENDON PRESS • OXFORD UNIVERSITY PRESS
1987

Oxford University Press

Oxford   New York   Toronto
Delhi   Bombay   Calcutta   Madras   Karachi
Petaling Jaya   Singapore   Hong Kong   Tokyo
Nairobi   Dar es Salaam   Cape Town
Melbourne   Auckland

and associated companies in
Beirut   Berlin   Ibadan   Nicosia

Library of Congress Cataloging-in-Publication Data

Naqvi, S. Mahmood.
  Precambrian geology of India.

  (Oxford monographs on geology and geophysics ; 6)
  Bibliography: p.
  Includes index.
  1. Geology, Stratigraphic—Pre-Cambrian. 2. Geology—
India. I. Rogers, John J. W. (John James William),
1930–   . II. Title. III. Series: Oxford monographs
on geology and geophysics ; no. 6.
QE653.N36   1987   551.7′1′0954   86-5420
ISBN 0-19-503653-0

9 8 7 6 5 4 3 2 1

Printed in the United States of America
on acid-free paper

PREFACE

Considerable progress has been made in the field of Precambrian geology during the past two decades. The quality and quantity of new data have improved, and conceptual models have become available for understanding Precambrian history. No models, however, have been as successful as the unifying concepts of sea floor spreading and plate tectonics applied to Phanerozoic history.

Until about 1960, most Precambrian rocks were studied for their petrologic and economic properties. Various types of information, including trace elements and isotopic data, were unavailable, and important features such as tectonothermal evolution patterns could not be determined. Thus, the more than two-thirds of earth history represented by the Precambrian was virtually unknown in comparison with the Phanerozoic.

A realization of the importance of the Precambrian came almost simultaneously with the development of new instruments and analytical techniques. These developments caused a search for areas of particular significance in which to study Precambrian history. One of these places is India, where many areas are well exposed and easily accessible and where extensive background maps and information were already available. Considerable work has been done there during the past two decades, both by Indian geologists and geologists from other countries.

Most of the new data on Indian Precambrian geology are not available in book form. Some of the best long publications remain the relatively old works of Pascoe (1950), Pichamuthu (1967), and Krishnan (1968); the stratigraphy of the Indian Precambrian was summarized by V. J. Gupta (1977) (all references in Chapter 1). Thus, this book has been written to collect in one place a summary of data, concepts, arguments, and opportunities to facilitate further research into the Precambrian geology of India. We have been able to summarize the Indian geological literature until about the end of 1984, including some references from 1985 and a few in press from 1986. Clearly, we have not been able to include all references and information, but we have tried to discuss the major issues and to construct a framework for further work.

January 1986   S. M. N.
                                  J. J. W. R.

ACKNOWLEDGMENTS

First and foremost we thank Rehana Naqvi and Barbara Rogers for their patience and assistance in various ways while we wrote this book. Drs. B. P. Radhakrishna, Hari Narain, K. Naha, and B. L. Narayana made many valuable suggestions on various chapters. Mr. D. V. Subba Rao, Dr. S. H. Jafri, and Mr. B. Govinda Rajulu helped to prepare the manuscript. Mr. P. N. Sharma and the drawing section of the National Geophysical Research Institute provided drafting assistance. Considerable typing assistance was provided by Ms. D. Copelan and Ms. Mary Crump, at the University of North Carolina.

S. M. Naqvi is grateful to Drs. B. P. Radhakrishna and Hari Narain for inspiration and to Prof. V. K. Gaur (Director, National Geophysical Research Institute, Hyderabad) and Dr. S. Varadarajan (Director-General, Council of Scientific and Industrial Research) for permission to publish this book. The book has been greatly improved by discussions with Drs. C. S. Pichamuthu, K. Naha, U. M. Raval, D. Mukhopadhyay, R. H. Sawkar, R. Srinivasan, V. Divakara Rao, and S. M. Hussain.

John Rogers thanks numerous colleagues who have improved his understanding of Indian geology, particularly Drs. T. Chacko, K. C. Condie, T. W. Donnelly, P. D. Fullager, G. N. Hanson, J. R. Monrad, V. Rama Murthy, R. C. Newton, and P. T. Stroh. Dr. Rama Murthy reviewed the entire manuscript and made valuable suggestions. Assistance with the bibliography was provided by Mr. P. Rogers.

Our own work on the geology of India has been greatly aided by support from the Indian Council of Scientific and Industrial Research and the U.S. National Science Foundation (grants INT78-17128 and EAR79-05723). We particularly appreciate the efforts at the National Science Foundation of Drs. J. F. Lance, E. R. Padovani, and O. Shinaishan, and Mr. Subramaniam (in Delhi).

CONTENTS

Precambrian Geology of India

1. INTRODUCTION

Precambrian rocks constitute most of the peninsula of India and occur in Himalayan and related mountainous regions along the northern part of the country. This book discusses these rocks in chapters subdivided by geologic and geographic area. This chapter is devoted to three topics: a brief survey of the geology of India, including the occurrence of Precambrian rocks, plus a bibliography of previous work that refers to the entire Indian Precambrian; a brief description of each of the seven cratons and other geologic subdivisions that are discussed in the remainder of the book; and a summary of the geophysical features of the Indian crust. The names of the Indian states are shown in Figure 1.1, along with a generalized geologic map of India and a proposed location of India in Gondwanaland.

GENERAL GEOLOGY OF INDIA
AND PREVIOUS WORK

India is divided into four geologic regions (Fig. 1.1). Two of the rock assemblages occur in the Indian peninsula, and the other two represent the Indo-Gangetic plain and the northern mountains. About two-thirds of the surface exposure of the peninsula consists of Precambrian rocks. The Precambrian suites have been proposed to extend through the entire range of Precambrian time, from 3,800 m.y. ago to the late Precambrian/early Paleozoic.

The Precambrian rocks of the peninsula are covered by Phanerozoic sedimentary suites and by the Deccan plateau basalts (Fig. 1.1). The sedimentary assemblages include rocks formed when India was still part of Gondwanaland, during the Paleozoic and much of the Mesozoic. These rocks are commonly referred to as "Gondwanas". They are mostly thin sequences, except in rift valleys, and are absent from much of interior India. Postrifting (late Mesozoic and Tertiary) sedimentary assemblages are mostly restricted to the continental margins, where they form thick sequences in some localized basins. The Deccan plateau basalts cover a large part of western and central India. They were formed during the Late Cretaceous to Early Tertiary and have been widely regarded as a volcanic consequence of the rifting of India from other parts of Gondwanaland.

The Indo-Gangetic plain forms a broad band of flat-lying sediments between the Precambrian rocks of the Indian peninsula and the highly deformed suites of the Himalayas. Surface exposures consist of very young sediments, and the ages and varieties of older rocks at depth are uncertain. Outcropping Precambrian rocks in the peninsula dip northward under the relatively thick sediments of the Indo-Gangetic plain and presumably form the basement for sedimentation. Thus, the sediments of the plain can be regarded as having accumulated on a downwarped portion of the Indian Precambrian shield. Approach of the Indian crust to the northward-directed subduction zone under the Himalayas presumably caused enhanced downwarp and greater thickness of sediment accumulation. Increased sedimentation was also caused by outpouring of clastic debris from the Himalayas during Late Tertiary and Recent time.

The Himalayas contain rocks of Precambrian to Tertiary age intensely deformed and metamorphosed during the Tertiary collision of the Indian peninsula with the Asiatic mainland. A northward-dipping boundary thrust belt separates the deformed ranges from the southern shield. Continuity between Precambrian and other assemblages of the Himalayas and various suites in the peninsula is difficult to establish because of the deformation and also because of cover by sediments of the intervening Indo-Gangetic plain.

Previous work

Information on the general geology of India has been provided by numerous investigators. We particularly call attention to books by Pascoe (1950), Wadia (1961), and Krishnan (1968) and the chapter by Pichamuthu (1967). Chatterjee (1974) published a book on igneous and metamorphic rocks in India and S. Ray (1976) published one on metamorphism. Mineralization was reviewed by Radhakrishna (1984).

Much of the discussion of Precambrian geology has centered around the presence of rocks of great diversity in metamorphic grade. In general, the term "high grade" refers to granulite-facies rocks, including the well-known Indian charnockites. "Low grade" refers to rocks of amphibolite or lower (including greenschist) facies.

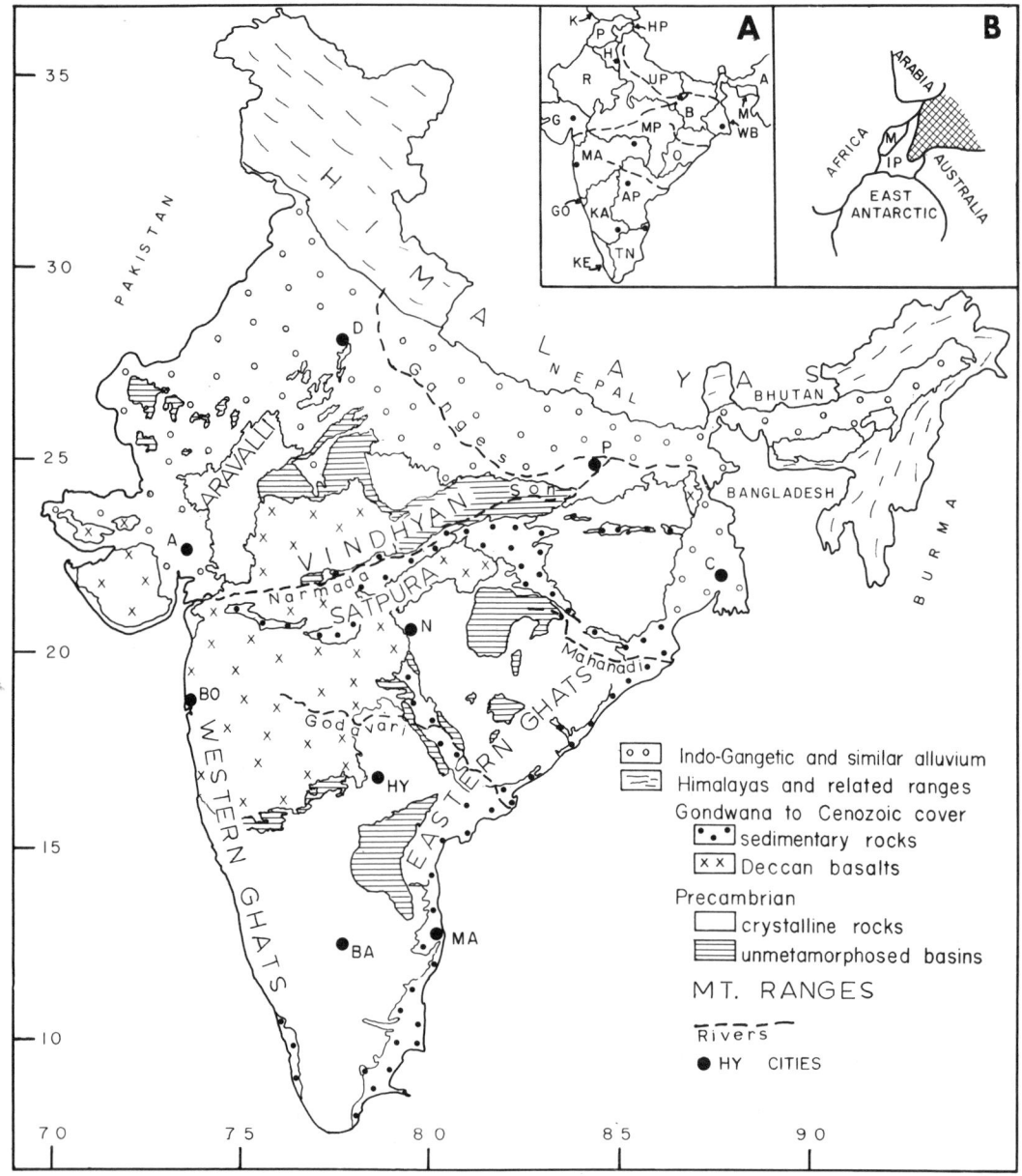

Fig. 1.1. Generalized geologic map of India. Locations on main map include D, Delhi; P, Patna; A, Ahmedabad; C, Calcutta; N, Nagpur; BO, Bombay; HY, Hyderabad; MA, Madras; BA, Bangalore.

Inset A shows the states of India: K, Kashmir; HP, Himachal Pradesh; P, Punjab; R, Rajasthan; UP, Uttar Pradesh; A, Assam; M, Meghalaya; B, Bihar; WB, West Bengal; MP, Madhya Pradesh; G, Gujarat; MA, Maharashtra; O, Orissa; AP, Andhra Pradesh; GO, Goa; KA, Karnataka; TN, Tamil Nadu; KE, Kerala.

Inset B shows position of India in Gondwanaland as proposed by Shields (1977). India is cross-hatched; M, Madagascar; IP, small plates now at various locations in Indian Ocean.

The Indian Precambrian terrain contains extensive suites of supracrustal rocks, both sedimentary and volcanic (summaries by V. J. Gupta, 1977; I. B. Singh, 1980). Iron formations are an important part of these assemblages in most areas (Pichamuthu, 1974), and manganese-rich rocks occur both as sedimentary (oxide) suites and

metamorphosed (silicate) assemblages (Roy, 1981).

CRATONS AND OTHER SUBDIVISIONS OF PRECAMBRIAN ROCKS
Figure 1.2 shows a subdivision of the Indian Precambrian shield into seven cratons. Each craton

represents one chapter in this book, and we have arbitrarily started with southern India and proceeded northward in the descriptions. In addition to the cratonic elements, Precambrian rocks also occur in the Himalayas, Meghalaya, and in rift valleys, all of which tectonic regions are described separately. The subdivision of cratons shown in Figure 1.2 is more detailed than that shown in Figure 1.3 (from Naqvi et al., 1974).

Naqvi et al. (1974; Fig. 1.3) separated the shield into three cratons (Dharwar, Singhbhum, and Aravalli) that had evolved about individual Archean nuclei. An essential coherence of the entire shield since the Late Archean or Early Proterozoic was proposed by Radhakrishna and Naqvi (1986) based on correlation of Early to Middle Proterozoic intracratonic sedimentary assemblages across much of the shield (Fig. 1.4). Further work has indicated the possibility of additional major structural joins (sutures?) between various Precambrian terrains. From the standpoint of writing this book, the larger number of subdivisions enables us to describe the various areas more comprehensively in individual chapters.

The seven cratons described in this book, and their respective chapter numbers, are Western Dharwar (2), Eastern Dharwar (3), Granulite or Southern Granulite, terrain (4), Eastern Ghats (5),

Fig. 1.2. Cratons and other subdivisions of India discussed in other chapters of this book plus area covered by Deccan plateau basalts. Locations include D, Delhi; P, Patna; A, Ahmedabad; C, Calcutta; N, Nagpur; BO, Bombay; HY, Hyderabad; MA, Madras; BA, Bangalore.

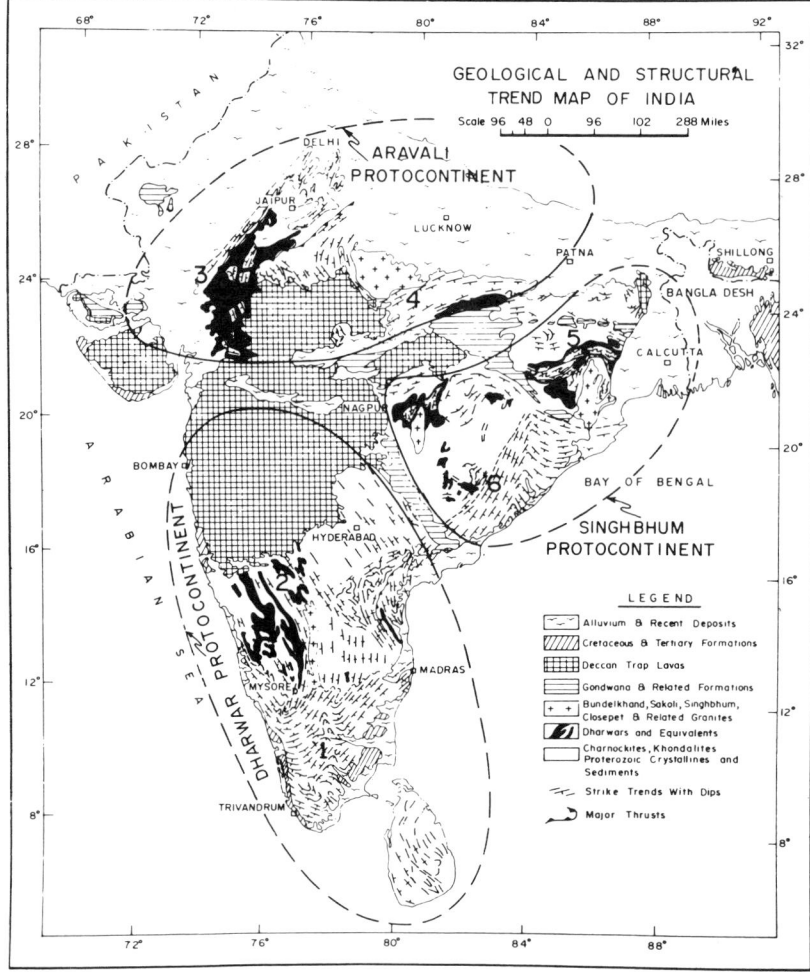

Fig. 1.3. Tectonic map of India (Naqvi et al., 1974).

Bhandara (6), Singhbhum (7), and Arvalli or Ar-avalli/Bundelkhand (9). Each of these geologic/geographic areas has characteristic features that distinguish it from other cratons. For example, the greenstone belts of India are best developed in the Western Dharwar craton, whereas potassic granites are best developed in the Eastern Dhar-war region. There are limited indications of differences in ages between different parts of the Indian shield. The characteristic features of each area, and the boundaries between them, are described in following chapters.

The general nature of the boundaries of (joins between) the various crustal terrains is uncertain. Most of the joins are along thrust zones and/or rift valleys, and at least two of the rifts appear to have formed along former thrust belts (Fig. 1.5; Ch. 10). The unresolved question is whether the joins are intracratonic thrusts and rifts, representing fractures of a coherent shield, or whether the

thrusts represent subduction-zone closure of ocean basins. Closure of ocean basins implies that the Indian shield formed by accretion of separate blocks. Rogers (1986) proposed that most of the joins were active about 1,500 m.y. ago and discussed evidence for and against accretion.

GEOPHYSICAL PROPERTIES

The Indian shield has crustal characteristics that are broadly typical of other Precambrian shield areas in the world. In detail, however, there are some interesting differences between India and other Precambrian terrains. This section discusses data obtained from investigations of gravity, seismic waves, magnetism, and heat flow. One aspect in which the Indian shield is unusual is the absolute velocity at which the plate containing India has moved, and we briefly correlate this characteristic with other properties of the shield.

Gravity data

The major information about gravity values in India has been presented in a series of maps prepared by the National Geophysical Research Institute (NGRI, 1975a, 1975b, 1975c, 1975d, 1975e). An explanatory brochure for these maps has also been published (NGRI, 1978). A generalized version of the free-air map is shown in Figure 1.6 (Verma, 1985).

A number of conclusions can be drawn from the gravity data. One is that the position of the Moho determined seismically is consistent with the gravity field (Kaila and Bhatia, 1981). Furthermore, on a broad scale, the free-air gravity anomalies are near zero, implying general isostatic compensation of the shield as a whole (Subrahmanyam and Verma, 1980; Verma and Subrahmanyam, 1984; Verma, 1985). Isostatic anomalies confirm regional compensation with

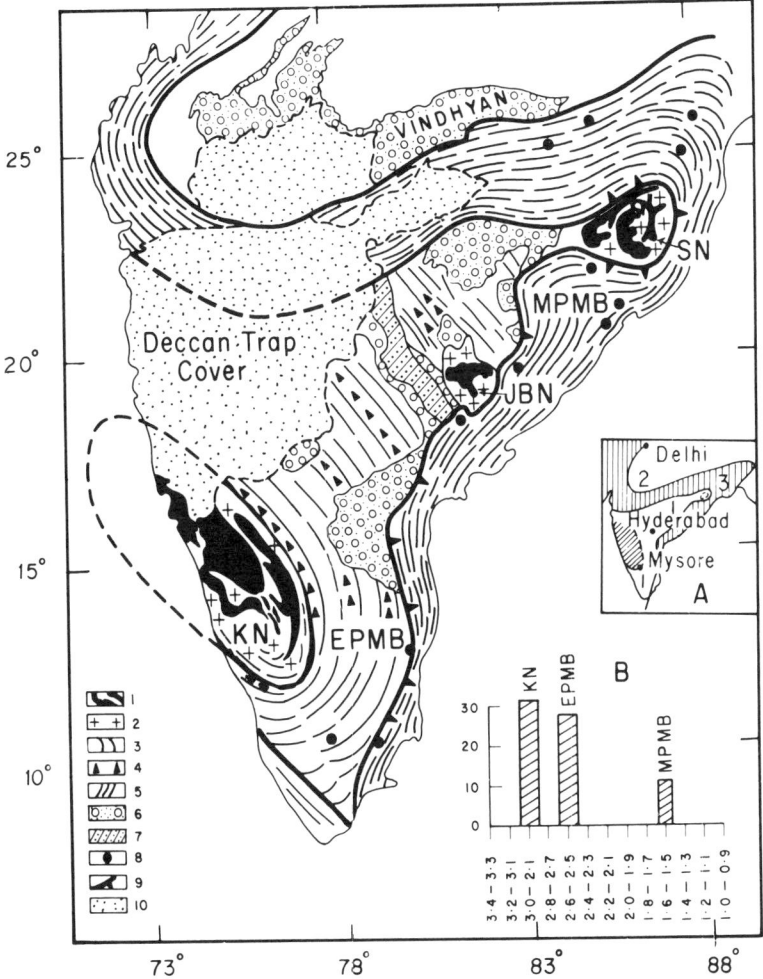

Fig. 1.4. Generalized geological map of Peninsular India (Radhakrishna and Naqvi, 1986). KN, Karnataka nucleus; JBN, Jeypore-Bastar nucleus; SN, Singhbhum nucleus; EPMB, Early Proterozoic Mobile Belt; MPMB, Middle Proterozoic Mobile Belt; 1, schist belts within nuclei; 2, tonalitic gneisses; 3, granodiorites, granodioritic gneisses, and granulites of EPMB; 4, K-granites in EPMB; 5, granulites and gneisses of MPMB; 6, Middle Proterozoic sedimentary basins; 7, Gondwana sediment of Godavari rift valley; 8, anorthosites emplaced along the EPMB-MPMB contact; 9, Eastern Ghat-Sukinda-Singhbhum thrust; 10, Deccan Trap cover.
 Inset A shows different crustal elements of Peninsular India: 1, Dharwar-Singhbhum protocontinent; 2, Bundelkhand protocontinent; 3, Middle Proterozoic mobile belts of Eastern Ghats, Satpura, and Delhi.
 Inset B shows three peaks of radiometric ages. Sixty-eight isochron ages include 36 in KN, 20 in EPMB, and 13 in MPMB. More than 90% of the radiometric ages in KN form a unimodal peak around 3,000 m.y. Similarly, 60% of radiometric ages yield a unimodal peak around 2,600 m.y. for the EPMB and about 70% yield a unimodal peak around 1,500 m.y. for the MPMB.

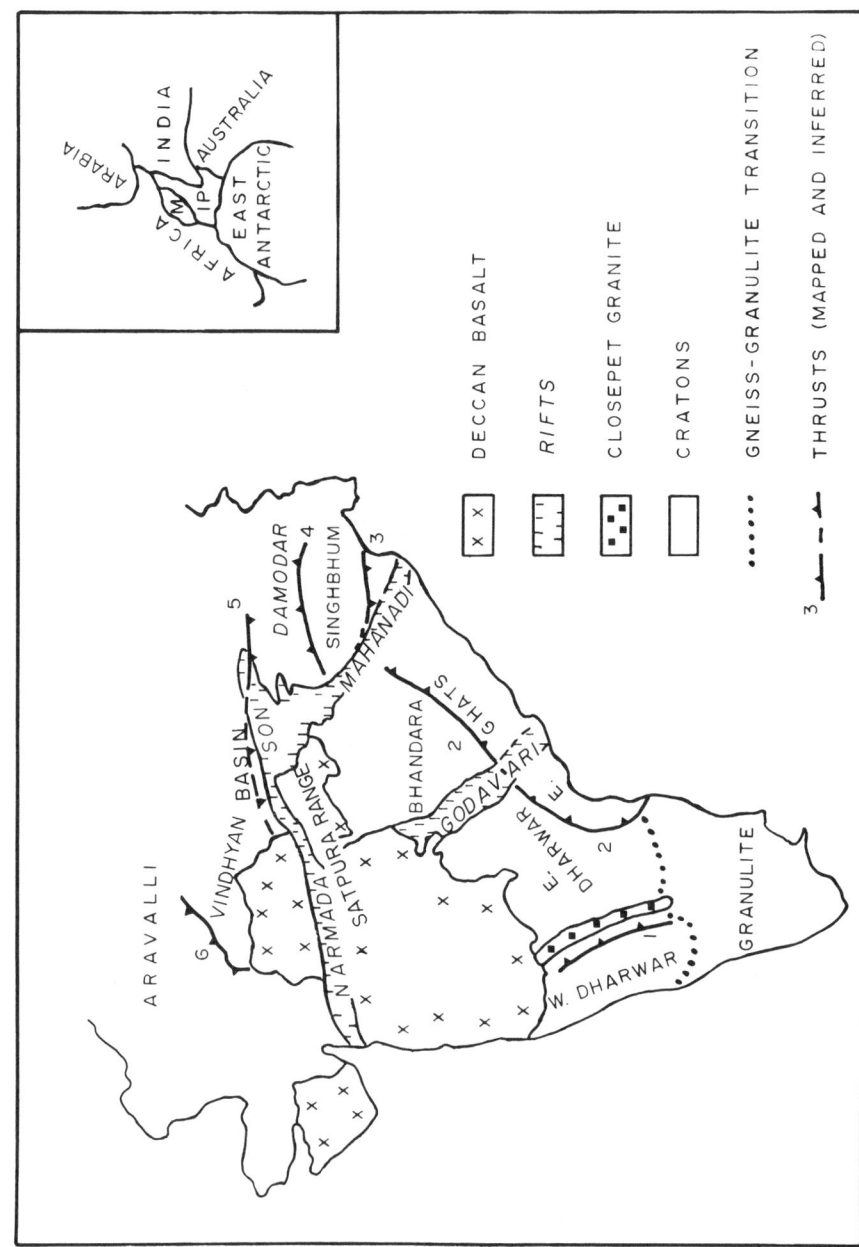

Fig. 1.5. Major structural features of India. Numbered thrusts include: 1, small thrust in Western Dharwar craton; 2, Eastern Ghats front; 3, Sukinda; 4, Singhbhum (Sopper Belt); 5, Son Valley thrust (inferred westward); 6, Great Boundary fault.

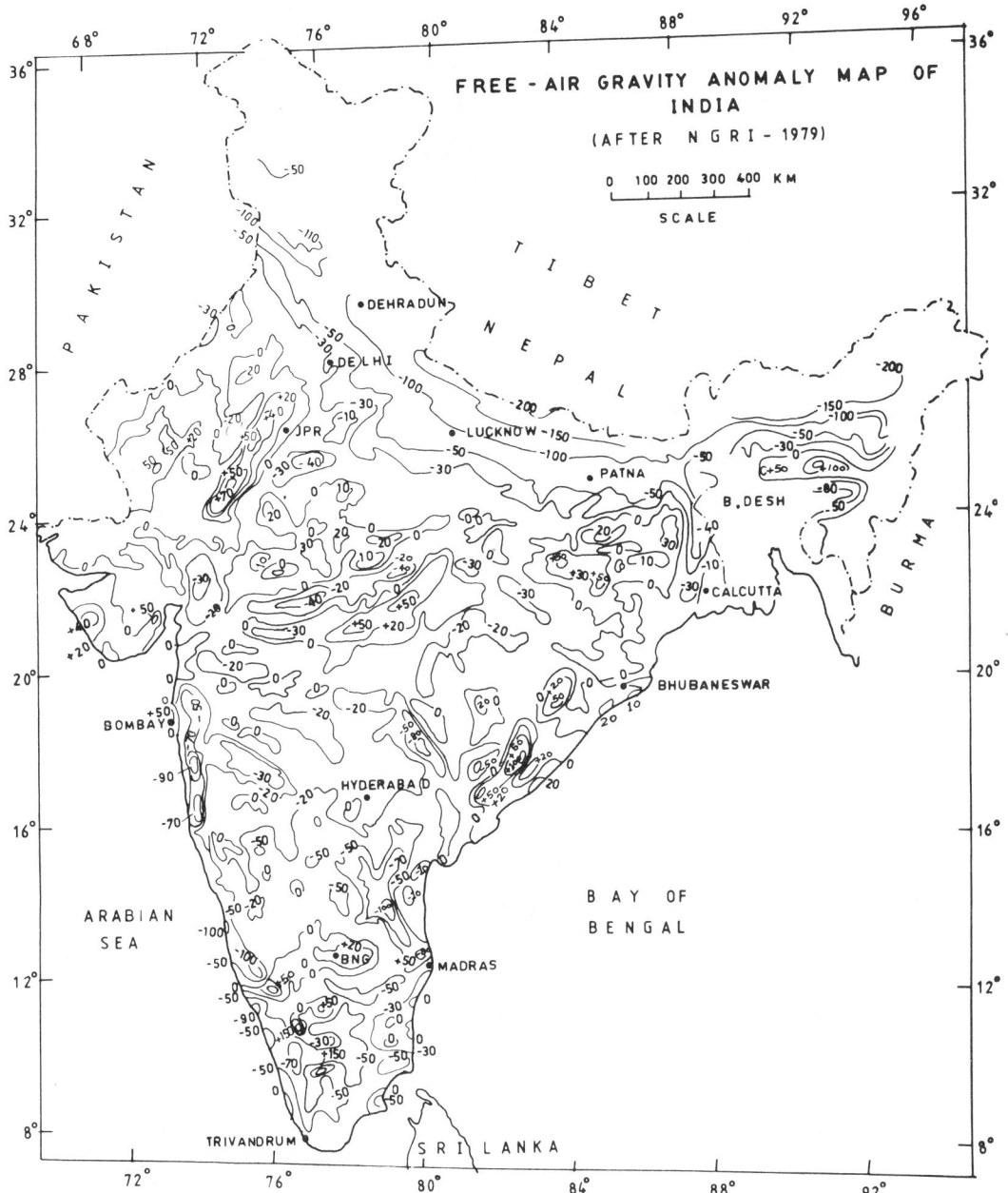

Fig. 1.6. Simplified free-air gravity anomaly map (Verma, 1985).

local imbalance (Aravamadhu, 1977). Some correlation can be made between gravity values and specific geologic features, but the relationship of gravity to surface geology is not clear everywhere.

Despite the relatively high (less negative) Bouguer values associated with high elevations, the overall Bouguer gravity of India is somewhat low relative to other continental areas (Qureshy and Warsi, 1980). This low gravity may be related to regional mass deficiency associated with the In-

dian Ocean gravity low discussed by Wagner et al. (1977; Fig. 1.7). It is possible that this mass deficiency is related to the rapid rate of movement of the Indian plate (discussed shortly), which may be accompanied by lateral movement of mantle material away from the path followed by India.

There appears to be very little relationship between Bouguer or free-air gravity values and major geologic features of the Indian shield. The only rift valley that is associated with an anomaly

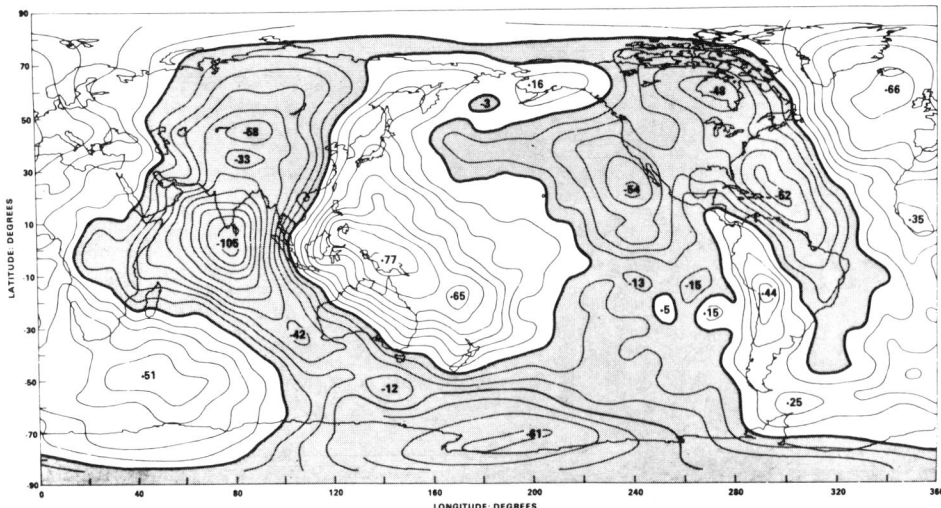

Fig. 1.7. World geoid map (Wagner et al., 1977). Numbers refer to heights of geoid in meters above (white areas) or below (shaded areas) the world average. Reproduced with permission from the Journal of Geophysical Research, v. 82, p. 911, copyright by the American Geophysical Union.

is the Godavari, with a negative Bouguer anomaly. A gravity gradient also occurs along the Eastern Ghats front. Other cratonic joins, however, have no effect on gravity values.

Seismic data

Information on crustal thicknesses and seismic wave velocities has been obtained from several lines of investigation. All studies agree that most of the Indian peninsula has a normal crustal thickness of 35 to 40 km. Reported thickness variations are generally within the margin of error of the measurements except for thinning to 30 to 35 km in the Indo-Gangetic plain and thickening to 70 to 80 km under the Himalayas. There is no indication of a deep root under the Satpura Range, but deep seismic sounding (DSS) studies suggest crustal thicknesses in the range of 40 to 45 km in the Eastern Ghats (Fig. 1.8).

Bhattacharya (1974, 1981) used surface waves to calculate P-wave velocities for a two-layered crust. Narain (1973) used body waves for the same purpose. Their data are summarized in Table 1.1, which shows a typically "granitic" upper half of the crust and a higher velocity in the lower crust. It is not clear whether these data indicate the presence of a Conrad discontinuity, but Narain referred to the value for the lower crust as P*, the refracted Conrad wave. P-wave velocities in the underlying upper mantle are in the range of 8.0 km/sec, which is normal for subcontinental mantle. D. D. Singh and Rastogi (1978) and Kharechko (1981) reported very small velocity variations in the crust.

Deep seismic sounding (Fig. 1.8) broadly confirms the presence of a two-layered crust (Kaila et al., 1979). P-wave velocities were reported for a traverse across the Eastern Ghats and the Eastern and Western Dharwar cratons (Table 1.1). The structural sections inferred by Kaila et al. contain a large number of near-vertical faults that extend through the whole crust. These faults, however, generally do not have a surface expression and have not been confirmed by other kinds of data. The interpretations of Kaila et al. are consistent with the proposals of Roy Chowdhury and Hargraves (1981).

TABLE 1.1. *Comparison of P-wave velocity estimates for India shield (in km/sec)*

Level	Narain (1973)	Bhattacharya (1981)	Kaila et al. (1979)
Upper crust (above 20 km)	5.7	5.8	6.0
Lower crust (below 20 km)	6.5	6.6	7.0
Upper mantle	8.0	8.2	

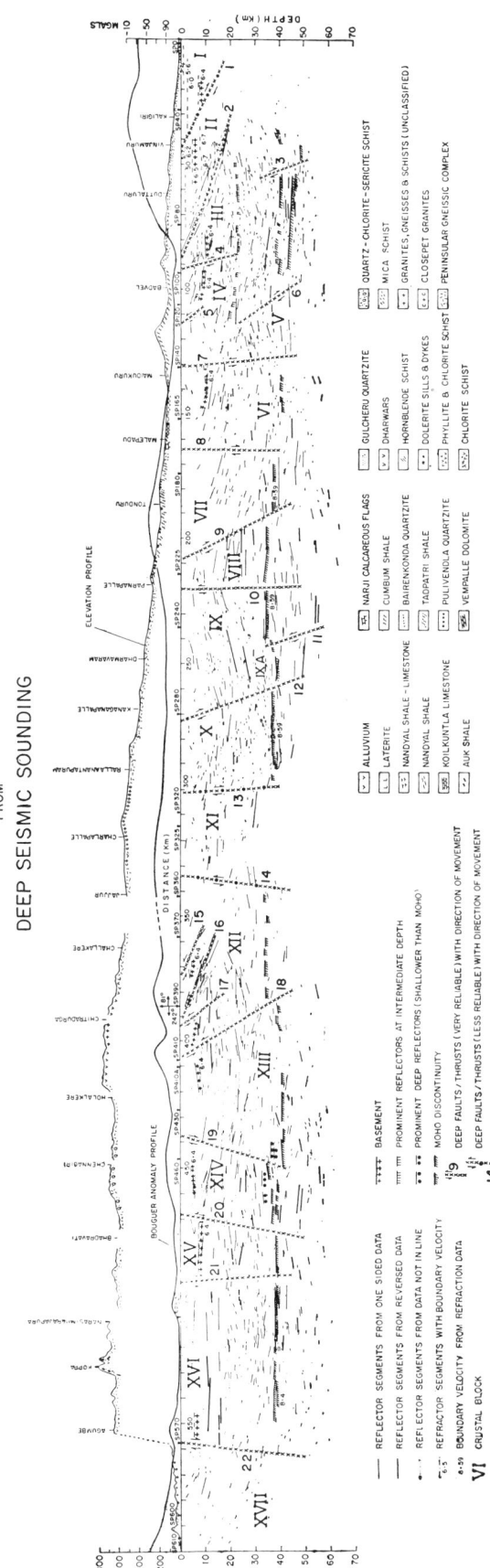

Fig. 1.8. Deep seismic sounding profile across Indian peninsula (Kaila et al., 1979). Profile is from Kavali, on the east coast, to Udipi, on the west coast, following an approximate east-west line. As located on Fig. 1.1, profile is from a point north of Madras to a point south of Bombay. Reproduced with permission from the Journal of the Geological Society of India, v. 20.

Inferences concerning mantle structure have been obtained largely from the Gauribidanur array, north of Bangalore (Berteussen et al., 1977; Ram and Mereu, 1977; Ram, 1979; Ram et al., 1981). Ram and Mereu (1977) found discontinuities at about 450 and 650 km under the shield and also noted that the 100- to 150-km low-velocity layer of oceanic regions was absent. Considerable homogeneity in the crust and upper mantle under the Gauribidanur array was determined by McCarthy et al. (1983).

Magnetic field

The magnetic field of India was discussed by Mishra (1984), and paleomagnetism was summarized by Hargraves and Bhalla (1983). The major variations are a comparative high magnetic intensity over southern India and a decrease toward the north. Mishra regarded the magnetic high as related to a shallow Moho under the Godavari rift. The decrease in magnetic intensity correlates with increase in heat flow northward in India (discussed shortly).

Heat flow

Heat flow measurements are not numerous in the Indian shield, but some generalizations can be drawn about both surface heat flow and crustal

heat production (M. L. Gupta, 1982). Sharma et al. (1980) summarized the related topic of paleo-uplift rates in India.

As a generalization, heat flow appears to increase northward in the Indian shield, although the number of measurements is too small for certainty (Negi and Pandey, 1976; R. N. Singh, 1984). Heat flow in the Dharwar area is about normal for old continental shields (approximately 1×10^{-6} cal/cm^2/sec, or 45 mW/m^2). Heat flow values of 1.5 to 2×10^{-6} cal/cm^2/sec in the northern part of the shield have been correlated with Moho temperatures as high as 850° to 900° (R. N. Singh and Negi, 1982). These high temperatures place the Curie temperature within the crust and prohibit the lower crust from contributing to magnetic intensity. Heat flow values in the southern part of the shield are consistent with normal Moho temperatures of about 550°, which places the Curie temperature at the Moho and permits the entire Indian crust in that area to contribute to magnetic intensity.

Average heat flow from the Indian shield (based on six points) appears to be slightly higher than for other Archean shields (Jessop and Lewis, 1978). If this difference is valid, then it can be explained by the observation that heat production in crustal rocks may be slightly higher in the average Archean rock of India than in similar rocks elsewhere (Rao et al., 1976).

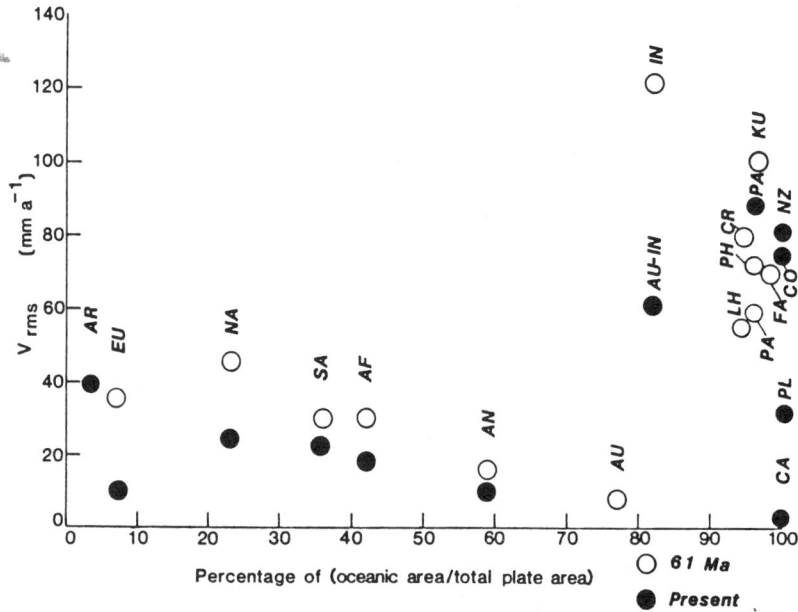

Fig. 1.9. Root mean square velocities with respect to hot spots for the Present and Early Tertiary as a function of the percentage of the plate area occupied by oceanic lithosphere (Jurdy and Gordon, 1984). AR, Arabian plate; EU, Europe; NA, North America; SA, South America; AF, Africa; AN, Antarctic; AU, Australia; AU-IN, Australia-India; IN, India; CR, Chatham rise; PH, Phoenix; LH, Lord Howe; PA, Pacific; KU, Kula; FA, Farallon; CA, Caribbean; CO, Cocos; PL, Philippine; NZ, Nazca. Reproduced with permission from the Journal of Geophysical Research, v. 89, p. 9931, copyright by the American Geophysical Union.

RATES OF MOVEMENT OF INDIAN PLATE

Absolute velocities of plate motions have been calculated in a variety of ways, commonly based on locations of magmatically active hot spots. A diagram from Jurdy and Gordon (1984) is shown in Figure 1.9, which compares the present and Paleocene plate velocities of most of the earth's plates. The principal conclusion from Figure 1.9 is that virtually all plates that contain continental material move relatively slowly in comparison with purely oceanic plates. The only exception is the Indian plate, which was the fastest-moving plate on the earth in the Paleocene and is still fast compared to other continental plates.

The significance of the absolute velocity of the Indian plate is not clear. Other geophysical parameters of India are also somewhat anomalous, including the gravity low south of the tip of peninsular India, the magnetic low in northern India, and the possibly high average heat flow and average crustal heat production of the shield. It is possible that these features are related, with the fundamental cause of the phenomena being the high heat production of the shield, resulting in high thermal gradients. If these gradients have persisted for large parts of geologic time, then an enhanced melting ("lubrication") might have taken place in the mantle, permitting movement more rapid than is normal for continental plates.

REFERENCES

Aravamadhu, P. S. (1977). Isostatic anomalies and their relation to tectonics in southern India. Geophys. Res. Bull. (Hyderabad), v. 15, no. 4, 83–91.

Berteussen, K. A., Husebye, E. S., Mereu, R. F., and Ram, A. (1977). Quantitative assessment of the crust-upper mantle heterogeneities beneath the Gauribidanur seismic array in southern India. Earth Planet. Sci. Lett., v. 37, 326–332.

Bhattacharya, S. N. (1974). The crust-mantle structure of the Indian peninsula from surface wave dispersion. Roy. Astron. Soc. London Geophys. J., v. 36, 273–283.

—— (1981). Observation and inversion of surface wave group velocities across central India. Seismol. Soc. Amer. Bull., v. 71, 1489–1501.

Chatterjee, S. C. (1974). Petrography of the igneous and metamorphic rocks of India. The Macmillan Co. of India, Madras, 559 p.

Gupta, M. L. (1982). Heat flow in the Indian peninsula; Its geological and geophysical implications. Tectonophysics, v. 83, 71–90.

Gupta, V. J. (1977). Indian Precambrian Stratigraphy. Hindustan Publ. Co., New Delhi, 333 p.

Hargraves, R. B., and Bhalla, M. S. (1983). Precambrian paleomagnetism in India through 1982; A review. In Precambrian of South India (ed. S. M. Naqvi and J. J. W. Rogers), Geol. Soc. India Mem. 4, 491–524.

Jessop, A. M., and Lewis, T. (1978). Heat flow and heat generation in the Superior province of the Canadian shield. Tectonophysics, v. 50, 55–77.

Jurdy, D. M., and Gordon, R. G. (1984). Global plate motions relative to the hot spots 64 to 56 Ma. J. Geophys. Res., v. 89, 9927–9936.

Kaila, K. L., and Bhatia, S. C. (1981). Gravity study along the Kavali-Udipi deep seismic sounding profile in the Indian peninsular shield: Some inferences about the origin of anorthosites and the Eastern Ghats orogeny. Tectonophysics, v. 79, 129–143.

——, Roy Chowdhury, K., Reddy, P. R., Krishna, V. K., Narain, H. Subbotin, S. I., Sollogub, V. B., Chekhunov, A. V., Kharechko, G. E., Lazarenko, M. A., and Ilchenko, T. V. (1979). Crustal structure along the Kavali-Udipi profile in the Indian peninsular shield from deep seismic sounding. Geol. Soc. India J., v. 20, 307–333.

Kharechko, G. E. (1981). Velccity irregularities in the upper crust beneath the Indian shield. Geophys. J., no. 3(4), 574–579.

Krishnan, M. S. (1968). Geology of India and Burma (5th ed.). Higginbotham's Ltd., Madras, 536 p.

McCarthy, S. M., Powell, C. A., and Rogers, J. J. W. (1983). A relative residual study of the southern peninsula of India using the Gauribidanur seismic array. In Precambrian of South India (ed. S. M. Naqvi and J. J. W. Rogers), Geol. Soc. India Mem. 4, 525–552.

Mishra, D. C. (1984). Long wavelength magnetic anomalies from the lithosphere: Indian shield and Himalaya. Tectonophysics, v. 105, 319–330.

Naqvi, S. M., Divakara Rao, V., and Narain, H. (1974). The protocontinental growth of the Indian shield and the antiquity of its rift valleys. Precamb. Res., v. 1, 345–398.

Narain, H. (1973). Crustal structure of the Indian subcontinent. Tectonophysics, v. 20, 249–260.

Negi, J. G., and Pandey, O. P. (1976). Correlation of heat flow and crustal topography in the Indian region. Roy. Astron. Soc. Geophys. J., v. 45, 201–217.

NGRI (1975a). Map of India showing gravity station distribution. Natl. Geophys. Res. Inst., Hyderabad, NGRI-GPH 1, 1 sheet, 1:5,000,000.

—— (1975b). Bouguer gravity anomaly map of India. Natl. Geophys. Res. Inst., Hyderabad, NGRI-GPH 2, 1 sheet, 1:5,000,000.

—— (1975c). Isostatic gravity anomaly map of India (Airy-Heiskanen, T = 30 km). Natl. Geophys. Res. Inst., Hyderabad, NGRI-GPH 3, 1 sheet, 1:5,000,000.

—— (1975d). Free air gravity anomaly map of

India. Natl. Geophys. Res. Inst., Hyderabad, NGRI-GPH 4, 1 sheet, 1:5,000,000.

———— (1975e). Bouguer gravity anomaly and regional geology map of India. Natl. Geophys. Res. Inst., Hyderabad, NGRI-GPH 5, 1 sheet, 1:5,000,000.

———— (1978). Gravity map series of India—Explanatory Brochure. Natl. Geophys. Res. Inst., Hyderabad, 13 p.

Pascoe, E. H. (1950). A Manual of the Geology of India and Burma, v. 1 (3rd ed.). Geol. Surv. India, Calcutta, 485 p.

Pichamuthu, C. S. (1967). The Precambrian of India. In The Precambrians (ed. K. Rankama), Wiley-Interscience, New York, v. 3, 1–96.

———— (1974). On the banded iron formation of Precambrian age in India. Geol. Soc. India J., v. 15, 1–30.

Qureshy, M. N., and Warsi, W. E. K. (1980). A Bouguer anomaly map of India and its relation to broad tectonic elements of the sub-continent. Roy. Astron. Soc. Geophys. J., v. 61, 235–242.

Radhakrishna, B. P. (1984). Crustal evolution and metallogeny—Evidence from the Indian shield. Geol. Soc. India J., v. 25, 617–640.

————, and Naqvi, S. M. (1986). Precambrian continental crust of India and its evolution. J. Geol., v. 94, 145–166.

Ram, A. (1979). A comparison of upper mantle travel-time branches using data from GBA, YKA and WRA seismic arrays. Ind. Soc. Earthquake Tech. Bull., v. 16, 153–158.

————, and Mereu, R. F. (1977). Lateral variation in upper mantle structure around India as obtained from Gauribidanur seismic array data. Roy. Astron. Soc. Geophys. J., v. 49, 87–113.

————, Yadav, L., and Singh, O. P. (1981). Seismic ray direction anomalies and relative residuals of earthquakes recorded at Gauribidanur array. Ind. Acad. Sci. Proc., Sect. A (Earth and Planet. Sci.), v. 90, 63–74.

Rao, R. U. M., Rao, G. V., and Narain, H. (1976). Radioactive heat generation and heat flow in the Indian shield. Earth Planet. Sci. Lett., v. 30, 57–64.

Ray, S. (1976). Metamorphism and Metamorphic Rocks of India. Jayanta Basu Publ., Calcutta, 156 p.

Rogers, J. J. W. (1986). The Dharwar craton and the assembly of peninsular India. J. Geol., v. 94, 129–144.

Roy, S. (1981). Manganese Deposits. Academic Press, London, 458 p.

Roy Chowdhury, K., and Hargraves, R. B. (1981). Deep seismic soundings in India and the origin of continental crust. Nature, v. 291, 648–650.

Sharma, K. K., Bal, D. K., Parshad, R., Nand Lal, and Nagpaul, K. K. (1980). Paleo-uplift and cooling rates from various orogenic belts of India, as revealed by radiometric ages. Tectonophysics, v. 70, 135–158.

Shields, O. (1977). A Gondwanaland reconstruction for the Indian Ocean. J. Geol., v. 85, 236–242.

Singh, D. D., and Rastogi, B. K. (1978). Crustal structure of the peninsular shield beneath Hyderabad (India) from the spectral characteristics of long-period waves. Tectonophysics, v. 51, 127–137.

Singh, I. B. (1980). Precambrian sedimentary sequences of India; Their peculiarities and comparison with modern sediments. Precamb. Res., v. 12, 411–436.

Singh, R. N. (1984). Thermal evolution of the Indian shield and its subjacent mantle. Tectonophysics, v. 105, 413–418.

————, and Negi, J. G. (1982). High Moho temperatures in the Indian shield. Tectonophysics, v. 82, 299–306.

Subrahmanyam, C., and Verma, R. K. (1980). The nature of free-air, Bouguer and isostatic anomalies in southern peninsular India. Tectonophysics, v. 69, 147–162.

Verma, R. K. (1985). Gravity Field, Seismicity and Tectonics of the Indian Peninsula and Himalayas. D. Reidel, Dordrecht, Netherlands, 213 p.

————, and Subrahmanyam, C. (1984). Gravity anomalies and the Indian lithosphere: Review and analysis of existing gravity data. Tectonophysics, v. 105, 141–161.

Wadia, D. N. (1961). Geology of India, 3rd ed. MacMillan Co., London, 536 p.

Wagner, C. A., Lorch, F. J., Brown, J. F., and Richardson, K. A. (1977). Improvements in the geopotential derived from satellite and surface data (GEM 7 and 8). J. Geophys. Res., v. 82, 901–914.

2. WESTERN DHARWAR CRATON

The Western Dharwar craton (Fig. 2.1) is bounded on the west by the Arabian Sea, on the north by overlap of the Deccan basalts, and on the south by transition into granulite-facies assemblages (Ch. 4). The eastern margin is more uncertain. The Eastern Dharwar craton (Ch. 3) contains abundant posttectonic granites, of which the westernmost major belt is the Closepet Granite (Fig. 2.1). Because of the prominence of the Closepet trend, and the scarcity of similar rocks to the west, the margin between the western and eastern parts of the craton may be the Closepet Granite outcrop.

Another possible designation of the eastern margin of the Western Dharwar craton is a major shear zone slightly west of the Closepet outcrop. Drury and Holt (1980) and Drury et al. (1984) recognized several "zones of high strain" (Fig. 2.1), the easternmost of which follows the eastern margin of the Chitradurga schist belt, particularly separating the Chitradurga and Javanahalli belts (Fig. 2.2). The fault has been recognized from deep seismic studies as a thrust that brings higher-grade rocks of the Javanahalli belt over lower-grade rocks of the Chitradurga belt (Kaila et al., 1979). Farther south, the same shear zone extends along the western side of the charnockite massif of the Billigirirangan Hills and terminates southward in the Moyar-Bhavani shear zone (Figs. 2.1 and 4.1). Viswanatha and Ramakrishnan (1981b) proposed that kyanite-bearing assemblages occur on the western side of the shear and cordierite-bearing assemblages (Radhakrishna, 1954) on the east. This difference implies a higher T/p metamorphic condition to the east than to the west.

For the purpose of this chapter, we have chosen the Closepet Granite as a tentative eastern margin for the Western Dharwar craton. We have, however, chosen to discuss the Closepet Granite along with similar rocks in Chapter 3, but all schist belts west of the Closepet are described in this chapter.

The Western Dharwar craton, with its well-known Dharwar greenstone (schist) belts, has attracted the attention of geologists for many years (recent summaries by Iyengar, 1976; Naqvi, 1976, 1978, 1981, 1982, 1983; Radhakrishna and Vasudev, 1977; Divakara Rao and Rama Rao, 1982; Pichamuthu and Srinivasan, 1983, 1984; Radhakrishna, 1983; Ramakrishnan and Viswanatha, 1983; Roy, 1983; Narain and Subrahmanyam, 1986; Radhakrishna and Naqvi, 1986; and Rogers, 1986). Major controversy has surrounded the relationship between the schist belts and the surrounding Peninsular Gneisses that constitute much of the craton. Three views have been most widely held: (1) the Dharwar greenstone/schist/supracrustal suite represents one major group/supergroup that is mostly younger than the Peninsular Gneiss, (2) the Dharwar assemblage represents one stratigraphic suite mostly older than the gneisses, and (3) the "Dharwar" supracrustal rocks represent a number of different suites whose ages overlap those of different members of the gneissic suite.

MAJOR STRUCTURES AND PROPOSED TILTING OF INDIA

The principal structural trends in the Western Dharwar craton are approximately north-south (Fig. 2.1). These trends are shown by the orientation of the major schist belts; the orientation of the Closepet Granite; and smaller-scale structures such as foliation in gneisses, schistose and other enclaves in gneisses, cleavage and schistosity in the greenschist belts, etc. (Mukhopadhyay, 1986). Some of these trends show a convexity toward the east (e.g., the Chitradurga belt and the Closepet Granite). This curvature may result from secondary structural processes, such as cross-folding at an angle to the N-S trend. Ultramafic to anorthositic bodies, possibly parts of layered intrusions, also cluster in an arcuate north-south trend approximately in the center of the craton (Nijagunappa and Naganna, 1983), which may indicate an upward arching along that zone.

Greenstone belts in the southern part of the craton tend to be small, engulfed in gneisses, and metamorphosed to amphibolite facies. Schist belts to the north are larger and occur mostly in greenschist facies. The northern belts, in general, also appear to be younger than the southern ones. As a first approximation, this distribution of schist belts can be explained by exposure of deeper levels of the craton farther south, with the deepest levels being exposed in the granulite terrain (Fig. 2.1).

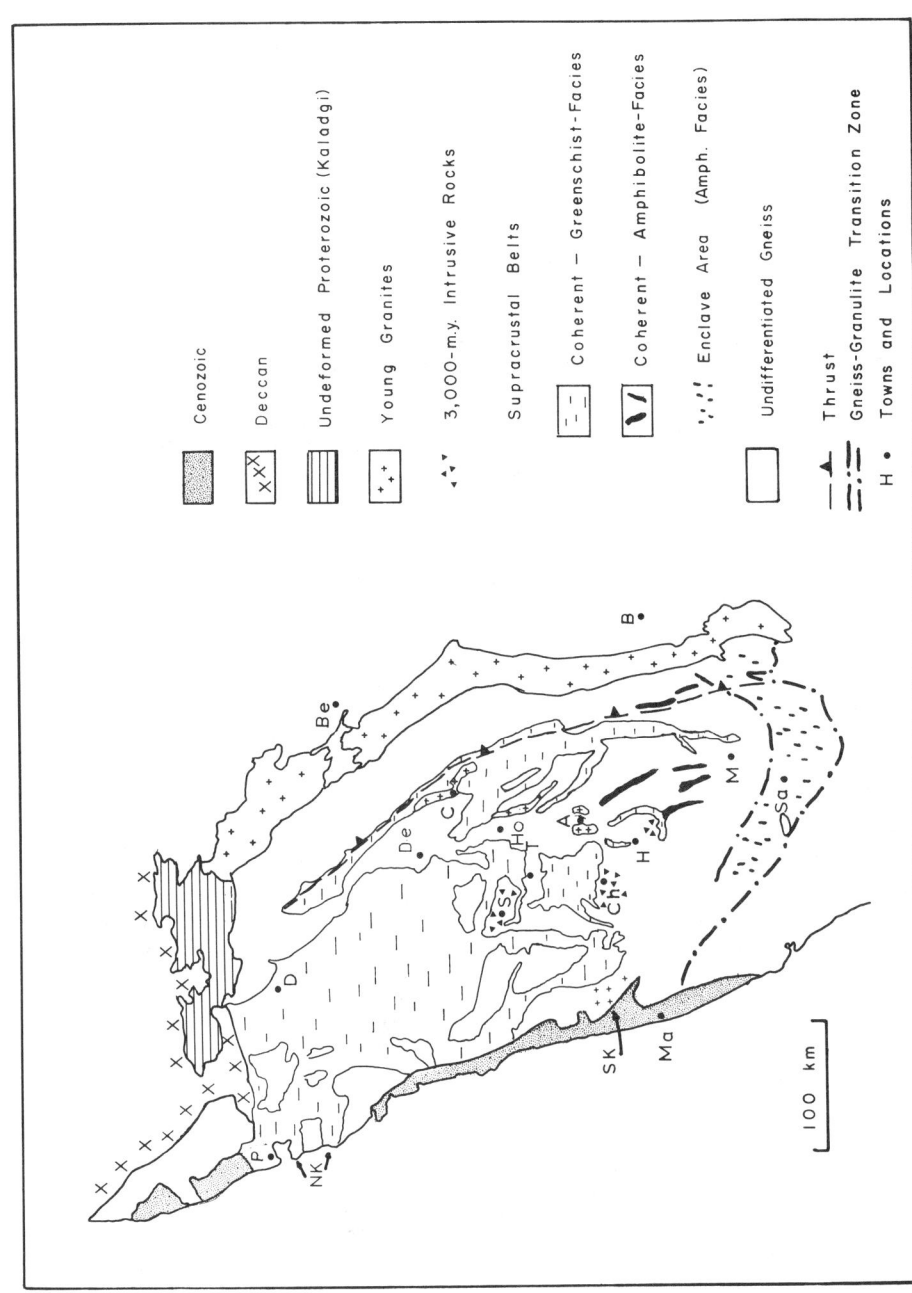

Cenozoic

Deccan

Undeformed Proterozoic (Kaladgi)

Young Granites

3,000-m.y. Intrusive Rocks

Supracrustal Belts

Coherent – Greenschist-Facies

Coherent – Amphibolite-Facies

Enclave Area (Amph. Facies)

Undifferentiated Gneiss

Thrust

Gneiss-Granulite Transition Zone

H • Towns and Locations

Fig. 2.1. Western Dharwar craton. P, Panjim; NK, North Kanara; SK, South Kanara; MA, Mangalore; D, Dharwar; BE, Bellary; De, Devangere; C, Chitradurga; HO, Hoskere; S, Shimoga; A, Arsikere; T, Tarikere; Ch, Chickamagalur; H, Hassan; B, Bangalore; M, Mysore; Sa, Sargur. Other locations mentioned in the text are Chamundi, at Mysore; Gundlupet and Doddakanaya, near Sargur; Halekote, southeast of Hassan; and Someshwar, near Dharwar.

16

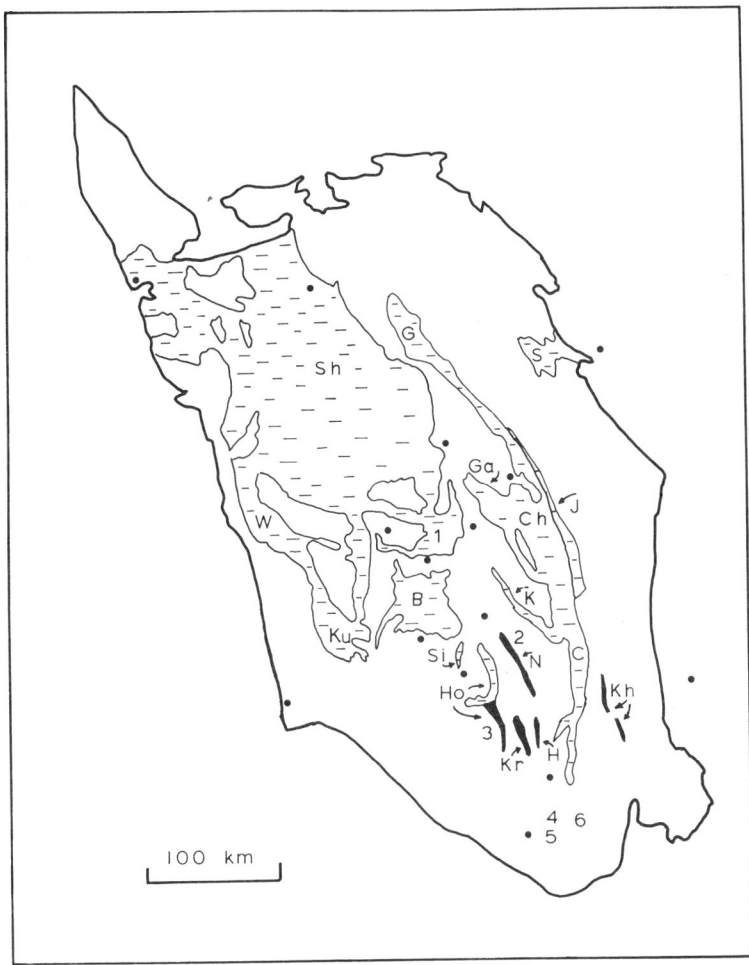

Fig. 2.2. Schist belts and anorthosites. Dots are towns named in Fig. 2.1. High-grade schist belts are in black; low-grade belts in lined pattern. Schist belts are Sh, Shimoga; G, Gadag; S, Sandur; Ga, Ghatti-Hosahalli; J, Javanahalli; Ch, Chitradurga; W, Western Ghats; B, Bababudan; K, Kibbenahalli; Ku, Kudremukh; Si, Sigegudda; C, Chiknay-akanhalli; Ho, Holenarasipur; N, Nuggihalli; Kr, Krishnarajpet; H, Hadnur; Kh, Kunigal-Hulliyurdurga. Anorthosite locations are 1, Masanikere; 2, Nuggihalli; 3, Holenarasipur; 4, Sindhuvalli; 5, Kurihundi; 6, Konkanhundi.

North-south change in metamorphic facies has been demonstrated in the schist belts (Raase et al., 1986). Geobarometers commonly indicate pressures in the range of 5 to 6 kb in the gneiss-granulite transition zone at the southern edge of the craton (Ch. 4). Measurements farther north include: an estimate of 2.6 to 3.8 kb and 570° ± 50° in the Sandur schist belt (Fig. 2.2), based on equilibrium in the system garnet-muscovite-biotite-andalusite-quartz-plagioclase (Harris and Jayaram, 1982); and 2 to 4 kb and 300° to 400° in schists in North Kanara (Fig. 2.1), based largely on breakdown of phengite and phengite-biotite equilibration (Sivaprakash, 1983). These lower pressures to the north of the transition zone can be interpreted in terms of a downward tilting of the craton to the north at an angle of 1° to 2°.

Variation in pressure in metamorphic assemblages is not necessarily accompanied by variation in temperature. For example, Harris and Jayaram (1982) reported equilibration conditions of about 5 kb and 700° in schistose enclaves near the southern part of the Closepet Granite based on the assemblages cordierite-biotite-garnet-quartz-plagioclase and cordierite-hypersthene-quartz-plagioclase. These conditions can be compared with pressures up to about 12 kb and temperatures of 700° to 800° in the granulite terrain farther south (Ch. 4). If the pressure and temperature determinations are correct, then much of southern India, including the Western Dharwar craton, has been affected by some process that established a near-isothermal condition throughout a large vertical range in the crust.

ROCK SUITES MAPPED

The geologic maps of the Western Dharwar craton (Figs. 2.1 and 2.2) show the distribution of eight different rock suites. These suites are discussed as follows.

Mafic schist belts

The term "Dharwar" has been in use for about 100 years since it was proposed by Foote (1889). The term originally applied to all mafic volcano-sedimentary "schist belts" within the sialic "Peninsular Gneisses." Recently, detailed investigations have led to more complex classifications of these assemblages and the development of numerous new stratigraphic terms (Swami Nath and Ramakrishnan, 1981). Table 2.1 shows three recent classifications. Some stratigraphic classifications assume that the various schist belts are fragments of a formerly continuous stratigraphic succession. On this basis, apparent lithologic similarities among isolated belts have a time-stratigraphic significance represented by a stratigraphic terminology applied to all belts in the craton. However, if the various belts developed in isolated basins, this assumption of continuity and synchroneity obviously cannot be made.

The three suites of schist belts shown in Figure 2.2 have been chosen in an effort to avoid any kind of genetic assumption. They are as follows:

1. *Enclaves of high-grade mafic rocks in gneisses in the southern part of the craton.* Metamorphic grade in the enclaves is amphibolite facies grading into granulite facies. Typical rock types include amphibolites, mafic granulites (commonly with both clino- and orthopyroxene), magnetite quartzites (or other iron formation), calc-silicate rocks, and minor quartzites with such varietal minerals as kyanite, sillimanite, and/or fuchsite. Ultramafic-mafic-anorthositic suites are common, but it is not clear that they are associated with the metasediments and later engulfed by the gneisses or whether they are simply a suite intrusive into some of the older gneisses. Enclaves range in size from widths of several kilometers and lengths of 10 km or more down to bands of less than a meter wide. They are almost invariably conformable with foliation in the gneisses (principally north-south). Intense deformation and disruption by the gneisses makes construction of a stratigraphic section virtually impossible. The enclaves have been referred to as "Sargurs."

2. *Coherent belts of high metamorphic grade*

TABLE 2.1. *Classification of supracrustal rocks in Western Dharwar craton*

Naqvi et al. (1980)	Pichamuthu and Srinivasan (1983)	Radhakrishna (1983)
Younger belts: Bababudan, Chitradurga, Shimoga, etc.; graywackes, quartzose sediments, shales, basalts to rhyolites, cherts, carbonates; greenschist facies	Chitradurga Subgroup: graywackes, conglomerates, silicic volcanic rocks, banded iron formations	Younger greenstones (Dharwar type): quartzose clastic sediments, carbonates, cherts, iron formations, basalts and rhyolites; includes Shimoga, Chitradurga and Sandur basins
Intermediate belt: Javanahalli; arkoses, amphibolites (metapelites), calc-silicates, metavolcanics; amphibolite facies	Dodguni Subgroup: quartz conglomerates and sandstones, metapelites, carbonates, manganese and iron formations	Ancient supracrustals (>3,000 m.y.): mafic/ultramafic igneous rocks, rare clastic sediments; sediments consist of volcanic debris or chemical precipitates; includes Holenarasipur, Nuggihalli, Krishnarajpet, etc. belts; Javanahalli belt is of same age but contains quartzose clastic sediments
Older belts: Nuggihalli, Krishnarajpet, southern part of Holenarasipur, etc.; ultramafic/anorthositic suites (mostly volcanic), metapelites, cherts and iron formations; no clastic quartz; amphibolite facies	Bababudan Subgroup: quartz conglomerates and sandstones, basalts and rhyodacites, graphitic metapelites, banded iron formations	
	Nuggihalli Subgroup: quartzites, metapelites, ultramafic/anorthositic intrusive rocks	Ancient supracrustals (Sargur type): occurs in transition zone to granulite facies at southern margin of craton; mafic/ultramafic/anorthositic suites, aluminous metapelites, fuchsite quartzites, iron and manganese cherts, graphitic schists

(amphibolite facies). Typical examples are the Nuggihalli and Krishnarajpet belts (Fig. 2.2). Other major features include a small outcrop area compared to that of the lower-rank belts, occurrence farther south in the craton than the low-rank belts, deformation of such intensity that lithologic units are disrupted and stratigraphic sections are difficult to establish, scarcity or absence of quartzose clastic sediment, and relative abundance of mafic and/or ultramafic volcanic rocks. Many authors have included belts in this category with the Sargur enclaves previously mentioned. We have separated the coherent and enclave suites because of uncertainty about the equivalence of the rock assemblages in the suites (e.g., Janardhan et al., 1979). These differences may be found in the relative proportions of rock types in the enclaves and in the larger belts and the possible differences between ultramafic/anorthositic and sedimentary assemblages in the two suites. There is little question that all of the high-grade belts have been intruded by silicic plutonic rocks at least as old as 3,000 m.y. It is possible that some of the belts are considerably older and have been intruded by older members of the Peninsular Gneiss complex (e.g., summary by Naqvi, 1981; discussion and reply by Drury and Naqvi, 1981).

3. *Belts of low metamorphic grade.* These belts are characterized by metamorphism predominantly of greenschist facies but locally up to epidote amphibolite or low amphibolite facies. Other major features include a large and coherent outcrop area, abundance of clastic and quartzose sediments, comparatively high proportion of sedimentary to volcanic rocks, conglomerates containing clasts of underlying gneisses, presence of biogenic structures, and relative absence of intrusion by gneisses. Typical examples (Fig. 2.2) are the Chitradurga, Bababudan, Shimoga, and Western Ghats belts. These belts have been deformed, in most cases in several episodes, but the deformation has not been so intense that it disrupted the essential coherence of lithologic suites, and stratigraphic sections can be constructed. In several places the belts lie unconformably on Peninsular Gneiss about 3,000 m.y. old. Many authors now use the term "Dharwar" for these belts, thus restricting its usage from former reference to all the mafic schist belts in the craton.

At least two schist belts in the Dharwar craton have proved to be controversial in any attempt at classification. The Javanahalli belt contains a suite that is predominantly clastic, like that of the low-grade belts, but at a higher grade of metamorphism. We have arbitrarily placed it with the low-grade belts in Figures. 2.1 and 2.2. The Holenarasipur belt has a clastic, low-grade assemblage toward the northern part and a high-grade assemblage with ultramafic rocks in the south. We have shown the belt as consisting of both of these assemblages in Figures. 2.1 and 2.2.

Ultramafic/mafic/anorthositic suites

Rocks of ultramafic, gabbroic, noritic, and anorthositic composition are common in the high-grade schist belts and elsewhere toward the southern part of the Western Dharwar craton. The problem in designating them as a mappable suite is whether these rocks represent one or more separate magmatic episodes or whether they are simply part of the igneous/sedimentary assemblage in schist belts. For example, deformed fragments of layered complexes occur in the high-grade schist belts, and also as enclaves in the southern, transitional zone. Rocks of this composition, however, also occur as volcanic products, such as komatiites, basalts, and minor anorthositic lavas (Naqvi and Hussein, 1979) in the coherent high-grade belts. After high-grade metamorphism and deformation, distinction between former volcanic rocks and fragments of layered complexes is difficult. A further complication is the occurrence of layered complexes intrusive into the gneisses after the major deformational episodes. One example is the Konkanhundi complex (Fig. 2.2), which has a relatively undeformed zonal structure (Ramakrishnan and Mallikarjuna, 1976).

Peninsular Gneiss

The term "Peninsular Gneiss" refers to a large, complex suite that constitutes much of the outcrop area of the Western Dharwar craton. Most of the rocks are tonalite-trondhjemite gneiss, but a wide variety of screens and inclusions of other rock types is present. Some of these conformable bands are undoubtedly engulfed fragments of older schist belts. Other bands may be metasedimentary and/or metavolcanic rocks formed on, or in, gneissic massifs previously isolated from the schist belts. The gneissic suite may include former sedimentary and volcanic rocks, former intrusive rocks subjected to gneissic metamorphism, and direct magmatic increments foliated during intrusion. These diverse components are conformably intermixed on such a small scale that they cannot be discriminated in Figure 2.1, and they are all mapped together. The only variant of the Peninsular Gneiss that is separated on Figure 2.1 is a suite of poorly foliated to massive, diapiric bodies with ages of about 3,000 to 2,900 m.y. (e.g., Halekote Trondhjemite and Chickmagalur Granite). Small trondhjemite bodies with similar features are not shown on the map.

Ages of the Peninsular Gneiss suite are about 3,400 to 3,000 m.y. Numerous ages of about 3,000 m.y. correspond to the apparent time of emplacement of the diapiric trondhjemites. We regard 3,000 m.y. as the time of stabilization (cratonizaton) of the Western Dharwar craton. Mafic schist belts engulfed by the 3,000-m.y.-old gneisses do not contain significant quartzose or other sialic debris, whereas those that rest unconformably on the 3,000-m.y. terrain are highly quartzose.

Granites

The Closepet Granite and most of its apparently related granitic suites occur primarily in the eastern part of the Dharwar craton (Ch. 3). Some small bodies of diapiric rocks with high concentrations of K feldspar and quartz are present in the Western Dharwar craton. These bodies are considerably more potassic than the trondhjemites that are part of the Peninsular Gneiss suite. Available ages are 2,300 m.y. for the Arsikere granite and 2,500 m.y. for the Chitradurga Granite. The petrologically similar Chamundi Granite has been dated as only about 800 m.y. old. The only alkaline complex in the Western Dharwar craton is the Kunduru Betta ring complex near the transition zone to granulite-facies rocks at the southern margin of the craton (Friend and Mahabaleswar, 1985).

Undeformed Proterozoic basins

Part of one Proterozoic basin crops out south of the Deccan overlap in the Western Dharwar craton. This area is referred to as the Kaladgi basin and is discussed in Chapter 3.

STRUCTURAL EVOLUTION OF THE CRATON

Most structural trends in the Western Dharwar craton are roughly north-south, as shown by major shear zones and the elongation of schist belts (Fig. 2.1). The same trends are also shown on smaller scales by foliation and cleavage in schists and gneisses. Much of the small-scale structure results from the tendency of first-generation axial planes of folds to have a north-south orientation throughout all major rock types in the craton (Naha et al., 1986). Folds tend to have steep plunges, and in many cases they are doubly plunging.

Most studies (referenced shortly) show at least two, commonly three, generations of folds within individual lithologic sequences. Evidence of an older deformational event is preserved in a few places in the amphibolite enclaves in the gneisses. The first, or first two, sets of folds are commonly tight and isoclinal; the youngest set tends to be more open (Plate 2.1a). Early fold sets may be coaxial (generally with north-south axes). The youngest fold set may have axes in different directions variable from area to area (Plate 2.1b). Intersection of the various fold axes yields dome-and-basin or other interference patterns. Steep plunges are characteristic of F1 folds, but refolding has resulted in dispersion of F1 axes from vertical to horizontal. Peak metamorphic conditions were generally synchronous with early stages of folding, commonly the first.

Rock suites of different ages show similar sequences of deformation. For example, the three episodes of deformation recorded in the pre-3,000-m.y. greenstone belts ("Sargurs") and in the younger sequences ("Dharwars") by Chadwick et al. (1978) show approximate parallelism of fabrics (Plate 2.1c). Similarly, Naha et al. (1986) reported parallelism and continuity of structural elements through gneisses, older greenstones, and younger greenstones. An older deformational period in the older sequences, including gneisses, is shown by folded pebbles of gneiss, quartzite, and banded iron formation with various foliation orientations in conglomerates in the younger schist belts (Plate 2.1d, e).

Many of the conclusions drawn by investigators in individual areas are highly controversial. The Chitradurga belt (described later) can be used as an example. Chadwick et al. (1981a) presented a rather simple history of the Chitradurga belt based on sedimentation and volcanism in a subsiding basin and only one major episode of folding. The folds developed in that episode are complex and doubly plunging, and structures have been accentuated by differential uplift and subsidence. This simple structure requires that easily recognized lithologic horizons commonly used for mapping purposes (such as banded iron formations) occur as separate units at various places in the stratigraphic sequence. A simple basin of this type is consistent with gravity models of as much as 10 km thickness of dense rocks in the Chitradurga belt (Subrahmanyam and Verma, 1982).

A more complex view of the Chitradurga belt has been proposed by most investigators, including Naqvi (1973), Mukhopadhyay et al. (1981), Mukhopadhyay and Ghosh (1983), Mukhopadhyay and Baral (1985), and Naha et al. (1986). The major structure of the belt can be viewed as a second-generation antiform refolding an earlier syncline. At least three generations of folding were recognized. First-generation folds are isoclinal and reclined, with axial plane schistosity parallel to bedding except at fold hinges (Plate 2.1a). The F2 folds are coaxial with F1 folds, open to isoclinal, accompanied by development of crenulation cleavage, and cause dispersion of the F1 axes. The F3 folds in the Chitradurga belt are broad warps on the limbs of both F1 and F2 folds. This complex structure requires banded iron formations to be repeated structurally at several places in the sequence and is consistent with gravity models that show less than 2.5 km thickness of mafic rock in the belt (Naqvi, 1973).

The north-south orientation of structures in the Chitradurga belt occurs throughout the Western Dharwar craton and affects rocks of widely differ-

Plate 2.1. (*a*) First generation (F₁) fold hinge showing the intersection of bedding and axial plane schistosity (Chitradurga belt). (*b*) Dome-and-basin structure developed by interference of F_1 and F_2 folds (Bababudan belt). (*c*) Three planes of folding in banded iron formation of Shimoga belt. Similar structures occur in gneisses and older greenstone belts (Plate 2.2e). (*d*) Gneissic pebbles in the conglomerate of the Chitradurga schist belt. One pen is parallel to the gneissosity, and one is parallel to schistosity in the conglomerate matrix. This figure demonstrates deformation of gneisses before incorporation into the conglomerate. (*e*) Folded quartzite pebble in conglomerate of Chitradurga schist belt. Pebble has undergone two deformations before deposition. Pencil indicates schistosity direction in the matrix of the conglomerate.

TABLE 2.2. *Radiometric ages*

Rock Type and Location (Reference)	Method	Age (in m.y.)	Initial $^{87}Sr/^{86}Sr$ or $^{238}U/^{204}Pb$ (μ)	Notes
Gneiss (including migmatites and conformable granitic layers)				
South Kanara (1)	$^{207}Pb/^{206}Pb$	3,174	8.99	Minimum K/Ar of minerals are about 3,100 m.y.
Kudremukh (2)	Rb/Sr	3,280 ± 230	0.7009 ± 0.0012	
North Mysore (3)	Rb/Sr	2,990 ± 120	0.707 ± 0.004	
Chitradurga area (4)	Rb/Sr $^{207}Pb/^{206}Pb$	2,970 ± 100 3,028 ± 28	0.7035 ± 0.0013 7.82	
Between Bababudan and Shimoga belts (5)	Rb/Sr	3,035 ± 60	0.7025 ± 0.001	
South of Bababudan belt (Chickmagalur) (4)	Rb/Sr	3,060 ± 160	0.7015 ± 0.0004	
	$^{207}Pb/^{206}Pb$	3,190 ± 100	8.00	
Between Holenarasipur and Nuggihalli belts (6)	Rb/Sr Rb/Sr	3,000 ± 100 3,020 ± 125	0.70323 ± 0.00064 0.70255 ± 0.00035	Metamorphic ages of earlier-formed rock
Near Holenarasipur belt (three suites) (7)	Rb/Sr	3,162 ± 61 3,139 ± 31 3,071 ± 67	0.7020 ± 0.0010 0.7015 ± 0.0003 0.7008 ± 0.0010	Low-Al gneiss
Near Holenarasipur (7)	Rb/Sr	2,959 ± 24	0.7019 ± 0.0002	High-Al gneiss
Mostly south of Holenarasipur belt (8,4)	Rb/Sr	3,358 ± 66	0.700 ± 0.0004	Ratio below mantle growth curve
	$^{207}Pb/^{206}Pb$	3,305 ± 13	8.00	
In transition zone (9)	Rb/Sr	2,850 ± 50	0.703 ± 0.003	
Around southern part of Closepet Granite (10)	Rb/Sr	3,010 ± 90	0.701 ± 0.001	Cordierite gneisses
Trondhjemites and related diapiric rocks				
Halekote (11)	Rb/Sr	2,977 ± 117	0.7016 ± 0.0005	Part of high-Al suite of Monrad (1983)
Chickmagalur Granite (4)	Rb/Sr $^{207}Pb/^{206}Pb$	3,080 ± 110 3,175 ± 45	0.7013 ± 0.0009 7.98	
Young granites				
Chitradurga (two suites) (12)	Rb/Sr	2,400 2,620		
Chitradurga (3)	Rb/Sr	2,475 ± 85	0.706 ± 0.004	Very large error of initial ratio
Chitradurga (4)	$^{207}Pb/^{206}Pb$	2,605 ± 18	7.68	
Closepet (12)	Rb/Sr	2,380 ± 35	0.7049 ± 0.0014	
Arsikere and Banavar (13)	Rb/Sr	2,310	0.705 ± 0.004	Only three samples
Chamundi (near Mysore) (12)	Rb/Sr	790 ± 60	0.7050 ± 0.0013	
South Kanara (1)	Rb/Sr	2,669 ± 60	0.7056 ± 0.0015	
Dikes in transition zone				
Dolerites (14)	Rb/Sr	2,420 ± 246	0.7012 ± 0.0010	
Alkaline rocks (14)	Rb/Sr	832 ± 40	0.7077 ± 0.0007	K/Ar ages of 805 to 810 m.y.
Greenstone belts				
Gneiss pebbles in Kaldurga conglomerate of Bababudan belt (15)	Rb/Sr	3,250 ± 150	0.702 ± 0.003	

TABLE 2.2. (*continued*)

Rock Type and Location (Reference)	Method	Age (in m.y.)	Initial $^{87}Sr/^{86}Sr$ or $^{238}U/^{204}Pb$ (μ)	Notes
Basalts of Chitradurga belt (two suites) (12)	Rb/Sr	2,345 ± 60 2,385 ± 3.0	0.7048 ± 0.0003 0.7048 ± 0.0003	Initial ratios are high and ages are probably low because of alteration
Basalts at Kudremukh (2)	Sm/Nd	3,020 ± 230	initial $^{143}Nd/^{144}Nd$ = 0.50878 ± 0.00022	Initial ratio is above most postulated mantle growth curves
Silicic volcanic rocks of Shimoga belt (4)	$^{207}Pb/^{206}Pb$	2,565 ± 28	7.46	Apparent crustal contribution

Sources
1. Balasubrahmanyan, 1978.
2. Drury et al., 1983.
3. Venkatasubramanian and Narayanaswamy, 1974c.
4. Taylor et al., 1984.
5. Rajagopalan et al., 1980.
6. Y. J. Bhaskar Rao et al., 1983.
7. Monrad, 1983.
8. Beckinsale et al., 1980.
9. Janardhan and Vidal, 1982.
10. Jayaram et al., 1976.
11. Stroh et al., 1983.
12. Crawford, 1969.
13. Venkatasubramanian and Narayanaswamy, 1974b.
14. Ikremuddin and Steuber, 1976.
15. Venkatasubramanian and Narayanaswamy, 1974a.

ent ages (Krishna Murthy, 1974, 1978; Chadwick et al., 1978; Roy and Biswas, 1979; Mukhopadhyay et al., 1981; Naha and Chatterjee, 1982). This continuing east-west compression has also been found in the Eastern Dharwar craton (including the Cuddapah basin) and in the Eastern Ghats, implying a long-term continuity of stress orientations over a very broad area.

The ages of deformational/metamorphic episodes in the craton are unclear. Pebbles of foliated gneiss from a terrain with ages of 3,000 to 3,400 m.y. occur in younger schist belts (Naqvi and Hussain, 1972), thus demonstrating one or more periods of gneissic metamorphism and deformation at least as old as 3,000 m.y. (Plate 2.1d). Janardhan and Srikantappa (1975) proposed that deformation in Sargur enclaves in the southern part of the craton was coincident with emplacement of the gneisses. Conversely, the parallelism and consistency of structures in older and younger greenstone belts and the 3,000-m.y. gneisses implies either that much of the deformation pattern now seen in all rock suites was imposed after deposition of the younger (Dharwar) greenschists or that this younger episode (or series of episodes) obliterated all older structures or formed from stresses with the same orientations as earlier ones. A principal argument against the idea that all structures in the Dharwar craton were imposed less than 3,000 m.y. ago is the scarcity of younger radiometric dates.

RADIOMETRIC AGE DETERMINATIONS

An increasing number of radiometric ages has become available in recent years. Table 2.2 lists ages grouped by rock type. We have eliminated Rb/Sr model ages, which assume an initial $^{87}Sr/^{86}Sr$ ratio, because the assumptions generally introduce errors that are unacceptably large. Where

available, we have given the authors' estimates of error (commonly two standard deviations). Large reported errors indicate scatter of data points or other unreliability. Additional information is given by Ramakrishnan et al. (1984). All of the information in Table 2.2 is based on whole-rock data unless otherwise specified.

The whole-rock data can be supplemented with a small amount of information on mineral ages. Venkatasubramanian et al. (1982) showed a number of vein galena samples on a uniform lead growth curve, ranging in age from about 900 to 3,400 m.y. Monrad (1983) reported Rb/Sr mineral isochrons of about 2,250 m.y. for gneisses near the Holenarasipur belt.

The geochronologic data in Table 2.2 yield a few firm conclusions. The most recognizable event is at 3,000 m.y. and represents a metamorphic age in most gneiss suites. Diapiric magmatic bodies (Chitradurga and Halekote) with the same age show low initial $^{87}Sr/^{86}Sr$ ratios, possibly implying mantle derivation and thereby sialic increment to the crust at that time. Ages of the foliated gneisses range from 3,000 to about 3,400 m.y. Some initial ratios in these gneisses are low and imply new increment to the crust; other ratios are high enough to require crustal residence time for some of the suites prior to metamorphism. Apparently much of the sialic terrain in the Western Dharwar craton evolved in the age range of 3,400 to 3,000 m.y. Extensive intrusion and resetting of isotopic systems did not occur after about 3,000 m.y. ago.

STRATIGRAPHY OF HIGH-GRADE BELTS (ENCLAVES)

With the exception of sparse contacts between different rock types preserved in some enclaves, stratigraphic relationships are impossible to establish in the high-grade fragments (Sargurs) in the southernmost part of the Western Dharwar craton (Janardhan and Srikantappa, 1975; Janardhan et al., 1978; Viswanatha and Ramakrishnan, 1981a; S. Viswanathan et al., 1981; Venkataramana, 1982). The best indication of the stratigraphy, therefore, is simply a list of rock types occurring in the Sargur suite. A map of part of the area is shown in Figure 2.3. Eight principal rock types are present.

Quartzites consist predominantly of fine-grained, interlocking quartz grains with a number of varietal minerals such as sillimanite, kyanite, garnet, corundum, fuchsite, cordierite, graphite, and paragonite.

Mica schists are commonly referred to as "metapelites." Mineralogy consists of some mixture of biotite, muscovite, kyanite, staurolite, garnet, corundum, and graphite. Rocks with high percentages of kyanite and staurolite are com-

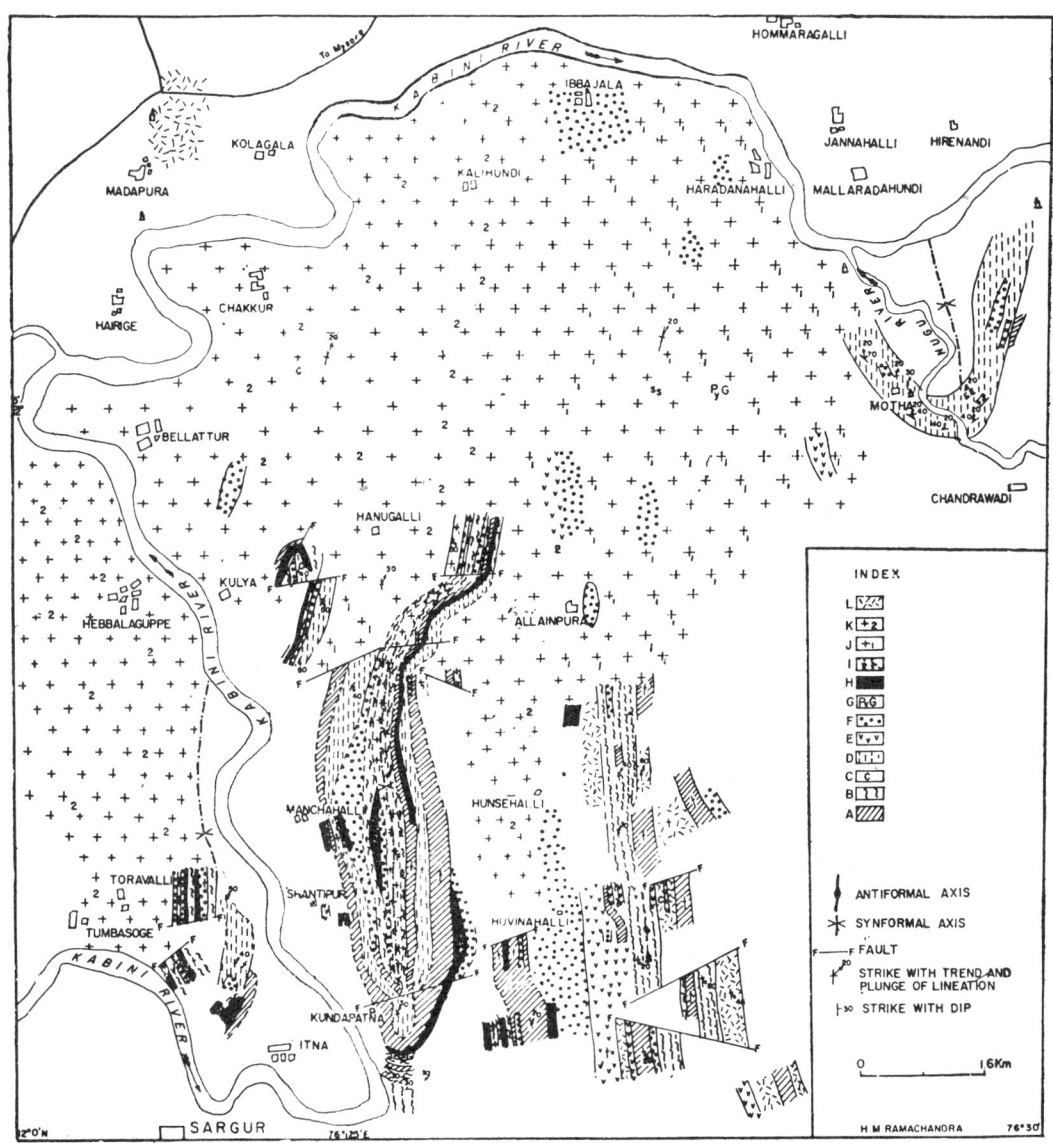

Fig. 2.3. Sargur enclaves (Janardhan et al., 1979). A, quartzite (fuchsite-bearing and commonly ferruginous); B, kyanite-sillimanite-graphite schist; C, calcareous rocks; D, banded magnetite quartzite; E, amphibolite/hornblendite; F, ultramafic rocks (serpentinite/pyroxenite/hornblendite); G, two-pyroxene granulite; H, pyribolite; I, kyanite-corundum-mica gneiss; J, gneiss; K, banded gneiss; L, pink gneiss/porphyritic granite. Reproduced with permission from the Journal of the Geological Society of India, v. 20, p. 67.

mon, indicating an excess of Al_2O_3 over total alkalies.

Calc-silicate rocks presumably are metamorphosed calcareous sediments. The major minerals are calcite, garnet, amphibole, diopside, anorthite, dolomite, and quartz. Magnesian varieties also contain talc, serpentine, and/or phlogopite.

Iron formations are largely quartzites rich in iron oxides and iron silicates. The major iron-bearing mineral is magnetite, and other varieties are grunerite, hedenbergite, cummingtonite, almandine, and Fe-rich orthopyroxene.

Manganese-rich rocks occur only locally, interbedded with quartzites and calc-silicate rocks (Janardhan et al., 1981). Principal manganiferous minerals are spessartite and Mn-bearing clino- and orthopyroxenes. Other major minerals are quartz and feldspars.

Amphibolites contain hornblende and plagioclase, commonly with abundant garnet (mostly almandine and pyrope). Clinopyroxene is present in some rocks. The suite is probably isochemical with the mafic granulites of the same terrain.

Mafic granulites consist primarily of plagioclase, ortho- and clinopyroxene, garnet, $\pm$ amphibole. Some of the granulites may be part of layered or other intrusive bodies. Conversely, some mafic granulites may represent higher-grade equivalents of paraamphibolites.

Ultramafic/anorthositic rocks are grouped together although it is not clear that they are all cogenetic. The high-grade enclaves include metamorphosed ultramafic assemblages with variable proportions of olivine (generally fo_{88-92}), serpentine, orthopyroxene, chromite, actinolite, anthophyllite, talc, carbonate, and chlorite. Dominant rock types include harzburgite and its serpentinized equivalent. Metapyroxenites are also abundant. Chromite pods occur in the ultramafic bodies. Some anorthosites show cumulate texture and are probably fragments of layered complexes; other suites (e.g., at Konkanhundi) may be massif anorthosites emplaced after major deformation. The ultramafic rocks generally show evidence of local cross-cutting and contact metamorphism of wall rocks. Thus, apparently the ultramafic/anorthositic rocks all represent intrusive suites. No definite evidence for komatiites or other ultramafic volcanic rocks has been found.

The very generalized "stratigraphy" of the high-grade enclaves is a succession of metasedimentary rocks (mica schist, quartzite, calc-silicate, iron formation, and manganese-rich rocks) and basaltic metavolcanic rocks (amphibolites, mafic granulites) intruded by ultramafic/anorthositic layered complexes. This suite was then disrupted by gneiss emplacement, deformation, and metamorphism. Some of the anorthosites, and possibly associated rocks, may have been emplaced toward the end of tectonic activity. Relative abundances of the nongneissic rock types are shown in Table 2.3.

STRATIGRAPHY OF HIGH-GRADE BELTS (COHERENT)

Stratigraphic relationships in high-grade greenstone belts are highly controversial. The major belts (Fig. 2.2) are Nuggihalli (Ramakrishnan, 1981; Jafri et al., 1983), Krishnarajpet (Divakara Rao et al., 1983), and the southern part of Holenarasipur (Naqvi et al., 1978b; Ganguli and Sen, 1980; Ramakrishnan and Viswanatha, 1981a; Hussain et al., 1982; Hussain and Naqvi, 1983). High-grade ("Sargur-type") rocks have also been described near the base of stratigraphic sequences in generally low-grade belts (Bababudan and

TABLE 2.3. *Proportions of rock types in supracrustal assemblages*

Rock Type	Sargur Enclaves	Holenarasipur Group	Javanahalli Group	Bababudan Group	Chitradurga Group
Ultramafic/anorthositic rocks	35	33–95	Minor	Minor	Minor
Mafic schists and granulites	15	20–40	Minor	60–70	20
Iron and manganese formations	5	Minor	2–3	10	2–3
Calc-silicates and carbonates	5		2–3	Minor	Abundant
Mica schists	35	2–36	70	20–25	10–20
Quartzite and cherts	5	2–3	4–6	Minor	2–3
Graywackes					70–80
Arkoses			10		

Abundances of rocks in enclaves are estimated from maps of Janardhan and Srikantappa (1975), Janardhan et al. (1979), and Viswanatha and Ramakrishnan (1981a). Abundances in Holenarasipur, Javanahalli, Bababudan, and Chitradurga suites are from Naqvi (1983).

Ghatti Hosahalli). Relative abundances of the various rock types in the belts are shown in Table 2.3.

The southern part of the Holenarasipur belt (Fig. 2.4; Table 2.4) provides a good example of the controversy concerning high-grade schist belts. The principal differences between interpretations center around two issues, the ultramafic rocks and the conglomerates. Hussain and Naqvi (1983) proposed that the majority of the ultra-

mafic suite is extrusive, including komatiite and anorthosite flows intercalated with more normal basalt; Ramakrishnan and Viswanatha (1981a), however, placed most of the ultramafic rocks in intrusive complexes. Chadwick et al. (1978) and Ramakrishnan and Viswanatha (1981a) regarded the Tattekere conglomerate to be an oligomict (quartz-pebble) clastic rock interbedded with other quartzose, including fuchsitic, clastic sediments. Hussain and Naqvi (1983) and Glikson (in

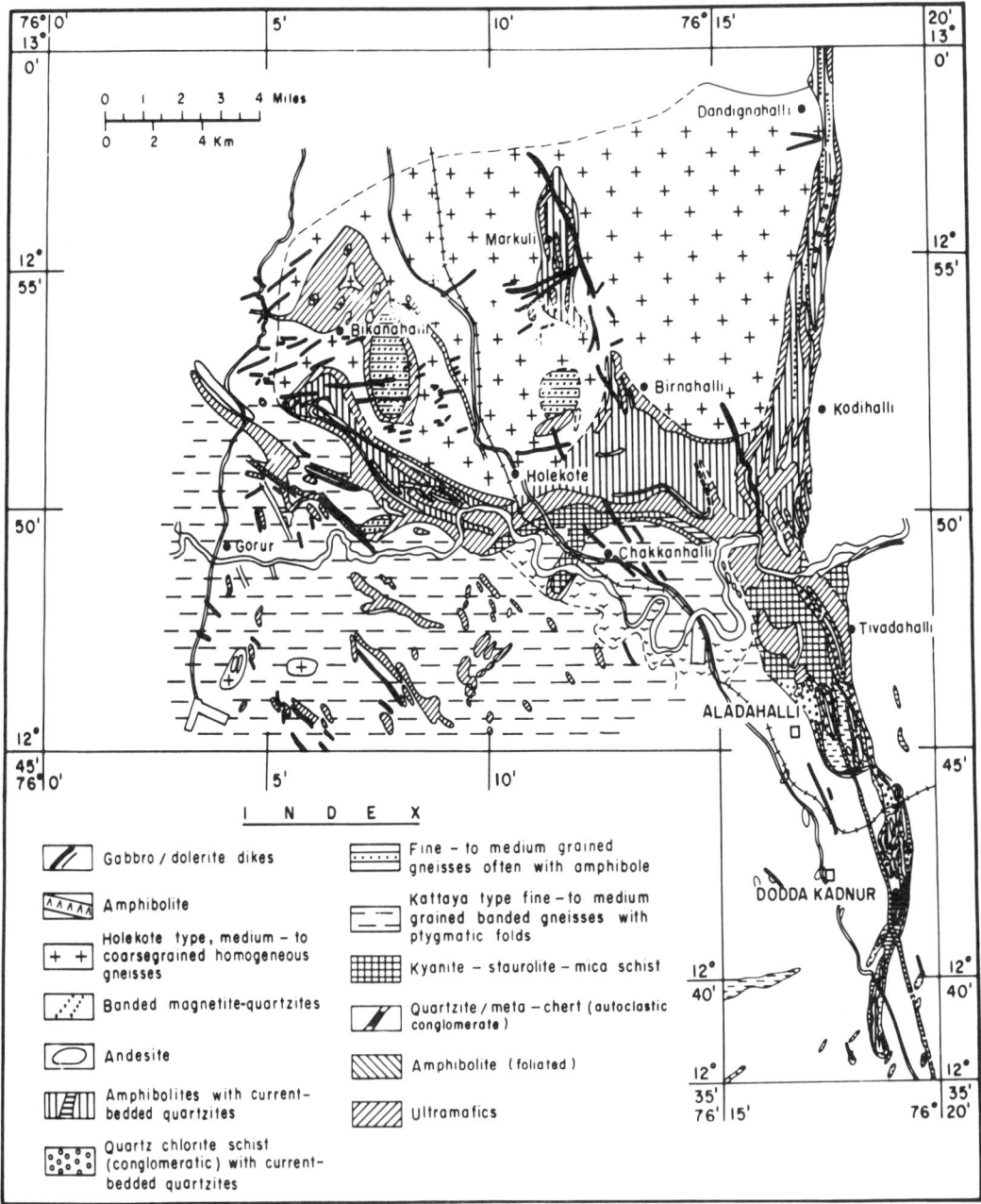

Fig. 2.4. Major part of Holenarasipur schist belt.

TABLE 2.4. *Stratigraphy of Holenarasipur belt*

Ramakrishnan and Viswanatha (1981a)		Hussain and Naqvi (1983)	
Dharwar Supergroup — Bababudan Group	Ironstone, polymictic conglomerate, and chlorite-biotite-garnet schist	Bababudan Group 3,200 m.y.	Banded magnetite quartzite
			Mafic intrusive rocks
	Metavolcanic rocks with quartzite, pelite, and ultramafic rocks		Polymictic conglomerate
			Alternating amphibolites and quartzites
	Kunkumna Hosur quartz pebble conglomerate		Metabasalts and metaandesites
	------------- Unconformity -------------		--------Peninsular Gneisses--------
3,000 m.y.	Development of Peninsular Gneiss	Holenarasipur Group 3,500 m.y.	High-Mg-Al mica schists, with kyanite, staurolite, garnet, chloritoid; associated fuchsite quartzite (recrystallized chert)
	------------- Unconformity -------------		Ultramafic/anorthositic volcanic and intrusive rocks
Sargur Group	Ultramafic/mafic intrusive complex		
	Iron formation (with garnet and grunerite)		
	Paraamphibolite and calc-silicate		
	Garnet-bearing, mafic intrusive (and extrusive?) rocks		
	Tattekere oligomictic conglomerate and associated detrital quartzites		
	Kyanite-staurolite-garnet-biotite-muscovite schist (metapelite)		

Glikson et al., 1979) regarded the conglomerate as tectonic (rodding of predominantly vein quartz) because of the extensive shearing and absence of any clearly sedimentary features; they considered the fuchsite quartzite to be recrystallized chert.

The differences in interpretation are obviously important. A scarcity of ultramafic flows and presence of clastic quartz permit the Holenarasipur belt to be interpreted as having formed on a continental crust. An absence of quartz and abundance of komatiites and related rocks permit the belt to be regarded as having formed on a mafic crust before segregation of continental sial.

PETROLOGY OF ROCKS IN ENCLAVE AND COHERENT HIGH-GRADE BELTS

This section describes the petrology and possible origin of the dominant rock types of the high-grade belts, including probable metasediments (quartzites and ironstones, mica schists, calc-silicate rocks) and probable metavolcanic rocks (amphibolites and mafic granulites). The problem of ultramafic and anorthositic rocks is discussed in a later section.

Quartzites and ironstones

Quartzose rocks are common in enclave suites but rare in coherent high-grade schist belts. They occur in the enclaves as thin layers associated with other presumed metasediments such as cal-careous rocks and aluminous mica schists (meta-pelites?). Iron-rich quartzites (ironstones) are quartz-magnetite or quartz-silicate rocks (containing minerals such as grunerite or hedenbergite). Manganese-rich rocks (containing manganiferous garnet, orthopyroxene, and clinopyroxene with quartz) have been described in one area of the gneiss-granulite transition zone (Janardhan et al., 1981). Varietal minerals in the quartzites include kyanite, sillimanite, fuchsite, and chromite.

There are basically two points of view regarding the origin of the quartzose rocks. One is that the iron- and manganese-rich quartzitic rocks are chemical precipitates, but fuchsitic, kyanitic, and sillimanitic varieties are detrital. The evidence for a detrital origin consists largely of their content of minor heavy minerals, primarily zircon, tourmaline, and rutile (Ramiengar et al., 1978b; Chadwick and others in Glikson et al., 1979; Raase et al., 1983). The presence of aluminous minerals such as kyanite or sillimanite signifies an argillaceous component in the sands.

If the quartzites are detrital, their Cr content has been redistributed during metamorphism. Raase et al. (1983) demonstrated equilibrium distribution of Cr between different mineral phases, which is highly unlikely for a clastic mineral assemblage. Furthermore, the Cr_2O_3 contents reported by Raase et al. for various minerals (up to 6% in kyanite) are too high to be characteristic of these minerals in any likely source rocks. Thus, even if the quartzites are primarily clastic, it is not certain that the chromium was derived from a clastic contribution. Naqvi et al. (1981) proposed enrichment of transition elements in rocks of the Western Dharwar craton by exhalation from the mantle. The principal evidence is the high transition-metal content and uniformity of Cr/Ni ratios in a variety of rocks types (including chert and limestone) in the older and younger schist belts (Table 2.5). The Cr/Ni ratios are similar to those postulated for a primitive mantle and could represent volcanogenesis during sedimentation or later metasomatism.

The second concept for the origin of the various quartzites is that they are cherty chemical precipitates, later recrystallized to quartz. Much of the evidence for this concept is simply the scarcity of quartz in other sedimentary rocks of the high-grade belts (Naqvi, 1978, 1981, 1983; Naqvi et al., 1981). It seems unlikely that clastic quartz from a sialic provenance would accumulate only in scattered quartzose lenses and not become a significant component of other sedimentary rock types. A second line of evidence in favor of chemical precipitation of rocks now occurring as quartzites is the $\delta^{18}O/^{16}O$ ratios of $+11.6$ and $+12.5$ in two samples (Naqvi et al., 1981); these ratios are higher than those typical of quartz in sialic igneous and metamorphic source rocks. Viswanathiah et al. (1979) described small bodies in these rocks in the enclaves (Sargur terrain) that are similar to proposed organic structures from Archean sediments in other areas; if they are organic, their presence supports a chemical or biochemical origin of the siliceous sediments.

Barite

Thin bands and lenses of barite (one meter or less in thickness) are associated with fuchsite quartzite

and quartz-arenite schist at one locality in the Ghatti Hosahalli belt (Radhakrishna and Sreenivasaiah, 1974). They are apparently syngenetic or diagenetic.

Mica schists

Mica schists represent a highly aluminous suite of rocks also characterized by high Mg, Al, and Cr contents (B. Bhaskar Rao and Mohanty, 1982; Naqvi et al., 1983c). The Al_2O_3 and Cr contents are closely correlated. The composition controls the presence of a mineral assemblage that includes biotite, muscovite, kyanite, staurolite, plagioclase, quartz, $\pm$ garnet, $\pm$ fuchsite, $\pm$ graphite, $\pm$ hornblende, $\pm$ corundum, $\pm$ chloritoid, $\pm$ chlorite. Their original sedimentary characteristics cannot be shown by any relic depositional features. Table 2.6 shows compositions of selected suites from both the coherent high-grade belts and the high-grade enclaves. The rare earth element patterns of the Holenarasipur mica schists are shown in Figure 2.5 (Naqvi et al., 1983a). Both positive and negative Eu anomalies are present. Patterns for staurolite and kyanite separates from the schists show high positive Eu anomalies (Eu/ Eu* = 5−10) in both minerals. Naqvi et al. attributed these characteristics to derivation of the sediments from a mafic and plagioclase-rich (partly anorthositic) source, with addition of elements such as Cr by volcanogenic or metasomatic processes.

Amphibolites and mafic granulites

Mafic suites in the enclaves may be different from those in the coherent belts. The mafic granulites in the enclaves contain plagioclase, clinopyroxene, orthopyroxene, small amounts of hornblende $\pm$ garnet. Amphibolites are predominantly pla-

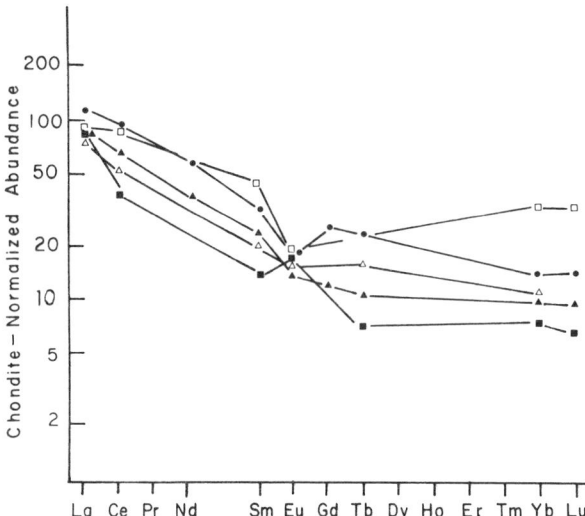

Fig. 2.5. REE in metasediments. Solid squares, Group I of Holenarasipur schist belt; open squares, Group II of Holenarasipur schist belt; open triangles, Group III of Holenarasipur schist belt; solid triangles, Ranibennur graywackes of Shimoga schist belt; solid circles, Ranibennar phyllityes of Shimoga schist belt. Holenarasipur data are from Naqvi et al. (1983c); Ranibennur data are from Naqvi (unpublished).

TABLE 2.5. *Abundances of transition and other elements in metasediments of schist belt*

	Limestone		Fuchsite Quartzite		Chert			Barite	Ironstone		
	Javana-halli Belt	Chitra-durga Belt	Javana-halli Belt	Ghatti Hosa-halli Belt	Pyritiferous (Chitra-durga Belt; Unmineralized)	Pyritiferous (Chitra-durga Belt; Mineralized)	Javana-halli Belt	Ghatti Hosa-halli Belt	Holenar-sipur Belt	Javana-halli Belt	Chitra-durga Belt
Number of Samples	6	3	9	2	3	1	4	6	5	6	2
ppm											
Cu	16	13	22	21	43	2625	19	17	403	42	41
Cr	35	83	230	414	152	45	269	225	221	278	185
Ni	11	36	83	87	144	25	99	72	129	86	167
Co	5	8	10	6	10	10	6	2	40	26	33
Zn	114	158	182	210	90	140	205	35	66	102	92
Pb	36	38	24	10	13	125	4	11	30	19	18
Rb	15	9	19	48	5	—	8	11	<5	19	<5
Sr	97	54	10	10	10	—	10	6700	<10	<5	<10
%											
Mn	0.270	0.270	0.161	0.186	0.220	—	0.175	0.106	0.256	0.242	0.240
Ti	—	—	—	—	0.09	—	—	—	0.068	0.234	0.182
P	0.35	0.71	0.085	0.06	—	—	—	—	0.025	0.026	0.062
Na	0.608	0.546	0.108	0.197	0.731	—	—	—	0.370	1.135	1.087
K	0.248	0.072	0.620	1.49	0.1	—	0.042	—	0.1	0.1	0.1
Ca	—	—	0.415	0.675	0.194	—	0.581	—	0.424	1.493	0.114
Mg	0.708	4.82	0.113	0.162	0.04	—	0.169	—	0.548	0.414	0.011

Abundances of Cu, Cr, Ni, Co, Zn, Pb, Rb, and Sr are in parts per million by weight. Abundances of other elements are in percent by weight. Prepared with permission from Geological Society of Australia Special Publication 7, p. 249.

TABLE 2.6. *Compositions of rocks in high-grade schist belts*

	1	2	3	4	5	6	7	8	9	10	11	12	13	14
SiO_2	54.4	51.7	65.0	56.4	76.6	51.0	48.9	51.6	50.0	53.6	47.2	38.7	46.7	46.4
TiO_2	3.6	0.91	0.45	0.82	0.44	0.82	0.93	0.35	0.69	0.73	0.98	0.08	0.30	0.39
Al_2O_3	29.2	16.3	15.7	27.8	11.4	14.0	11.3	7.1	15.3	11.9	15.4	1.2	27.0	7.0
Fe_2O_3	2.6	11.2	4.8	9.0	4.1	3.0	1.5	10.7	12.2	2.1	2.6	6.7	0.91	7.5
FeO	1.9					8.5	11.3			8.3	7.9	2.1	2.9	3.7
MnO	0.01	0.25	0.10			0.19	0.19	0.22	0.18	0.28	0.06	0.09	0.19	0.16
MgO	5.1	6.0	2.7	1.4	0.81	7.5	12.0	16.1	8.1	7.3	12.6	46.3	3.3	21.6
CaO	0.37	2.2	1.6	0.24	0.19	9.8	9.8	11.7	11.5	11.8	7.8	1.7	15.9	7.2
Na_2O	0.76	1.9	2.3	0.74	0.61	2.1	2.1	0.7	2.1	2.2	4.2	0.60	2.4	1.1
K_2O	1.1	2.5	3.0	2.7	0.86	0.45	0.24	0.04	0.06	0.17	0.34	0.06	0.08	0.07
Rb			130	96	9	6	3			5				
Sr						171	170			202				
Y														
Zr						92		42	43					
V													196	
Cr		975	321	936	138	212	117	1950	253	449		2000	269	1790
Ni		444	144			132	420	587	176	192			120	
Th		1.4	8		9									
U		0.76	3		1.3									

All abundances of major elements are in weight percent of oxides. Where no entry is made in the FeO position, the value at Fe_2O_3 is total iron calculated as ferric oxide. All abundances of trace elements are in parts per million by weight. Values are taken directly from quoted sources and not recalculated to 100%.

Table 2.6. Column References

1. Mica schist in Sargur area (Janardhan et al., 1978).
2. Group 1 mica schist in Holenarasipur belt (Naqvi et al., 1983a, 1983b).
3. Group 2 mica schist in Holenarasipur belt (Naqvi et al., 1983a, 1983b).
4. Group 3 mica schist in Holenarasipur belt (Naqvi et al., 1983a, 1983b).
5. Group 4 mica schist in Holenarasipur belt (Naqvi et al., 1983a, 1983b).
6. Amphibolite in enclave in transition zone (Janardhan et al., 1978).
7. Mafic granulite in enclave in transition zone (Janardhan et al., 1978).
8. High-Mg basalt in Holenarasipur belt (Drury et al., 1983).
9. Tholeiitic basalt in Holenarasipur belt (Drury et al., 1983).
10. Amphibolite in Krishnarajpet belt (Divakara Rao et al., 1983).
11. Hullahalli anorthosite in transition zone (Janardhan and Ramachandra, 1978).
12. Peridotite in enclave in transition zone (Divakara Rao et al., 1975).
13. Olivine-normative anorthosite in Holenarasipur belt (Jafri and Saxena, 1981).
14. Spinifex-textured komatiite in Nuggihalli belt (Jafri et al., 1983).
Additional compositions are available in Hussain et al. (1982) and B. Bhaskar Rao and Vereebhadrappa (1983).

gioclase-hornblende rocks. Their compositions are shown in Table 2.6. Both amphibolites and mafic granulites may represent original basalts metamorphosed to different facies. In high-grade metamorphic terrains, however, it is difficult to distinguish eruptive basalts from gabbroic rocks of fragmented layered complexes, and some of the amphibolites and granulites may be parts of such complexes.

Some rocks in the coherent belts are clearly volcanic. For example, some amphibolites show relic pillow structures, and evidence of metasediments converted to amphibolites has not been found on a major scale (Satyanarayana et al., 1974). Some suites have komatiitic or other ultramafic compositions (e.g., high-Mg basalts in the Holenarasipur belt; Table 2.6). Typical REE patterns are shown in Figure 2.6.

Mafic and ultramafic rocks in the Holenarasipur belt were studied by Hussain et al. (1982). Many of the rocks show spinifex texture, indicative of volcanic origin. Hussain et al. regarded those rocks that exhibit cumulate texture as the basal parts of thick lava flows rather than parts of intrusive layered complexes. Komatiitic compositions (Table 2.6) are preserved with little modification in many rocks, but scatter in the abundances of some elements, particularly Ca, indicates significant postconsolidation metasomatism. A bimodality of MgO contents, with few values in the range of 20% to 30%, indicates generation of at least two distinct magma batches.

Drury (1983a) studied metabasaltic rocks from both high- and low-grade schist belts in the southern part of the Western and Eastern Dharwar cratons. Most of his samples were from rocks classified as the Bababudan Group (Table 2.1). Drury proposed that basaltic rocks from all belts form a single population, indicating either evolution of all suites from a single magmatic event or evolu-

tion of different suites by similar processes. The high MgO content of some rocks was proposed to be the result of large percentages of partial melting of rising mantle diapirs. Drury suggested that vapor transfer accounted for the light REE depletion in some high-Mg basalts (Fig. 2.6) but that removal of mafic minerals accounted for light REE enrichment in more fractionated rocks. Development of negative Eu anomalies by plagioclase fractionation occurred only during development of andesites and dacites. Removal of a spinel phase accounted for depletion of Ti relative to other basalts. Variability of Zr/Y was explained as a further result of vapor-phase transfer. Drury generally regarded the basaltic suite as similar to basalts in back-arc basins above modern subduction zones, with the differences between Dharwar mafic rocks and modern ones attributed partly to a thinner crust in the Archean.

Ultramafic/anorthositic suites

The southern part of the Western Dharwar craton contains a number of outcrops of metamorphic rocks with an ultramafic composition, including pyroxenites, norites, gabbros, and anorthosites. Many of these occurrences are within identifiable schist belts, but some occur simply as bodies within the general gneissic terrain. Many of the rocks form layered intrusive complexes or fragments of former complexes. Other rocks show relic spinifex texture and pillow structures, thus indicating an extrusive origin. Where age relationships can be determined, the layered complexes appear to intrude schist belts prior to major metamorphism and gneiss emplacement, but mafic/ultramafic volcanic suites appear to be the oldest members of schist belts. Compositions of various rocks are shown in Table 2.6.

Ultramafic rocks consist of small bodies of un-

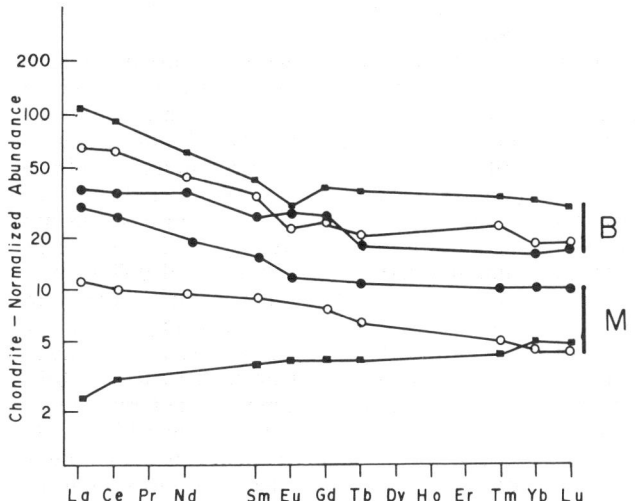

Fig. 2.6. REE in metavolcanic rocks. B samples are from Bababudan belt (Y. J. Bhaskar Rao and Drury, 1982): solid circles, Lingadahalli basalt; open circles, Santavery basalt; solid squares, Santavery andesite. M samples are high-Mg basalts (Drury, 1983a): solid circles, Kudremukh (9.2% MgO); open circles, Sigegudda (22.0% MgO); solid squares, Holenarasipur (25.9% MgO).

altered dunites, harzburgites, serpentines, and tremolite-actinolite schists ± talc ± chlorite. Chromite pods and bands are common in the ultramafic rocks, and cumulate texture has been reported in the Nuggihalli belt (Jafri et al., 1983; Nijagunappa and Naganna, 1983). In the high-grade enclaves and other parts of the gneiss-granulite transition zone, the percentage of anorthosite is relatively high (as much as 50% in some complexes). In the coherent schist belts, however, anorthosite is a very minor component in comparison with metaultramafic rocks (Ramakrishnan et al., 1978; Venkataramana et al., 1984). Spinifex texture and pillow structure occur in the ultramafic rocks of several of the coherent schist belts (Naqvi, 1981; Chalokwu and Sood, 1983; Divakara Rao et al., 1983; Hussain and Naqvi, 1983; Jafri et al., 1983) but not in enclave suites.

At least one comparatively undeformed layered intrusive body is present in the area. The Konkanhundi massif (Fig. 2.2) has an oval-shaped outcrop area of about 50 km^2, consists mostly of layered gabbros with less abundant norites and anorthosites, and has abrupt contacts with country rock, although it does not show chilled margins (Ramakrishnan and Mallikarjuna, 1976). Ultramafic rocks are uncommon at Konkanhundi. The body does not show granulite-facies metamorphism (e.g., development of garnets) and may be younger than complexes that have been fragmented by emplacement of the gneisses.

Some highly fragmented outcrops show evidence of being former layered intrusions. Criteria listed by Srikantappa et al. (1980) at Sinduvalli, in the gneiss-granulite transition zone, include layered masses of spinel-bearing dunite, harzburgite, peridotite, and pyroxenite; cumulate texture of the chromite; chromite grains surrounding and occluding olivine grains; and low Al and high total Fe content of the chromites. Anorthosites may be somewhat less abundant in layered complexes in the gneiss-granulite transition zone than they are farther south in the Granulite terrain (Janardhan and Ramachandra, 1978). In addition to the greater abundance of anorthosite, more southerly suites show more calcic plagioclase, presence of garnet, and the occurrence of chromite in ultramafic rocks instead of in anorthosites.

Anorthositic flows in the Holenarasipur belt have been proposed by Naqvi and Hussain (1979), with further chemical data from Drury et al. (1978). The flows form sheets up to 20 m thick, are very fine grained, are interlayered with spinifex-textured komatiites, and contain chromite only in veinlets (not in layers). The mineralogy of these anorthositic sheets is plagioclase (An$_{92-95}$), hornblende, ± garnet, ± quartz. There is no evidence of differentiation of the anorthosite from the surrounding ultramafic rocks. Jafri and Saxena (1981) described an anorthositic dike in the Nuggihalli belt. Titaniferous magnetite layers, some rich in vanadium, are intercalated with the ultramafic/mafic assemblages of some coherent belts (e.g., Nuggihalli; Vasudev and Srinivasan, 1979).

Tremolite/actinolite schists of the coherent belts range from ultramafic (Table 2.6) to basaltic. Rocks of basaltic composition tend to form tremolite/actinolite assemblages when subjected to the conditions that produce serpentine in more ultramafic rocks. A komatiitic nature of the rocks before metamorphism is commonly indicated by REE patterns (Drury, 1981, 1983a).

In summary, we can make the following tentative statements. Layered complexes and fragments of layered complexes are clearly present in the gneiss-granulite transition zone. It is not clear that similar complexes also occur in the coherent schist belts to the north. Conversely, ultramafic and related volcanic rocks are present in the coherent belts. Their presence in the transition zone to the south has not been proven. Associated mafic rocks definitely include basaltic flows in the coherent belts and gabbros/norites in the transition zone. The occurrence of flows in the transition zone and major intrusions in the coherent belts has not been proven.

COMPARISON OF COHERENT HIGH-GRADE BELTS, HIGH-GRADE ENCLAVES, AND SUITES IN GRANULITE TERRAINS

The rock types in the coherent belts, the enclaves in the amphibolite terrain, and the mafic suites in the Granulite terrain (Ch. 4) show many similarities. It is possible to regard the three suites as essentially identical (e.g., Swami Nath and Ramakrishnan, 1981), which indicates a rather simple evolutionary history for southern India. In this history, suites of similar rocks formed throughout the area and are now exposed at different levels, and at different degrees of metamorphism, in the various terrains.

A contrary view is that the nature of the mafic and supracrustal suites is different in different areas. This conclusion is based on two principal observations. The first is that ultramafic and anorthositic rocks may be predominantly volcanic in the coherent belts and predominantly intrusive in the enclave suite, with anorthosites constituting a much higher percentage of the suite in the Granulite terrain. A second set of differences is that the sedimentary suite in the coherent amphibolite-facies belts is dominated by mica schists (metapelites?), whereas these rocks are less abundant in the enclaves than in the coherent belts, and the suite is very minor in the Granulite terrain. Conversely, the highly siliceous khondalites (Ch. 4) are far more abundant in the Granulite terrain than any equivalent siliceous suite in the amphibolite-facies area. Rogers (1986) attributed these differences to evolution of the old supracrustal suites in a stable platformal environment in the Granulite terrain and in an unstable, more ensimatic environment in the Western Dharwar craton.

STRATIGRAPHY OF LOW-GRADE GREENSCHIST BELTS

Radhakrishna (1983) proposed subdivision of the low-grade schist belts into three basins: Chitradurga, Shimoga (including Bababudan, Kudremukh, and North and South Kanara areas), and Sandur (Fig. 2.2). Present exposures apparently conform to the former limits of the basins. Generalized stratigraphic terminology for the belts is shown in Table 2.1. Proposed age relationships between the various schist belts and the gneissic

terrain were summarized by Naqvi (1983; Fig. 2.7). We have chosen to discuss the stratigraphy of five belts in which there has been considerable work and controversy: Chitradurga, Bababudan, Shimoga, Javanahalli, and the northern part of the Holenarasipur belt (Fig. 2.2). Several other belts are mentioned briefly.

Chitradurga belt

The geology of the 450-km-long Chitradurga belt (Fig. 2.8) is highly controversial; (see Special Issue of the Journal of the Geological Society of India, 1985, v. 26, no. 8). A variety of terms has been applied to different parts of the belt, including Chitaldrug, for the central part; Gadag, for the northern part; Kibbenhalli for a northwesterly trending arm on the western side; and Chikkanayakanhalli, for the southern part. Crawford

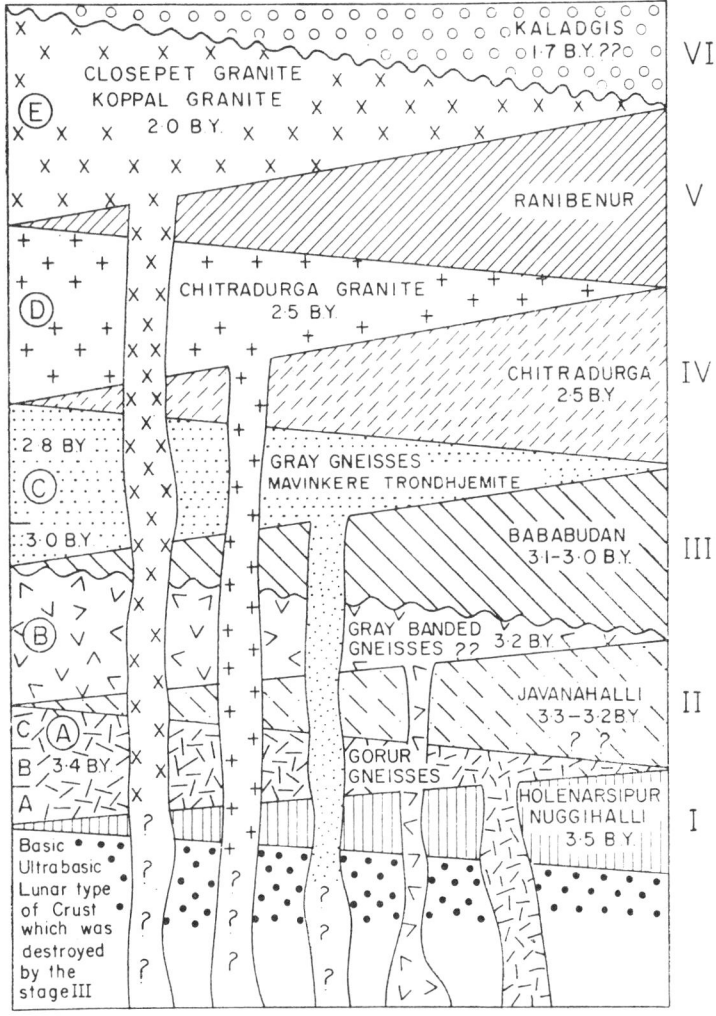

Fig. 2.7. Age relationships of gneisses and schists (Naqvi, 1983).

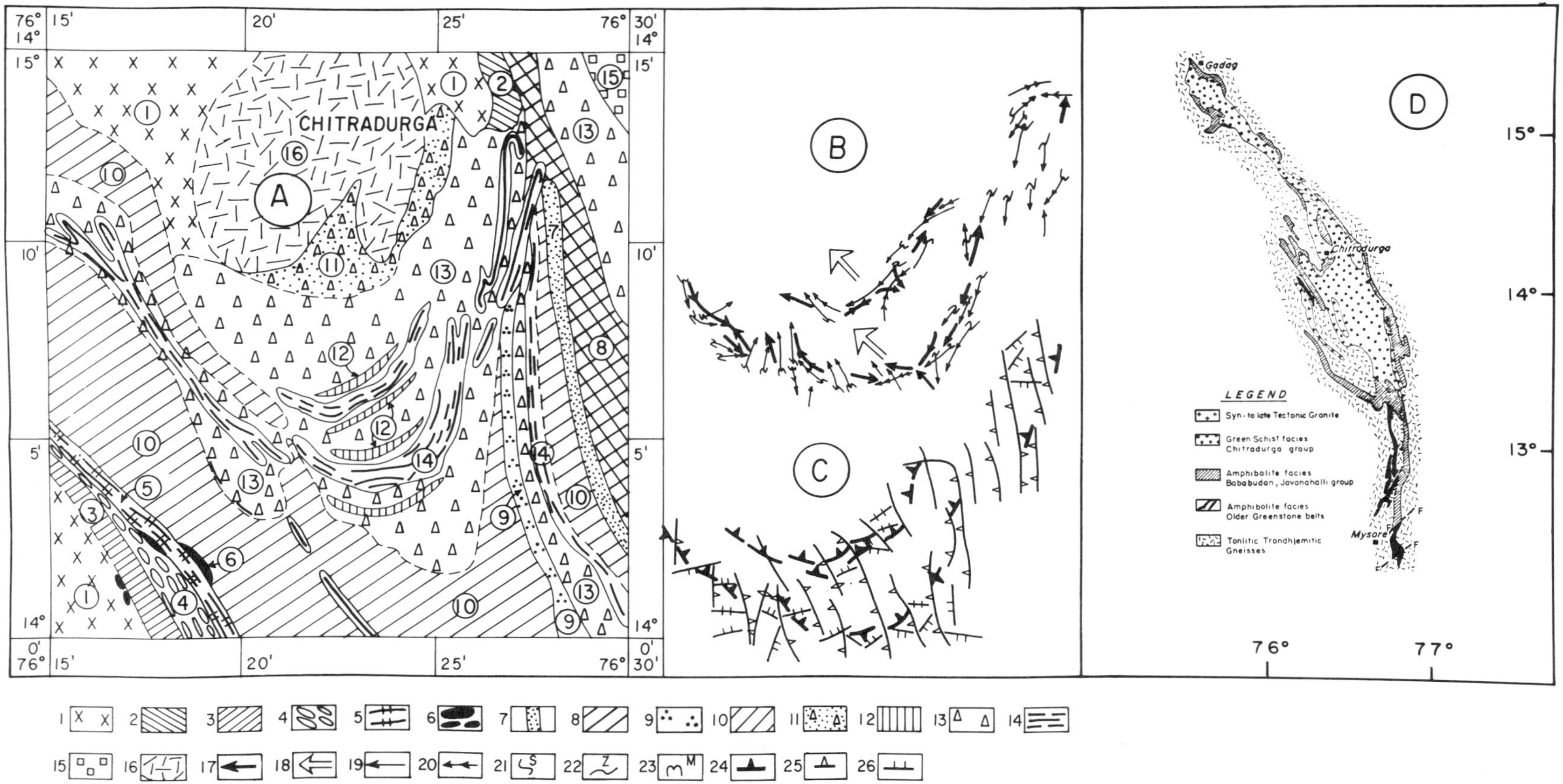

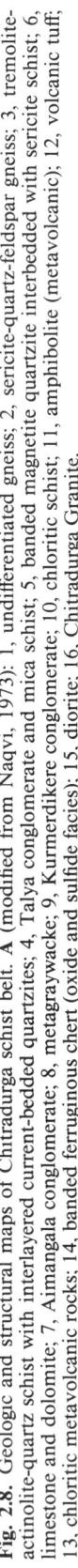

Fig. 2.8. Geologic and structural maps of Chitradurga schist belt. **A** (modified from Naqvi, 1973): 1, undifferentiated gneiss; 2, sericite-quartz-feldspar gneiss; 3, tremolite-actinolite-quartz schist with interlayered current-bedded quartzites; 4, Talya conglomerate and mica schist; 5, banded magnetite quartzite interbedded with sericite schist; 6, limestone and dolomite; 7, Aimangala conglomerate; 8, metagraywacke; 9, Kurmerdikere conglomerate; 10, chloritic schist; 11, amphibolite (metavolcanic); 12, volcanic tuff; 13, chloritic metavolcanic rocks; 14, banded ferruginous chert (oxide and sulfide facies); 15, diorite; 16, Chitradurga Granite. **B** (redrawn from Mukhopadhyay and Baral, 1985): trend map of linear structures: 17, fold axes of first phase; 18, elongation lineation of second phase; 19, fold axes and other linear structures of second phase; 20, fold axes of third phase; 21, 22, 23, shapes of minor folds. **C** (redrawn from Mukhopadhyay and Baral, 1985): trend map of secondary planar structures: 24, axial planes of first phase; 25, axial planes of second phase; 26, axial plane cleavage of third phase. **D** (prepared by Naqvi): Map of entire Chitradurga belt.
Parts B and C reproduced with permission from the Journal of the Geological Society of India, v. 26, pp. 550 and 551.

(1969) measured Rb/Sr whole-rock ages of 2,345 m.y. and 2,385 m.y. on basalts in the belt (Table 2.2). These ages are probably lower than true ages, which are presumably older than the age of intrusion of the posttectonic Chitradurga Granite, dated at about 2,500 m.y. (Table 2.2).

Within the belt itself there is no assemblage of older ("Sargur-type") rocks such as ultramafic flows or layered igneous complexes. Rocks of higher metamorphic grade occur in the Javanahalli belt to the east and in other smaller belts in the area. One of these smaller belts is Ghatti Hosahalli, west of Chitradurga, which contains komatiite (Narayana and Naqvi, 1980; Chalokwu and Sood, 1983). Seshadri et al. (1981) regarded these more mafic rocks as a lower stratigraphic level to the Chitradurga suite. Komatiites have also been described in the Kibbenahalli part of the belt (Srikantia and Bose, 1985).

Mukhopadhyay et al. (1981) described stratigraphic differences between the western, central, and eastern parts of the belt (Table 2.7). Drury (1983b) also proposed differences between the eastern and western sides of the belt, with the eastern side being in thrust contact with an overlying block containing the Javanahalli belt and the western side being at least partly a depositional unconformity. Drury explained some of the complexity of the belt by a series of listric, east-dipping, normal and thrust faults. Chadwick et al. (1981a) and Seshadri et al. (1981) regarded the belt as a simply deformed, thick, volcano-sedimentary accumulation. Naqvi (1973) and Mukhopadhyay and Ghosh (1983) stated that several of the layers of banded iron formation proposed by Chadwick et al. (1981a) as separate sedimentary horizons are actually refolded parts of the same horizon. A stratigraphic section based on

this more complex structure is shown in Table 2.7.

Radhakrishna (1983) summarized stratigraphic relationships of iron formations in the Chitradurga basin. Magnetite-quartz, with iron silicates, is common around the basin margin and on structural highs within the belt. Somewhat more rapidly subsiding areas accumulated an iron oxide-carbonate suite with chert and some manganese oxides. Deep parts of the basin (toward the south-central part) contain pyrite associated with chert and carbonaceous shale. The oxide and sulfide facies of the banded iron formations were recognized as two separate stratigraphic horizons by Naqvi (1967).

Conglomerates appear at various positions in the Chitradurga section (Naqvi et al., 1978a). With the exception of the Neralakette quartz pebble conglomerate, all of them are polymictic, containing fragments of the older gneiss-greenstone terrain.

Bababudan belt

The stratigraphy of the Bababudan belt is somewhat less controversial than that of the Chitradurga. The basal sequence contains a quartz-pebble conglomerate and other siliceous sediments. The quartzites commonly show horizontal axial planes and both upright and inverted current bedding (Plate 2.2a). Similar rock types occur upward in the section intermixed with graywackes, carbonate rocks, cherts, and polymictic conglomerates. Both siliceous and mafic volcanic rocks also occur at various stratigraphic levels, and some of them show columnar jointing. Kyanite-bearing mica schists and metamorphosed ultramafic rocks are present locally.

TABLE 2.7. *Stratigraphy of Chitradurga belt (Mukhopadhyay et al., 1981)*

		Western Margin	Central Belt	Southern Belt
Chitradurga Group			Phyllites and schists (metapelites) and graywackes	Phyllites and schists (metapelites) and graywackes; includes Kurmerdikere and Aimangala conglomerates, ferruginous quartzite, and volcanic rocks
		Marble Banded ferruginous quartzite Talya conglomerate		Basic and minor acid volcanic rocks alternating with banded ferruginous quartzite and associated phyllite
		------------------------------------Unconformity ?------------------------------------		
Bababudan Group		Metabasic rocks and intercalated quartzite and schist Neralakette conglomerate		
		------------------------------------Unconformity ?------------------------------------		
		Peninsular Gneiss		

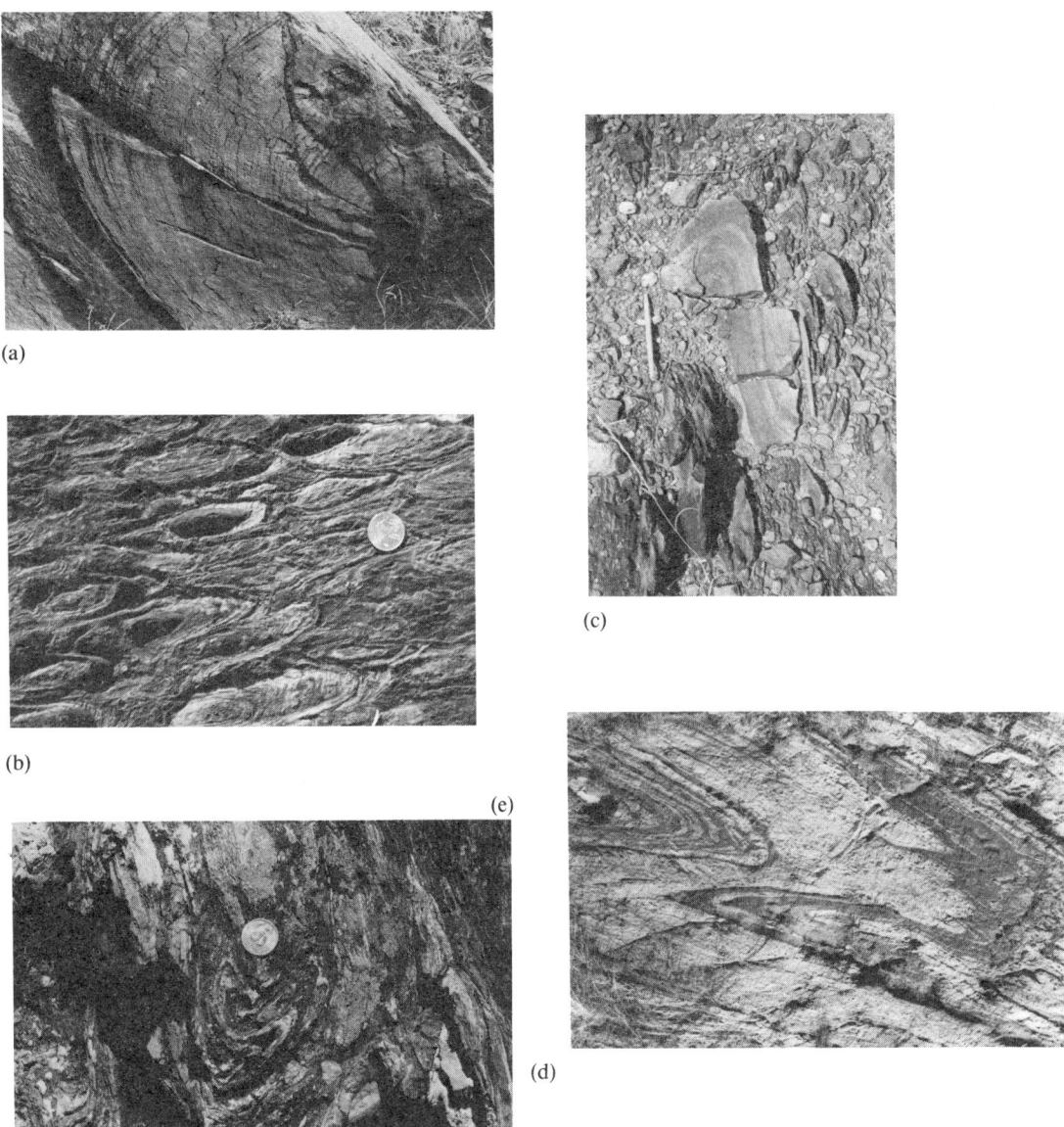

Plate 2.2. (*a*) Hinge of F₁ fold with horizontal axial plane in Bababudan schist belt, 11 km north of Chickmagalur. Inversion of current bedding is seen in the overturned limb. (*b*) Stromatolites from Sandur schist belt, showing earliest evidence of life in Western Dharwar craton. The Sandur schist belt has been correlated with rocks of the Bababudan Group (classification of Radhakrishna, 1983). Correlative rocks in the Kudremukh belt have been dated as 2,900 m.y. old. (*c*) Folded pebble of banded iron formation from polymictic conglomerate of Chitradurga schist belt. Folding occurred before deposition and represents a pre-F₁ event. Fractures within the pebbles were formed by brittle deformation at the time of development of schistosity in the matrix of the conglomerate. (*d*) Isoclinal, reclined F₁ folds in Peninsular Gneisses near Mysore. This photograph shows similarity of structures in gneisses and schist belts of all ages. (*e*) Type-3 interference pattern in supracrustal rocks of Sargur area, indicating similarity of structures in older greenstone belts, gneisses, and younger greenstone belts [see Plate 2.1 (c)].

The Bababudan belt has been regarded as older than much of the Chitradurga (e.g., Viswanatha and Ramakrishnan, 1981c), although the correlations are controversial. The basal rocks of the Bababudan belt are considered to be younger than the Chickmagalur Granite of 3,000 m.y. age by Chadwick et al. (1981b, 1985) and Taylor et al. (1984). Naha et al. (1986), however, found evidence of intrusion of at least parts of the Chickmagalur Granite into basal members of the belt.

Venkatasubramanian and Narayanaswamy (1974a) dated gneiss pebbles in the Kaldurga conglomerate of the belt as being about 3,250 m.y. old (Table 2.2).

Shimoga belt

The Shimoga belt is a comparatively unmetamorphosed sequence of greenschist-facies rocks resting on older gneiss and mafic/ultramafic enclaves (Harinadha Babu et al., 1981). The graywackes and other rock types of the Shimoga belt are commonly referred to as the Ranibennur Group (Iyengar, 1976; Naqvi, 1981). Intrusions of gabbro/anorthosite and ultramafic rocks occur in the belt, and volcanic rocks include basalt, andesite, and dacite in an apparent calcalkaline assemblage (Ziauddin et al., 1978). Manganese-rich horizons were described by Krishna Rao et al. (1982). Some conglomerates in the Shimoga belt are referred to as equivalents of the Kaldurga conglomerate of the Bababudan belt. A study of the Kaldurga and other conglomerates in the Shimoga belt, south of gneiss outcrops in the Tarikere Valley (Fig. 2.1), showed that gneiss pebbles in the conglomerate appear to contain less Ti and Mg than the presumably underlying gneiss (Ranganna et al., 1981).

Northern part of the Holenarasipur belt

Low-grade rocks occur in the northern part of the Holenarasipur belt, where they are stratigraphically above the high-grade suite to the south. The upper sequence consists primarily of mafic volcanic rocks (now amphibolites) and siliceous sediments (sandy and argillaceous) (Table 2.4; Fig. 2.4). Current-bedded quartzites occur only in the northern part of the belt, where they rest directly on ultramafic schists with which they show parallel deformational patterns. Naqvi (1981) and Ramakrishnan and Viswanatha (1981a) both considered the nothern sequence of the Holenarasipur belt to be equivalent to rocks of the Bababudan Group (Table 2.1).

As an example of different interpretations of rock types and the stratigraphic consequence, we mention the controversy over the Kunkumna Hosur conglomerate. Ramakrishnan and Viswanatha (1981a) interpreted this conglomerate as representative of an angular unconformity that separates older Sargur sequences from younger rocks of the Dharwar Supergroup (Table 2.1). The angular unconformity was regarded as marking the emplacement of Peninsular Gneisses about 3,000 m.y. ago. Naqvi (1981) and Naha et al. (1986), however, did not find evidence of an (angular) unconformity in this area and showed evidence that the Kunkumna Hosur conglomerate is a tectonic feature, formed by deformation and quartz veining. Their evidence is based on complete gradation of the older sequence of rocks, beneath the conglomerate, upward into apparently conglomeratic material.

Javanahalli belt

The Javanahalli belt is on the eastern side of the Chitradurga belt, with the two belts possibly separated by a major thrust (up on the east side). Metamorphism is generally to amphibolite grade, but the volcanic/sedimentary assemblage is more siliceous than in high-grade belts farther south. Seshadri et al. (1981) regarded the Javanahalli belt as an older part of the Chitradurga sequence. Naqvi et al. (1980) and Narayana et al. (1983) considered the Javanahalli belt to be intermediate in rock type, degree of metamorphism, and possibly age between low-grade belts, such as Chitradurga, and high-grade belts, such as Nuggihalli.

The rock types in the Javanahalli belt are as follows (Narayana et al., 1983): metaarkoses (paragneisses) recognized by an absence of inclusions, lack of gneissic banding, lack of migmatite, and interbedding with calc-silicate rocks and quartzites; quartzites, with or without fuchsite, magnetite, and epidote; calc-silicate rocks and limestones, containing such metamorphic index minerals as diopside, actinolite, garnet, and clinozoisite; amphibolites that are largely schistose, interbedded with quartzites and calc-silicate rocks, and grade into banded iron formations; tremolite/actinolite schists (ultramafic rocks) as scattered lenses stratigraphically below the fuchsite quartzites; and cordierite-anthophyllite-garnet rocks, which seem to have formed by a volcanogenic process (Narayana et al., 1982).

Other belts

Schists occur along the western coast of Karnataka (in the North Kanara area) and north to Goa, but no particular name has been assigned to them. Possible southern correlatives have been invaded by the South Kanara batholith (Balasubrahmanyan et al., 1976; Balasubrahmanyan, 1978), with an age of about 2,700 m.y., and komatiites have been reported in the area (Dessai and Deshpande, 1979). Sivaprakash (1983) determined metamorphic conditions of 300° to 400°C and 2 to 4 kb in the northern part of the area, which indicates greenschist facies. Many horizons are manganiferous (Sawkar, 1980).

The Kudremukh (Western Ghats) belt contains an apparently older assemblage of ultramafic, anorthositic, and mafic rocks (layered complexes and komatiites) with mica schists overlain by graywackes, carbonates, and mafic to acid volcanic assemblages (Krishna Rao et al., 1977; Drury, 1981; Ramakrishnan and Harinadha Babu, 1981). Mafic metavolcanic rocks near the base cf the belt yielded an Sm-Nd data of about 3,000 m.y. (Table 2.2).

The Sandur schist belt (Bellary District) has a lithologic assemblage broadly similar to those of other low-grade belts (Table 2.8; Krishna Murthy, 1974). A. Roy and Biswas (1979) showed metamorphic conditions ranging from greenschist facies in the center of the belt to amphibolite facies on the margins. Harris and Jayaram (1982) determined metamorphic equilibration conditions of 2.6 to 3.8 kb and 570° ± 50°. Krishna Murthy (1974) showed that several generations of folding in the Sandur belt were consistent with continued east-west compression (Fig. 2.9). Murthy and Krishna Reddy (1984) reported stromatolites (Plate 2.2b).

The Sigegudda belt is quite small (Fig. 2.2). Ramakrishnan and Viswanatha (1981b) and Viswanatha et al. (1982) described an older, partly ultramafic, sequence separated by structural unconformity from a younger sequence. Naha et al. (1986) found that both the older and younger sequences have similar deformation patterns and attributed the apparent angular discordance to interaction of bedding planes with foliation.

PETROLOGY OF ROCKS IN LOW-GRADE GREENSCHIST BELTS

Some of the rock types in the low-grade belts are similar to those described previously for the high-grade belts, and they are not discussed further in any detail. For example, minor anorthositic and ultramafic suites are similar to those in the belts designated as high grade.

Metavolcanic rocks

Most of the volcanic rocks in the low-grade greenschist belts are mafic, although significant amounts of silicic rocks occur locally. Mafic flows constitute about 10% to 20% of the assemblage in the low-grade schist belts, whereas mafic and ultramafic volcanic rocks constitute nearly 50% (or more) of the coherent high-grade belts (Table 2.3). Both pillow structures and columnar structures have been found in different areas, and eruption apparently occurred under a variety of conditions. Welded tuffs occur in the Chitradurga belt (Philip, 1977). Massive, coarse-grained amphibolites in the Holenarasipur and Javanahalli belts are generally regarded as metavolcanic, whereas schistose, finer-grained amphibolites may be me-

tasediments (Narayana et al., 1983). The metabasalts now commonly consist of a mineral assemblage of amphibole and plagioclase ± quartz, ± garnet, ± epidote as major mineral phases.

Compositions of various presumed metabasalts are shown in Table 2.9. Discriminant function analysis by Yellur and Nair (1978) placed different samples as ocean floor basalts, low-K tholeiites, and calc-alkali basalts. The wide diversity of presumed environments may indicate the dominant control of posteruptive processes on the compositions of the rocks. Rare earth element patterns for samples of the Bababudan suite (Y. J. Bhaskar Rao and Drury, 1982; Fig. 2.6) and the Chitradurga suite (Murali et al., 1979) show moderate light REE enrichment.

Several conclusions have been drawn from petrologic studies of the basalts. Drury (1983a) emphasized the coherent relationship between the basaltic and more siliceous rocks in the low-grade schist belts and the mafic and ultramafic (high-Mg) rocks that occur primarily in the high-grade belts. He regarded the more siliceous magmas as products of crystal fractionation by removal of pyroxene, spinel, and plagioclase from the high-Mg magmas. Naqvi (1972) and Naqvi and Hussain (1973a, 1973b) proposed that basalts of the Chitradurga belt had been derived from mantle enriched in Cr and related elements relative to comparable modern mantle. The volcanic suite shows no evidence of crustal contamination.

The mafic and more silicic volcanic rocks of the Bababudan belt (Y. J. Bhaskar Rao and Naqvi, 1978; Y. J. Bhaskar Rao and Drury, 1982) show a flat heavy REE pattern and moderate light REE enrichment, probably indicating a mantle source enriched in lithophile elements. Y. J. Bhaskar Rao and Drury proposed that an increase in negative Eu anomaly with increasing whole-rock SiO_2 content indicated plagioclase fractionation in the more siliceous melts. Fractionation of the more mafic melts, however, was restricted to removal of olivine and pyroxene. Very similar conclusions were drawn by Drury (1981) for evolution of basalts in the Kudremukh belt.

Iron and manganese rocks

The iron- and manganese-rich rocks of the low-grade schist belts appear to be compositionally

---

TABLE 2.8. *Stratigraphy of Sandur schist belt (after A. Roy and Biswas, 1979)*

Upper Unit
Bimodal (mafic-silicic) volcanic rocks interbedded with phyllites and graywackes
Bimodal volcanic association interbedded with banded iron formation, conglomerate, fuchsite quartzite, graywacke, and metapelite (mica schist and phyllite)
Metasedimentary rocks consisting of former sandstones and dolomitic limestones; with volcanic rocks
Lower Unit
Mafic volcanic rocks with minor quartzite and quartz-mica schist.

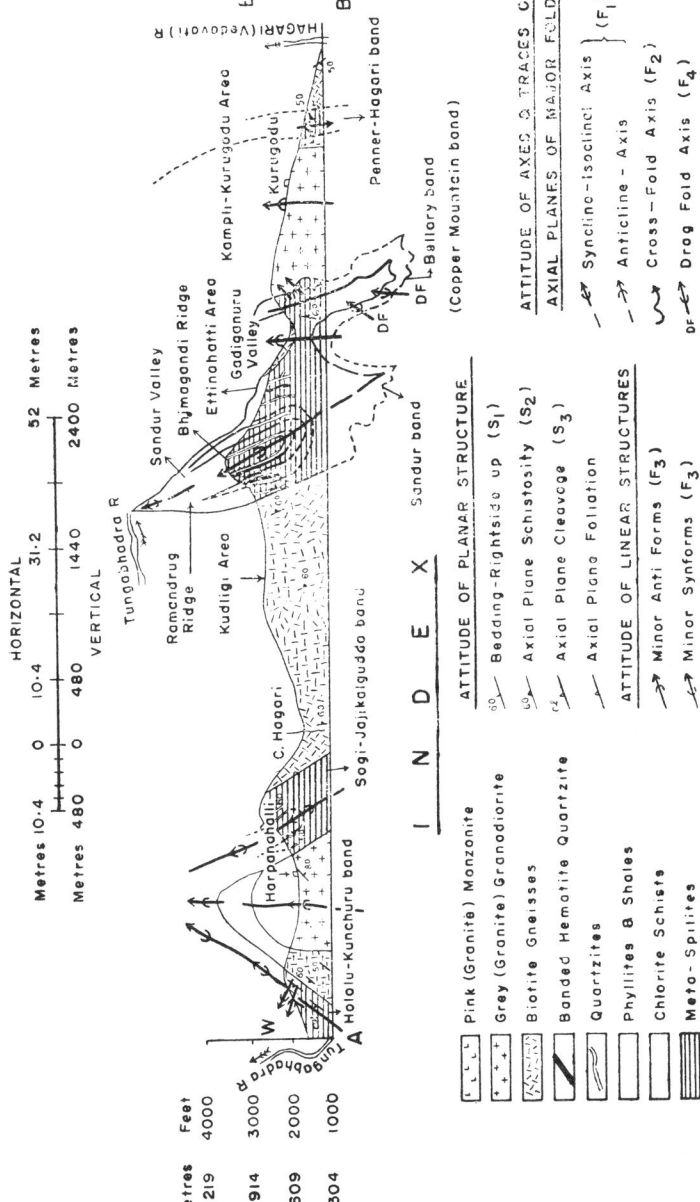

Fig. 2.9. Cross-section of Bellary District, across Sandur schist belt (Krishna Murthy, 1974). Reproduced with permission from the Journal of the Geological Society of India, v. 15, p. 43.

TABLE 2.9. *Compositions of rocks in low-grade schist belts*

	1	2	3	4	5	6	7	8	9	10	11	12	13	14	15	16	17	18
SiO_2	51.6	50.8	51.9	52.9	50.6	51.8	51.6	42.2	41.6	48.1	67.3	70.9	61.8	60.2	61.9	61.3	72.3	72.7
TiO_2	0.86	0.47	0.55	1.6	1.2	1.4	0.82	0.49	0.21	0.52	0.48	0.59	0.83	0.88	0.40	0.55	0.53	0.24
Al_2O_3	13.4	15.7	13.7	14.5	15.2	14.2	14.3	3.7	3.3	5.9	14.6	12.7	13.8	14.6	14.6	14.4	11.0	11.8
Fe_2O_3	4.6	2.2	2.0	2.4	2.5	2.1	11.7	12.8	8.0	1.3	6.4	6.7	7.1	1.2	5.1	8.0	4.9	3.1
FeO	9.0	7.4	6.3	9.2	8.5	8.8			3.5	8.6				8.4				
MnO	0.15	0.16	0.27	0.36	0.37	0.38	0.17	0.15	0.18	0.18	0.03	0.05	0.26	0.41	0.17	0.18		
MgO	7.6	6.6	8.5	5.5	6.9	5.8	7.5	29.8	32.0	18.4	1.8	1.8	4.1	4.0	3.6	2.5	2.3	1.1
CaO	10.4	12.9	11.5	8.8	2.5	8.6	8.8	1.5	1.1	14.1	2.0	1.7	4.6	4.5	5.7	3.9	4.7	3.2
Na_2O	2.8	2.7	2.9	2.7	3.6	2.9	4.5		0.30	0.70	0.91	0.84	4.1	3.0	4.2	3.4	3.8	2.0
K_2O	0.22	0.44	0.59	0.65	0.31	0.62	0.08		0.10	0.40	2.1	2.1	1.0	0.81				
Rb		10														70		
Sr		130														210		
Y							94								19			
Zr							94						48		34	119		
V															69			
Cr	280	439			84	28	241	2908			42	75	162		141	202	74	84
Ni	91	211			63	40	110	2305			95	142	58		70	93		
Th																		
U																		

All abundances of major elements are in weight percent of oxides. Where no entry is made in the FeO position, the value at Fe_2O_3 is total iron calculated as ferric oxide. All abundances of trace elements are in parts per million by weight. Values are taken directly from quoted sources and not recalculated to 100%.

Table 2.9. Column References
1. Amphibolite in northern Holenarasipur belt (Hussain and Naqvi, 1983).
2. Massive amphibolite in Javanahalli belt (Narayana et al., 1983).
3. Voblapur basalts in Chitradurga belt (Mukhopadhyay et al., 1981).
4. Kondli basalts in Chitradurga belt (Mukhopadhyay et al., 1981).
5. Olivine-normative basalt in Bababudan belt (Y. J. Bhaskar Rao and Naqvi, 1978).
6. Quartz-normative basalt in Bababudan belt (Y. J. Bhaskar Rao and Naqvi, 1978).
7. Tholeiitic basalt in Kudremukh belt (Drury, 1983a).
8. Ultrabasic basalt in Kudremukh belt (Drury, 1983a).
9. Spinifex-textured komatiite in Ghatti Hosahalli belt (Narayana and Naqvi, 1980).
10. Spinifex-textured komatiite in Sanguem area, Goa (Dessai and Deshpande, 1979).
11. Oxide-facies schist in Chitradurga belt (Naqvi, 1978).
12. Sulfide-facies schist in Chitradurga belt (Naqvi, 1978).
13. Matrix of conglomerate in Chitradurga belt (Naqvi et al., 1978a).
14. Matrix of conglomerate in Shimoga belt (Naqvi et al., 1978a).
15. Graywacke in Chitradurga belt (Naqvi, 1983).
16. Graywacke in Shimoga belt (Naqvi, 1983).
17. Na-rich metaarkose in Javanahalli belt (Naqvi, 1983).
18. K-rich metaarkose in Javanahalli belt (Naqvi, 1983).

similar to those in high-grade belts (Naqvi et al., 1981). High iron and manganese contents occur in quartzites/cherts and in metapelites.

Mukhopadhyay et al. (1980) summarized the mineralogy of banded ferruginous quartzites and argillites in the Bababudan belt. Magnetite, quartz, and clays appear to be the primary minerals in these rocks. Hematite formed secondarily, although the exact time of formation is unclear. Metamorphic minerals in the quartzites include cummingtonite, grunerite, magnesioriebeckite, and acmite; minerals formed in the argillites also include stilpnomelane, minnesotaite, and chlorite. Siderite is present in some argillites. Pyritic chert and black shale are presumably deep-water facies of the iron-oxide-bearing rocks.

Manganese deposits in the Shimoga, Chitradurga, and similar belts are interbedded in phyllite/chert sequences. Primary minerals are mostly pyrolusite and cryptomelane. Secondary mobilization, presumably near-surface, has been extensive, forming both veinlets and conformable concentrations. There is apparently no relationship to volcanic rocks, indicating a sedimentary, rather than volcanogenic, origin.

Chert and carbonate

Thin beds of chert and carbonate rocks occur sporadically through the low-grade schist belts. Most of the carbonates are dolomitic limestones (Devaraju and Anantha Murthy, 1984), although some pure limestone is present. Recrystallization has formed fine-grained quartzites from much of the chert and given a coarser crystalline texture to the limestones. Distinction between clastic quartzites (unless they are conglomeratic) and recrystallized chert is difficult.

The question of organic contribution to siliceous and calcareous rocks is always controversial. Suresh (1982) isolated apparent organic (algal?) remains in limestones of the Dodguni Chert in the Chitradurga belt. Stromatolitic structures have been found in the same area by Mukhopadhyay and Ghosh (1983). Murthy and Krishna Reddy (1984) described stromatolites in the Sandur belt, and microbiota have also been obtained from these rocks. Kumar et al. (1983) found $\delta^{13}C/^{12}C$ ratios from $+0.3$ to -1.3 parts per thousand, possibly indicative of organic activity.

Micaceous schists

Schistose rocks consisting of quartz, minor feldspar, and high abundances of biotite, muscovite, and/or chlorite are generally regarded as metapelites (Table 2.9). The major schist belts, such as Chitradurga, have thick sequences of these fine-grained rocks, although the greatest thicknesses of sedimentary accumulation seem to consist of graywackes. Both positive and negative Eu anomalies are shown on REE patterns (Fig. 2.5). The abundance of micaceous schists indicates extensive development of clays in the source areas of the schist belts.

Quartzites

Quartzites in the low-grade belts are clearly detrital and exhibit structures such as current bedding (Srinivasan and Ojakangas, 1986). Quartzites in both high- and low-grade schist belts have high contents of ferrous metals, such as iron (magnetite or pyrite), manganese (generally as an oxide), and chromium (fuchsite). The persistent abundance of these elements, and their dispersed distribution in the quartzites, has been explained previously as a result of chemical, including volcanogenic, processes. Sedimentational conditions for the accumulation of pure quartzite appear to have been limited in the Archean, as shown by the abundance of argillaceous matrices in the quartz-pebble conglomerates.

A study by Argast and Donnelly (1983) of quartzites from the Javanahalli belt showed that the K/H ratio in the silica-illite mixture is higher than is possible for micas equilibrated with modern ocean water. They explained this observation as the result of mixing of detrital illite with chemically precipitated amorphous silica and reaction of the assemblage with sea water having a very acidic pH (about 6.3). Argast and Donnelly also pointed out that the extreme variability of Cr contents in the quartzites (from 3 to 1083 ppm in their samples) is difficult to explain by increment of Cr-bearing detrital mica. Thus, the Cr was probably introduced by syn- or postdepositional reactions.

Conglomerates

Conglomerates are extremely useful for tectonic and stratigraphic analysis, but different varieties must be carefully distinguished. Three types whose presence leads to widely different genetic interpretations are as follows: (1) polymictic conglomerate (pebbles of various rock types), in which the pebbles are in contact and support each other, a prime indication of accumulation of coarse debris on an erosion surface; (2) polymictic conglomerate in which the pebbles are isolated in a fine-grained matrix and do not support each other, indicating simultaneous deposition of pebbles and matrix, as from a turbidity flow; and (3) oligomictic conglomerate (pebbles of one kind). Many oligomictic conglomerates contain only quartz pebbles, which could indicate extreme weathering in a source area or on an erosional surface. A quartz matrix would strengthen this interpretation by showing a source terrain devoid of anything but quartz debris. Conglomerates with a similar appearance, however, can also be formed by tectonic processes. An example would be a shear zone in which vein quartz deposition was interrupted by shear movements that formed quartz rods and fragments with superficial resemblance to sedimentary pebbles.

Considerable controversy exists about conglomerates in the Dharwar schist belts. A classification is shown in Table 2.10. Contrary views on several of these units have been sumarized by various papers in Swami Nath and Ramakrishman (1981) and by Ramakarishnan and Viswanatha (1983). The polymictic conglomerates of the schist belts have a clay matrix and generally contain fragments of all underlying rock types (Naqvi and Hussain, 1972; Naqvi et al., 1978a; Ranganna et al., 1981), including debris from underlying rocks within the schist belt in addition to those from surrounding terrains (Plate 2.2c). Oligomictic sedimentary conglomerates are rare.

Examples of compositions of conglomerate matrices are shown in Table 2.9. The relatively high Na/K ratios in the matrices are consistent with the high ratio of trondhjemites to granites in the pebbles and indicates a scarcity of potassic granites in the craton at the time the conglomerates were deposited. The Na was presumably derived from plagioclase in the source. Commonly there is no exact correspondence between the composition of the gneissic pebbles and the composition of apparent source rocks. One major observation is that the average composition of pebbles reported by Naqvi et al. (1978a) is less potassic than

that of underlying Peninsular Gneiss, possibly because of K-metasomatism of the gneiss after the schist belts were formed.

Graywackes

Graywackes constitute thick sections of many of the low-grade schist belts and are completely absent from the high-grade belts. Their average composition is about 40% quartz, 40% rock fragments and micaceous matrix, 10% plagioclase, 5% K-feldspar, and 5% chert and quartzite. They were commonly deposited late in the evolutionary history of the sedimentary basins and were clearly derived by erosion of the gneissic terrain (including older mafic belts).

Compositions of typical graywacke suites are shown in Table 2.9 and an REE pattern is shown in Figure 2.5. The graywackes show either no Eu anomalies or positive Eu anomalies, whereas adjacent shales show negative ones. This difference may be a result of diagenetic absorption of cations from sea water by clays, which dilutes the effect of the positive Eu anomalies shown by detrital plagioclase. The composition of the graywackes is consistent with derivation of clastic material from a source terrain consisting of tonalite and mafic/ultramafic rocks (Naqvi, 1978, 1982).

Argast and Donnelly (1986) recognized both texturally mature and texturally immature (graywacke-type) sandstones in the Chitradurga belt. The components of the sandstones were attrib-

TABLE 2.10. *Classification of conglomerates in schist belts (Naqvi et al., 1978a)*

Conglomerates of Dharwar greenschists

Tectonic (in greenstone belts with basic-ultrabasic base)   Sedimentary (in greenschists of geosynclinal piles with orthoquartzite-base)

Autoclastic   Pyroclastic

Tattekere (Holenarasipur   Peddapalli (?)
  greenstone belt)   (Kolar greenstone belt)

Disrupted framework   Contact framework
                                                      Basal conglomerates

                                                      Chickmagalur (Bababudan
                                                        schist belt)

Graded bedded (Polymictic)   Ungraded bedded

Kurmerdikere,
Aimangala,
Talya (Chitradurga schist belt)   Oligomictic   Polymictic

Benikoppa,   Chinnagiri (Shimoga schist belt)   Peddapalli (?) (Kolar greenstone belt)
Bagewadi (Gadag schist belt)

Kalvarangana Betta,   Holalur, Bikonahalli (Shimoga schist belt)
Ubrani (Shimoga schist belt)

Kaldurga (Bababudan schist belt)

Prepared with permission from Canadian Journal of Earth Sciences, v. 15, p. 1088.

uted to erosion of a dominantly sialic (tonalitic) crust with some mafic rocks. The high abundance of ferromagnesian minerals in some samples was attributed to exposure of mafic rocks adjacent to the depositional basins.

Presumed metasediments of the Javanahalli belt

The Javanahalli belt is generally in amphibolite facies but has a higher proportion of clastic sediments than high-grade belts farther south (Narayana et al., 1982, 1983). Among the apparent sedimentary suites are paragneisses (metaarkoses) and paraamphibolites (schistose rocks probably formed from calcareous sediments). Although no sedimentary textures have been preserved in these rocks, their origin is shown by interlayering with, and gradation into, other rocks that are clearly metasediments (quartzites, calc-silicates, and iron formations). Compositions (Table 2.9) indicate derivation from a source similar to that of the low-grade schist belts. The REE patterns of the paragneisses generally show negative Eu anomalies, presumably inherited from silicic source rocks (Naqvi et al., 1983c; Narayana et al., 1983). The Javanahalli paraamphibolites are apparently mostly mafic pyroclastic or epiclastic material admixed with shales and limestones.

COMPARISON OF SEDIMENTARY ROCKS AND PROCESSES IN LOW-GRADE AND HIGH-GRADE BELTS

The major differences in the properties of sedimentary suites between the older (high-grade) and younger (low-grade) belts are as follows: sediments of very high Mg and Al contents (the kyanite-staurolite-bearing mica schists of high-grade belts) generally do not occur in low-grade sequences; coarse clastic sediments are much more abundant in younger (low-grade) rocks, particularly as shown by graywacke abundances; unquestionable clastic quartz occurs only in younger suites; iron formations are all oxides in older belts, but some sulfide suites occur in younger belts. Even if the quartzites of the high-grade belts are clastic, their abundance is so small in comparison with the quartz of the graywackes that there must have been extensive increase in availability of gneissic rocks as sedimentary provenance for the younger belts. These features lead to the conclusion that younger, low-grade, schist belts formed in environments that provided upland massifs for erosion and deep intervening basins with diverse water depths and configurations. The proportion of coarse debris was consequently high.

PENINSULAR GNEISS

Similar to all other gneissic complexes in the world, the Peninsular Gneiss of India is extremely difficult to investigate. The major problem is the great diversity of the suite, with variability on a scale ranging from hand specimens to mountain ranges (Pichamuthu, 1976). In reconnaissance studies, map units commonly comprise rocks formed by different processes at different times, and even small samples may contain a mixture of suites of different origin.

In Figure 2.1, we have attempted to distinguish only two parts of the Peninsular Gneiss: the major part of the gneiss, mostly with complex banding and evidence of multiple deformation, and a suite of much less deformed, in places nearly massive, rocks of broadly similar composition to the banded rocks. These more massive, trondhjemitic bodies are only a small part of the total gneiss outcrop area.

Structures and ages of the gneisses have been reviewed in earlier sections of this chapter. Briefly, the typical gneiss records at least three episodes of deformation, with older ones perhaps obliterated. Early deformational pulses were apparently caused by east-west compression. The later compressions were either east-west or, less commonly, from other directions and caused open refolding of the older isoclines. Fold axes range from steep to horizontal, and axial planes vary from reclined to vertical as a result of superposed folding (Plate 2.2d, e). The resultant interference pattern of the different folds yields complex outcrops. Interference patterns in the gneisses are identical to those in the supracrustal suites of all ages (Plate 2.2e).

The gneisses record whole-rock ages ranging from about 3,400 to 3,000 m.y. (Table 2.2). Where dated, massive trondhjemitic bodies show ages of about 3,000 m.y. (Halekote, Stroh et al., 1983; and Chickmagalur, Taylor et al., 1984). Initial $^{87}Sr/^{86}Sr$ ratios in most rocks are slightly higher than a presumed mantle growth curve. Apparently, the gneisses were equilibrated from material that had been derived from the mantle only a short time before equilibration and with only slightly higher Rb/Sr ratios than the mantle. This equilibration could either represent a remelting event or a metamorphic resetting of the isotopic system.

Mineralogy and composition

The general mineralogy of the Peninsular Gneisses is very simple. All rocks contain quartz, plagioclase, and biotite in variable proportions. Most rocks also contain small amounts of K-feldspar, and muscovite or hornblende is present in some samples. The rocks are, therefore, the tonalitic/trondhjemitic "gray gneiss" of many Archean terrains.

Texturally, the gneisses are typically granoblastic and appear to have reached metamorphic equilibrium. Some areas are migmatized; except for a high abundance of migmatite near the granulite transition zone, the complete distribution of migmatitic vs. nonmigmatitic gneisses has not been mapped. Pegmatite lenses and veins are abundant in some areas; some are concordant to the gneissic layering, but many are discordant. There is a possibility that K-rich (pink) pegmatites are more common near the Closepet Granite than elsewhere in the craton, where albite-quartz pegmatites predominate.

Various investigators have attempted different mineralogical and/or compositional classifications of the gneisses. Y. J. Bhaskar Rao et al. (1983) recognized the following varieties of gneiss in areas east of the Holenarasipur schist belt: (1) leucocratic migmatite variable from tonalite to granite—these rocks are interlayered with ultramafic/mafic schists and with alumina- and silica-rich schists in a "trimodal" association, (2) a younger suite of polydeformed "grey gneisses" that intrudes the trimodal suite—the grey gneisses occur as two phases of different mafic mineral content that are deformed together, and (3) late-tectonic granites with minor mafic/ultramafic hypabyssal rocks—the granites occur as sheets only a few meters thick, in places forming complex stockworks.

A study by Naqvi et al. (1983b) used the following classification of gneisses: (1) quartzofeldspathic gneisses, consisting of quartz, K-feldspar, plagioclase, muscovite, and biotite and corresponding to the leucocratic migmatite of Y. J. Bhaskar Rao et al. (1983); (2) banded migmatitic gneisses, which consist of quartz, K-feldspar, plagioclase, biotite, small amounts of amphibole, range in composition from tonalite to granite, and commonly intrude the quartzofeldspathic gneisses; (3) paragneisses, consisting of quartz, feldspar, biotite, muscovite, ± amphibole and aluminous varietal minerals (kyanite and sillimanite)—the paragneisses are interbedded locally with calc-silicate rocks and do not show intrusive relationships with, or contain inclusions of, other rock types; and (4) comparatively massive trondhjemite-granodiorite-granite plutons that intrude other members of the gneiss suite and range in size from small dikes and sills to large batholiths.

Monrad (1983) used a predominantly compositional discrimination of two gneiss suites (Table 2.11): a low-alumina (Hassan) suite, with $Al_2O_3 <$ 13%, that is consistently banded and polydeformed; and a high-alumina (Marginal) suite, with $Al_2O_3 > 15\%$, that ranges from banded to nearly massive. In outcrops where relationships can be determined, the high-alumina suite intrudes the low-alumina one. Monrad regarded the high-alumina suite as a set of intrusions "marginal" to the Holenarasipur schist belt, where his studies were conducted.

The low-alumina suites of Monrad (1983) yielded ages ranging from 3,100 to 3,200 m.y., with initial $^{87}Sr/^{86}Sr$ of 0.7008 to 0.7020 (Table 2.2). The high-alumina rocks, which include the Halekote Trondhjemite (discussed in the next section), have an age of about 3,000 m.y. and initial $^{87}Sr/^{86}Sr$ ratio of 0.7019. Representative compositions of low- and high-alumina suites are shown in Table 2.11. Figure 2.10 shows REE diagrams of various samples of the low- and high-alumina suites.

Janardhan and Srikantappa (1975) and Janardhan et al. (1978) recognized different gneiss suites in the gneiss-granulite transition zone. The terminology is similar to that used by Y. J. Bhaskar Rao et al. (1983) and Naqvi et al. (1983b).

Most of the mafic (metasedimentary and metaigneous) enclaves and inclusions in the gneisses are regarded as fragments of the high-grade schist belts engulfed by emplacement of the gneiss. It is not clear whether the siliceous pelitic members of the gneiss suite are also fragments of schist belts. Metamorphosed quartz-clay rocks do not occur in recognized high-grade schist belts, except for very minor fuchsite-, kyanite-, or sillimanite-bearing quartzite. It is possible, therefore, that the pelitic parts of the gneisses were not originally deposited in schist belts but represent sediments engulfed from another, more platformal, environment.

3,000-M.Y. INTRUSIVES

As mentioned in the preceding section, some areas of the gneissic terrain are comparatively massive, showing fewer episodes of deformation and at least partly cross-cutting other structures. These bodies range in size from small dikes and sills to plutons with diameters of many kilometers. The larger ones are shown separately, to the extent they are known, on Figure 2.1 as "3,000-m.y. intrusive rocks."

Most bodies apparently related to the 3,000-m.y. event are trondhjemitic. They show two important differences from younger granites discussed in the next section: the trondhjemites show some deformation, particularly open folding, in contrast with the almost completely massive younger bodies; and the trondhjemites have gradational margins into the wall rocks, whereas the granites are sharply cross-cutting. As the terms imply, the K/Na ratios in the granites are much higher than in the older trondhjemites.

Mineralogically, the trondhjemites are a simple mixture of quartz, plagioclase, biotite, K-feldspar, ± hornblende in variable proportions. The composition of the Halekote Trondhjemite is shown in Table 2.11, and a pattern of REE abundances in Figure 2.10. The REE pattern is partly explainable by equilibration of the trondhjemite melt with residual garnet. The composition of the Chickmagalur Granite (granodiorite) of the same age is also shown in Table 2.11.

TABLE 2.11. *Compositions of gneisses*

	1	2	3	4	5	6	7	8	9	10	11	12
SiO_2	73.3	72.8	72.7	71.3	73.8	75.2	77.7	69.5	72.3	70.1	72.0	70.5
TiO_2	0.07	0.23	0.09	0.26	0.22	0.42	0.11	0.34	0.28	1.1	0.11	0.36
Al_2O_3	13.7	13.2	15.1	14.3	11.9	13.0	13.0	16.0	13.1	11.4	16.8	14.4
Fe_2O_3	0.93	0.80	0.91	1.7	4.5	2.3	1.1	2.5	1.0	2.8	0.86	2.3
FeO	1.5	1.4		1.7					1.5	2.5		
MnO	0.10	0.09	0.02	0.13	0.15				0.06	0.06		0.03
MgO	0.86	0.71	0.20	0.86	1.2	0.48	0.11	0.54	1.9	2.6	0.24	0.67
CaO	2.2	1.5	1.9	2.5	2.0	3.6	0.79	2.3	2.6	3.6	1.6	2.1
Na_2O	4.8	4.2	6.3	4.2	3.8	4.7	3.9	5.0	4.1	2.7	5.7	4.5
K_2O	1.6	3.9	2.3	2.5	1.4	0.19	2.6	2.2	0.80	1.2	2.5	2.8
Rb	82	125				1	38	76	18	13	71	81
Sr	332	267				103	160	375	118	110	720	401
Y						54	130	25	6		14	17
Zr						223	208	198			124	206
V												
Cr						20	20	20	51	10	5	
Ni									29	34		
Th						8.4	13.3		9.1		4.2	
U						2.3	2.4		3.2		1.9	

All abundances of major elements are in weight percent of oxides. Where no entry is made in the FeO position, the value at Fe_2O_3 is total iron calculated as ferric oxide. All abundances of trace elements are in parts per million by weight. Values are taken directly from quoted sources and not recalculated to 100%.

Table 2.11. Column References
1. Trondhjemitic gneiss near Channarayapatna (Y. J. Bhaskar Rao et al., 1983).
2. Granitic gneiss near Channarayapatna (Y. J. Bhaskar Rao et al., 1983).
3. Quartzofeldspathic gneiss (Naqvi et al., 1983b).
4. Banded migmatitic gneiss (Naqvi et al., 1983b).
5. Paragneiss (Naqvi et al., 1983b).
6. Low-Rb suite of low-alumina (Hassan) gneiss near Holenarasipur belt (Monrad, 1983).
7. High-SiO_2 suite of low-alumina (Hassan) gneiss near Holenarasipur belt (Monrad, 1983).
8. High-alumina (Marginal) gneiss near Holenarasipur belt (Monrad, 1983).
9. Quartzofeldspathic gneiss in transition zone (Janardhan et al., 1978).
10. Biotite-hornblende gneiss in transition zone (Janardhan et al., 1978).
11. Halekote Trondhjemite (Stroh et al., 1983).
12. Chickmagalur Granite (Taylor et al., 1984).

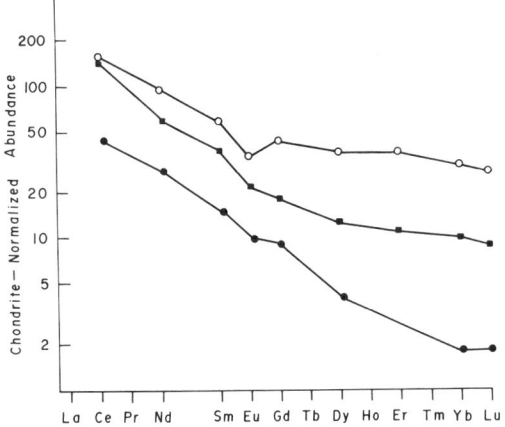

Fig. 2.10. REE in silicic rocks. Open circles, low-alumina gneiss (Monrad, 1983); solid squares, high-alumina gneiss (Monrad, 1983); solid circles, Halekote Trondhjemite (Stroh et al., 1983).

TECTONIC SIGNIFICANCE OF 3,000-M.Y. EVENT

The intrusion of the trondhjemite plutons marks an important time in the evolution of the Western Dharwar craton. The 3,000-m.y.-old plutons clearly invade the high-grade schist belts, but debris eroded from the plutons occurs in the low-grade belts. In a few places, the sedimentary sequences of the low-grade belts can be shown to rest unconformably on the 3,000-m.y.-old rocks. Furthermore, there have been no measurements of whole-rock ages substantially younger than 3,000 m.y. in any parts of the gneissic terrain investigated thus far. All igneous activity after this time was restricted to intrusion of diapiric granites and dikes of various compositions, none of which affected the whole-rock isotopic systems of the gneisses.

Based on the preceding evidence, the 3,000-m.y. event separates a period of continual instability, deformation, and intrusion from a younger time of quiescent sedimentation and diapiric intrusion. Deformational activity did not cease at 3,000 m.y., however. As discussed earlier, the post–3,000-m.y. schist belts were isoclinally folded during polycyclic deformation. Furthermore, great thicknesses of volcanic and sedimentary rocks may have accumulated in the depositional basins (although the total thicknesses is controversial, based on interpretations of structure and stratigraphy).

The types of activity that occurred before and after 3,000 m.y. are distinguished in three ways. One is that younger compressive deformation was conducted at a sufficiently low temperature that whole-rock isotopic systems were not reset, thus leaving 3,000 m.y. and older ages characteristic of all gneisses. A second distinction is that intrusion after 3,000 m.y. was localized and diapiric, instead of widely pervasive in the gneisses. The

third distinction is that evidence of faulting can be found only for events younger than 3,000 m.y. The principal evidence for fault movements consists of the presence of uplifted massifs that provided a sialic provenance for clastic sediments and the occurrence of rift basins in which Late Archean sediments were deposited. The localization of these various processes could have been possible only if the craton had become a rigid or semirigid block. This 3,000-m.y. age, therefore, should be regarded as the age at which the Western Dharwar area became a craton.

ORIGINS OF GNEISSES

The materials that now constitute the gneissic terrain could have been derived originally in three ways: as sediments metamorphosed to gneisses, as igneous rocks (extrusive or intrusive) later metamorphosed to gneisses, and by direct igneous emplacement of magmas without extensive further metamorphism. The magmas involved in these processes could have been derived either from the mantle or crust, with or without extensive differentiation.

After initial formation, the gneisses could have been subjected to a variety of processes that altered their original condition. Chemical changes can be classified in the following ways: (1) impoverishment in some elements by local melting or extraction of mineral components into a fluid phase; (2) enrichment in specific elements, presumably mostly lithophilic ones, by fluids or magmas that invaded local areas; (3) pervasive addition of new components to the gneisses, presumably through pore fluids; and (4) pervasive chemical alteration without visible effects in outcrop or thin section. Examples of these various processes of initial formation and later modification in the Peninsular Gneiss are discussed, where possible, in the remainder of this section.

Metasediments

Gneisses that have been thoroughly deformed rarely retain primary sedimentary structures and textures. Therefore, the best indication of a sedimentary parent is generally compositional. Because so few igneous rocks are significantly peraluminous, the presence of aluminous minerals such as kyanite and sillimanite may indicate a sedimentary component in a gneiss. Pelitic and paragneisses have been reported in the Western Dharwar craton (e.g., Janardhan et al., 1978; Naqvi et al., 1983b) based on the presence of aluminous minerals and association with other metasediments. Analyses of these rocks, however, generally do not show high Al_2O_3 contents (Table 2.11). Furthermore, these assemblages are not widely reported, and the metasedimentary component of the Peninsular Gneiss suite apparently is very small (Jayaram et al., 1984).

Metamorphosed igneous rocks

If the metasedimentary component of the gneisses is small, then most of the polyphase-deformed rocks must be meta-igneous. Proof of this mode of formation is difficult to obtain if rocks have been subjected to metasomatic compositional changes. One attempt to demonstrate an igneous origin for a gneiss suite was made by Monrad (1983), who showed that low-alumina gneisses near the Holenarasipur belt plot in the eutectic region of the quartz-albite-orthoclase phase diagram (Fig. 2.11) at moderate anorthite contents and $P_{H2O} = P_{total}$ of less than 3 kb. These observations are consistent with equilibration of a trondhjemite/tonalite melt at depths of less than 10 km and subsequent solidification, either at the same depth or after rapid upward emplacement. The observation that the REE patterns (Fig. 2.10) show that garnet was not an equilibrating phase supports this conclusion of equilibration at a shallow depth. An igneous origin for most gneisses was proposed by Jayaram et al. (1984) on the basis of REE and other element distributions.

Another attempt to demonstrate an igneous origin of the gneisses was based on studies of zircon morphology. T. V. Viswanathan (1968) showed that zircons in gneisses near Bangalore (in the Eastern Dharwar craton) are primarily euhedral to subhedral; only 23% of zircons were rounded, and only 9% show overgrowths, leading to the

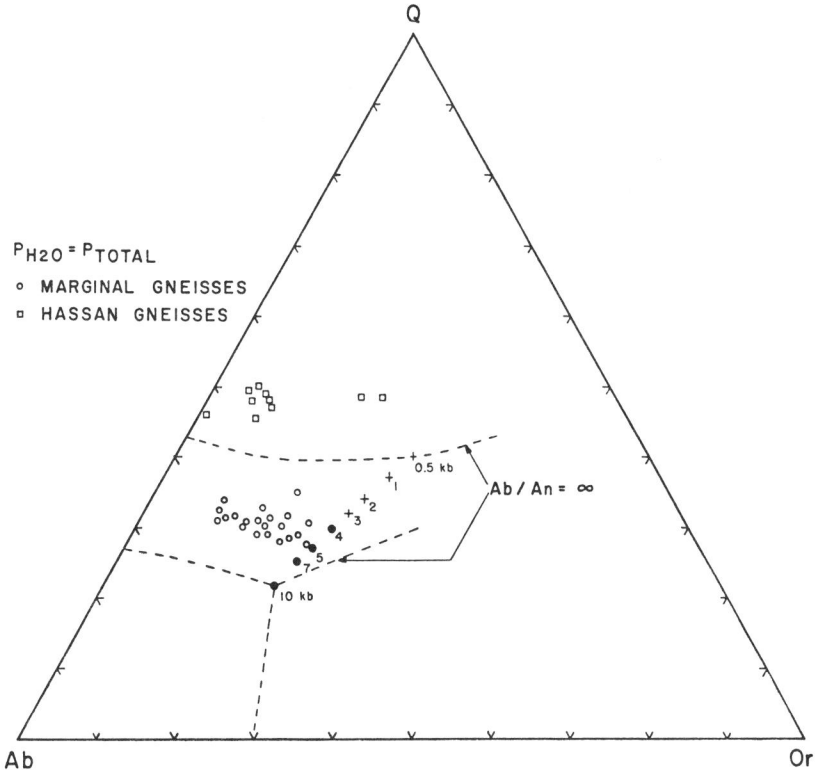

Fig. 2.11. Quartz-albite-orthoclase diagram for Peninsular Gneisses (Monrad, 1983). Projection is from An. Cotectic field boundaries are at pure Ab composition, appropriate for the gneisses plotted. Field boundaries are shown at 0.5 and 10 kb water pressure, and minima points are shown at intermediate water pressures.

conclusion that the gneisses were primarily derived from pre-existing igneous rocks.

Direct igneous intrusion

Comparatively massive small bodies of tonalite/trondhjemite, granodiorite, and granite occur widely throughout the gneissic suite. Their source is unclear; possibly they were melted from older gneisses or amphibolites in the greenstone belts. The total proportion of such rocks in the gneiss terrain is also unclear; including major bodies and minor dikes and sills, possibly about 10% to 20% of the gneiss outcrop is composed of these plutons.

Depleted residues

Obvious depleted residues have not been analyzed in the Peninsular Gneiss terrain, with the possible exception of the low-Rb suite of Monrad (1983) (Table 2.11). Textural and structural evidence for the process is also absent. For example, the abundant pegmatites in the gneisses generally do not have margins of biotite-hornblende layers, indicating a residue of rock from which the pegmatites were anatectically derived.

Enrichment in specific elements by intrusion, pegmatization, and veining

Pegmatitic veins, late granites, and similar rocks are abundant throughout the craton. Their abundance is included in the 10% to 20% just estimated for the total increment of late intrusive material to the gneiss suite, but we emphasize the lack of precise evidence for this estimate. Naqvi et al. (1983b) concluded that extensive pegmatization occurred primarily in the center of gneissic areas between schist belts. These areas may represent antiforms more conducive to fluid penetration than the synforms containing schist belts. Although pegmatites have not been specifically dated, it is unlikely that they could have formed in abundance younger than 3,000 m.y. ago, the last date at which isotopic whole-rock systems were reequilibrated in the gneisses. Presumably they were synchronous with the 3,000-m.y. event.

Pervasive enrichment in new components (neosome development)

Neosomes would be recognized as areas of transformation, with partial or total obliteration of preexisting structures. Reports of this phenomenon over areas of any significant size are lacking for most of the Peninsular Gneiss except in the gneiss-granulite transition zone. Small areas of pegmatization and migmatization probably contain wall-rock gneisses that have been affected by this process.

Pervasive metasomatism without megascopic effects

Two types of investigations, one direct and one indirect, indicate extensive metasomatism of the gneisses after their initial formation. The direct evidence is based on the distribution of uranium in the gneisses and diapiric trondhjemites (Callahan and Rogers, 1986). The U and Th contents of the diapiric trondhjemites are extremely low for any type of intrusive igneous rock, and conversely, the U contents of the gneisses are very high (about 3 ppm) relative to other Archean gneisses. This high U content gives the gneisses Th/U ratios that are too low for magmatic rocks. Presumably the U, and possibly the Th, are metasomatic additions to the gneisses, and the most likely time of metasomatism is 3,000 m.y. ago, when the trondhjemites were still hot and repelling fluids.

An indirect type of evidence for metasomatism is comparison of compositions of gneisses with compositions of gneiss pebbles in the conglomerates of low-grade schist belts (e.g., Naqvi et al., 1978a; Ranganna et al., 1981). Results of the various studies are inconsistent. As a generalization, however, there appears to be a higher content of lithophilic elements (such as potassium) in the gneisses than in the pebbles. If this difference is valid, then one likely explanation is metasomatism of the gneisses after erosion of pebbles from them into the depositional basins of the schist belts. The reason why this proposed metasomatism affected the gneisses and not the pebbles or other rocks in the schist belts could be that the metasomatizing fluids were effectively isolated from the pebbles by the matrices of the conglomerates. This type of metasomatism must have occurred after development of the low-grade schist belts, possibly during granulite metamorphism to the south 2,500 m.y. ago.

Summary of origin of gneisses

The following conclusions for the evolution of the Peninsular Gneiss suite are summarized from Rogers et al. (1986). Most of the gneiss is metamorphosed former igneous rocks that have been deformed in several episodes. Metasediments in the gneisses are not abundant, either as discrete blocks or as assimilated screens in the igneous rocks. Extensive veining, pegmatization, migmatization, and late intrusion have occurred, particularly in the (antiformal) centers of broad gneiss terrains. This material appears to be a new increment to the gneiss rather than fluids derived by local anatexis and constitutes perhaps 10% to 20% of the gneiss; a most likely age of formation is 3,000 m.y. ago. Lithophilic metasomatism, shown particularly by U, probably occurred at the same time. After 3,000 m.y., deformation and metasomatism occurred at sufficiently low temperatures that isotopic systems were not affected. Intrusion was localized, rather than pervasive. Lithophilic elements may have been added to the gneisses at this young time, possibly about 2,500 m.y. ago during granulite formation in the lower crust.

GRANITES AND DIKES

Igneous activity in the Western Dharwar craton younger than 3,000 m.y. is not associated with compressive deformation or metamorphism. Both the granites and the dikes (generally mafic) were apparently emplaced into cold, rigid crust.

The major granites of the Dharwar craton are the Closepet massif and related rocks. Because most of these bodies are in the eastern part of the craton, they are discussed in Chapter 3. Some plutons, however, occur to the west of the Closepet outcrop, the major studied ones being at Arsikere and Banavar, Chitradurga (Chitaldrug), Hoskere, and in South Kanara (Fig. 2.1). Lithologically similar, but undated and less well-known, plutons

occur elsewhere in the craton (Gokul and Srinivasan, 1976; Krishna Murthy, 1978). The proportion of the Western Dharwar craton occupied by these granites is less than 5%.

Bodies that have been dated yield the following ages (Table 2.2): at Chitradurga—2,605 m.y. by Pb/Pb methods and 2,475 m.y. by Rb/Sr methods, at Arsikere—2,310 m.y. by Rb/Sr methods, and in South Kanara—2,669 m.y. by Rb/Sr methods. A granite with approximately the same lithology occurs at Chamundi Hill (Babu, 1956) but has an age of only about 800 m.y. (Table 2.2; Table 2.12). Dike suites are generally undated but, in most cases, appear to be younger than the granites.

Granites

The granites have a number of common characteristics. Contacts are discordant and abrupt on a large scale, although gradational over short distances on a small scale. Rocks are generally massive, except for some porphyritic phases that

TABLE 2.12. *Compositions of young magmatic suites*

	1	2	3	4	5	6	7
SiO_2	72.9	70.9	72.7	68.8	47.8	48.5	52.4
TiO_2		1.2	0.19	0.19	1.4	1.0	1.4
Al_2O_3	14.7	12.6	13.7	14.7	15.4	14.1	14.9
Fe_2O_3	1.4	0.45	1.5	0.94	3.8	3.3	2.9
FeO	1.3	1.3		1.5	8.9	9.1	8.1
MnO	0.37	0.28	0.03	0.06	0.10	0.05	0.02
MgO	0.48	3.9	0.32	1.6	5.9	10.1	7.1
CaO	0.84	2.3	1.3	2.9	10.4	9.0	7.4
Na_2O	4.6	3.6	3.7	4.4	3.6	2.4	3.1
K_2O	4.1	0.84	4.2	3.7	1.0	0.67	0.66
Rb			209	150			
Sr			205	200			
Y			31		14	16	18
Zr			165	1800	82	82	98
V							
Cr					247	294	277
Ni					205	276	165
Th							
U							

All abundances of major elements are in weight percent of oxides. Where no entry is made in the FeO position, the value at Fe_2O_3 is total iron calculated as ferric oxide. All abundances of trace elements are in parts per million by weight. Values are taken directly from quoted sources and not recalculated to 100%.

Table 2.12. Column References
1. Arsikere Granite (Narayana Das, 1970).
2. Hoskere (Hosdurga) Granite (Divakara Rao and Quershy, 1969).
3. Chitradurga Granite (Taylor et al., 1984).
4. Porphyritic pink Chamundi Granite (Babu, 1956; Babu and Shankar Das, 1977).
5. Nepheline-normative basalt dike near Chitradurga (Naqvi et al., 1974).
6. Olivine-normative basalt dike near Chitradurga (Naqvi et al., 1974).
7. Quartz-normative basalt dike near Chitradurga (Naqvi et al., 1974).

show flow foliation. The rocks are petrographically true granites, with a high quartz content and a high ratio of K-feldspar to plagioclase. Varietal minerals are mostly biotite with some amphibole. Alkali minerals, such as aegerine or riebeckite, are generally absent. Textures range from fine grained to porphyritic and pegmatitic, and the same body may have more than one kind of texture (Narayana Das, 1970). Compositions are shown in Table 2.12. The origin of the late granites is not clear, but it seems likely that granites such as Arsikere and Chitradurga formed by anatexis of older crustal rocks.

Dikes

Mafic dikes are common throughout the Western Dharwar craton. They have a variety of orientations, although the preferred direction is approximately north-south. Most of the dikes have a typical tholeiitic mineralogy, including plagioclase, pyroxene, and either olivine or quartz. All variations occur from tholeiites to minor alkali olivine basalts (Naqvi et al., 1974; Table 2.12). Alkaline dikes occur at a few places in the craton, principally in the gneiss-granulite transition zone, where they tend to be younger than the tholeiitic dikes.

ECONOMIC DEPOSITS

The economic deposits of the Dharwar craton have been summarized by Radhakrishna (1976). Sedimentary iron and manganese ores occur in several of the low-grade schist belts. Chromium (as chromite) and nickel (as sulfide) occur in some ultramafic rocks of the high-grade belts. Titaniferous magnetite deposits are locally vanadium bearing (Ramiengar et al., 1978a; Vasudev and Srinivasan, 1979). Copper occurs with some ultramafic rocks (Radhakrishna et al., 1973). Gold is also present.

Most of the commercial deposits in the Western Dharwar craton are sulfides (Vasudev, 1983). Stratiform deposits occur in schist belts, and hydrothermal (vein-type) deposits occur both in the schist belts and the gneisses and granites. The principal metal is copper (Radhakrishna, 1967). Lead and zinc are also present, commonly associated with the copper ores.

The principal ore deposit is in the area of Ingaldhal, in the Chitradurga belt just south of Chitradurga (Fig. 2.2; Naqvi et al., 1976, 1977; Mookherjee and Philip, 1979). Vasudev (1983) listed four principal types of ores at Ingaldhal: stratiform massive base-metal sulfides, bedded pyrite or pyrite-chert rhythmites, massive sulfide veins, and fracture-controlled drusy quartz veins carrying sulfides of Pb-Cu-As. All ores occur in volcaniclastic and chemical sediments of the upper part of the Jogimardi volcanic sequence. The ores have been deformed during folding of the schist belt. A volcanogenic origin is likely for at least some of the stratiform deposits. Isotopic studies at Ingaldhal have been made by Menon et al. (1981), Seccombe et al. (1981) and Venkatasubramanian et al. (1981). Many $\delta^{34}S/^{32}S$ ratios are approximately zero, indicating mantle derivation of the mineralizing fluids, but some are as high as ± 10, indicating a different source or a large variation in temperature and composition of precipitating fluids.

REFERENCES

Argast, S., and Donnelly, T. W. (1983). Javanahalli quartzites: Evidence for sedimentary mica and implications for the chemistry of Archean ocean water. In Precambrian of South India (ed. S. M. Naqvi and J. J. W. Rogers), Geol. Soc. India Mem. 4, 158–168.

——— and ——— (1986). Compositions and sources of metasediments in the Upper Dharwar Supergroup, South India. J. Geol., v. 94, 215–232.

Babu, S. K. (1956). Petrochemistry of the granites and associated rocks of Chamundi Hill, Mysore State. Madras Univ. J., v. 26B, 533–551.

———, and Shankar Das, M. (1977). Trace element geochemistry of Chamundi granites, Mysore. Chayanica Geologica, v. 3, 166–189.

Balasubrahmanyan, M. N. (1978). Geochronology and geochemistry of Archaean tonalitic gneisses and granites of South Kanara District, Karnataka State, India. In Archaean Geochemistry; Proc. First Internat. Symp. on Archaean Geochem.: The Origin and Evolution of Archaean Continental Crust (ed. B. F. Windley and S. M. Naqvi), Elsevier, Amsterdam, v. 1, 59–77.

———, Ramamohan Rao, G., and Venkataraman, S. (1976). The gneissic complex and Kanara batholith, Mysore State. Geol. Surv. India Misc. Publ. 23, Part 2, 363–373.

Beckinsale, R. D., Drury, S. A., and Holt, R. W. (1980). 3,360-My-old gneisses from the South Indian craton. Nature, v. 283, 469–470.

Bhaskar Rao, B., and Mohanty, A. K. (1982). Petrochemistry and P-T conditions of metamorphism of the metasediments from the southern part of the Holenarasipur schist belt, Hassan District, Karnataka. Ind. Geol. Assoc. Bull., v. 15, 59–64.

———, and Veerabhadrappa, S. M. (1983). Geochemistry and genesis of Mg-rich metavolcanic rocks from the southern part of the Holenarasipur schist belt, Karnataka, South India. Ind. Geol. Assoc. Bull., v. 16, 21–32.

Bhaskar Rao, Y. J., and Drury, S. A. (1982). Incompatible trace element geochemistry of Archaean metavolcanic rocks from the Bababudan volcanic-sedimentary belt, Karnataka. Geol. Soc. India J., v. 23, 1–12.

——, and Naqvi, S. M. (1978). Geochemistry of metavolcanics from the Bababudan schist belt; A late Archaean-early Proterozoic volcano-sedimentary pile from India. In Archaean Geochemistry; Proc. First Internat. Symp. on Archaean Geochem.: The Origin and Evolution of Archaean Continental Crust (ed. B. F. Windley and S. M. Naqvi), Elsevier, Amsterdam, 325–341.

——, Beck, W., Rama Murthy, V., Nirmal Charan, S., and Naqvi, S. M. (1983). Geology, geochemistry, and age of metamorphism of Archaean gray gneisses around Channarayapatna, Hassan District, Karnataka, South India. In Precambrian of South India (ed. S. M. Naqvi and J. J. W. Rogers), Geol. Soc. India Mem. 4, 309–328.

Callahan, E. J., and Rogers, J. J. W. (1987). Thorium and uranium contents of gneisses and trondhjemites in the Western Dharwar craton, India. Can. J. Earth Sci., in press.

Chadwick, B., Ramakrishnan, M., Viswanatha, M. N., and Sreenivasa Murthy, V. (1978). Structural studies in the Archaean Sargur and Dharwar supracrustal rocks of the Karnataka craton. Geol. Soc. India J., v. 19, 531–542.

——, ——, ——, —— (1981a). The stratigraphy and structure of the Chitradurga region: An illustration of cover-basement interaction in the late Archaean evolution of the Karnataka craton, southern India. Precamb. Res., v. 16, 31–54.

——, ——, ——, ——(1981b). Structural and metamorphic relations between Sargur and Dharwar supracrustal rocks and Peninsular Gneiss in central Karnataka. Geol. Soc. India J., v. 22, 557–569.

——, ——, and —— (1985). Bababudan—A Late Archaean intracratonic volcano-sedimentary basin, Karnataka, South India: Part I, Stratigraphy and basin development; Part II, Structure. Geol. Soc. India J., v. 26; Part I, p. 769–801; Part II, p. 802–821.

Chalokwu, C. I., and Sood, M. K. (1983). Komatiites from southern India. Ind. Geol. Assoc. Bull., v. 16, 1–12.

Crawford, A. R. (1969). Reconnaissance Rb-Sr dating of the Precambrian rocks of southern peninsular India. Geol. Soc. India J., v. 10, 117–166.

Dessai, A. G., and Deshpande, G. G. (1979). Komatiites from Sanguem area, Goa, India. Neues Jahrb. Mineral. Abh., v. 135, 209–220.

Devaraju, T. C., and Anantha Murthy, K. S. (1984). Carbonates of Chiknayakanhalli schist belt, Karnataka. Geol. Soc. India J., v. 25, 162–174.

Divakara Rao, V., and Qureshy, M. N. (1969). Major element geochemistry of parts of the Closepet Granite pluton, Mysore State, India. Geophys. Res. Bull. (Hyderabad), v. 7, 145–157.

——, and Rama Rao, P. (1982). Granitic activity and crustal growth in the Indian shield. Precamb. Res., v. 16, 257–271.

——, Satyanarayana, K., Naqvi, S. M., and Hussain, S. M. (1975). Geochemistry of Dharwar ultramafics and the Archaean mantle. Lithos, v. 8, 77–91.

——, Rama Rao, P., Govil, P. K., and Balaram, V. (1983). Geology and geochemistry of the Krishnarajpet schist belt: A greenstone belt of the Dharwar craton, India. In Precambrian of South India (ed. S. M. Naqvi and J. J. W. Rogers), Geol. Soc. India Mem. 4, 293–308.

Drury, S. A. (1981). Geochemistry of Archaean metavolcanic rocks from the Kudremukh area, Karnataka. Geol. Soc. India J., v. 22, 405–416.

—— (1983a). The petrogenesis and setting of Archaean metavolcanics from Karnataka State, South India. Geochim. Cosmochim. Acta, v. 47, 317–329.

—— (1983b). A regional tectonic study of the Archaean Chitradurga greenstone belt, Karnataka, based on LANDSAT interpretation. Geol. Soc. India J., v. 24, 167–184.

——, and Holt, R. W. (1980). The tectonic framework of the South Indian craton: A reconnaissance involving LANDSAT imagery. Tectonophysics, v. 65, T1-T15.

——, and Naqvi, S. M. (1981). The oldest supracrustals of the Dharwar craton, India; Comment and reply. Geol. Soc. India. J., v. 23, 99–100.

——, ——, and Hussain, S. M. (1978). REE distributions in basaltic anorthosites from the Holenarasipur greenstone belt, Karnataka, South India. In Archaean Geochemistry; Proc. First Internat. Symp. on Archaean Geochemistry: The Origin and Evolution of Archaean Continental Crust (ed. B. F. Windley and S. M. Naqvi), Elsevier, Amsterdam, 363–374.

——, Holt, R. W., Van Calsteren, P. C., and Beckinsale, R. D. (1983). Sm-Nd and Rb-Sr ages for Archaean rocks in western Karnataka, South India. Geol. Soc. India J., v. 24, 454–459.

——, Harris, N. B. W., Holt, R. W., Reeves-Smith, G. J., and Wightman, R. T. (1984). Precambrian tectonics and crustal evolution in South India. J. Geol., v. 92, 3–20.

Foote, R. B. (1889). The "Dharwar System," the chief auriferous series in South India. Geol. Surv. India Records, v. 2, no. 1, 17–39.

Friend, C. R. L., and Mahabaleswar, B. (1985). The Kunduru Betta ring intrusion, Malavalli, Karnataka, South India. Geol. Soc. India J., v. 26, 73–83.

Ganguli, S. K., and Sen, S. (1980). A superposed

linear belt on a greenstone sequence in a portion of the Holenarasipur schist belt, Karnataka. Geol., Min. Metall. Soc. India Quart. J., v. 52, 89–106.

Glikson, A. Y., Chadwick, B., Ramakrishnan, M., Viswanatha, M. N., and Srinivasa Murthy, V. (1979). Foundation of the Sargur Group; Comment and reply. Geol. Soc. India J., v. 20, 248–255.

Gokul, A. R., and Srinivasan, M. D. (1976). Chandranath Granite, Goa. Geol. Surv. India Records, v. 107, Part 2, 38–45.

Harinadha Babu, P., Ponnuswamy, M., and Krishnamurthy, K. V. (1981). Shimoga belt. In Early Precambrian Supracrustals of Southern Karnataka (ed. J. Swami Nath and M. Ramakrishnan), Geol. Surv. India Mem. 112, 199–218.

Harris, N. B. W., and Jayaram, S. (1982). Metamorphism of cordierite gneisses from the Bangalore region of the Indian craton. Lithos, v. 15, 89–98.

Hussain, S. M., and Naqvi, S. M. (1983). Geological, geophysical, and geochemical studies over the Holenarasipur schist belt, Dharwar craton, India. In Precambrian of South India (ed. S. M. Naqvi and J. J. W. Rogers), Geol. Soc. India Mem. 4, 73–94.

———, ———, and Ghaneshwar Rao, T. (1982). Geochemistry and significance of mafic-ultramafic rocks from the southern part of the Holenarasipur schist belt, Karnataka. Geol. Soc. India J., v. 23, 19–31.

Ikremuddin, M., and Steuber, A. M. (1976). Rb-Sr ages of Precambrian dolerite and alkaline dikes, southeast Mysore State, India. Lithos, v. 9, 235–241.

Iyengar, S. V. P. (1976). The stratigraphy, structure and correlation of the Dharwar Supergroup. Geol. Surv. India Misc. Publ. 23, Part 2, 415–455.

Jafri, S. H., and Saxena, R. (1981). Geochemistry of anorthositic dyke from Nuggihalli schist belt of Dharwar craton, Karnataka, India. Geol. Soc. India J., v. 22, 85–91.

———, Khan, M., Ahmad, S. M., and Saxena, R. (1983). Geology and geochemistry of Nuggihalli schist belt, Dharwar craton, Karnataka, India. In Precambrian of South India (ed. S. M. Naqvi and J. J. W. Rogers), Geol. Soc. India Mem. 4, 110–120.

Janardhan, A. S., and Ramachandra, H. M. (1978). Anorthosites from Hullahalli, Mysore District, Karnataka State. Geol. Soc. India J., v. 19, 277–280.

———, and Srikantappa, C. (1975). Geology of the northern part of the Sargur schist belt between Mavinahalli and Dodkanya, Mysore District, Karnataka. Ind. Mineralogist, v. 16, 66–75.

———, and Vidal, P. (1982). Rb-Sr dating of the Gundlupet gneisses. Geol. Soc. India J., v. 23, 578–580.

———, Srikantappa, C., and Ramachandra, H. M. (1978). The Sargur schist complex—An Archaean high-grade terrain in southern India. In Archaean Geochemistry; Proc. First Internat. Symp. on Archaean Geochem.: The Origin and Evolution of Archaean Continental Crust (ed. B. F. Windley and S. M. Naqvi), Elsevier, Amsterdam, v. 1, 127–149.

———, Ramachandra, H. M., and Ravindra Kumar, G. V. (1979). Structural history of Sargur supracrustals and associated gneisses southwest of Mysore, Karnataka. Geol. Soc. India J., v. 20, 61–72.

———, Shadakshara Swamy, N., and Ravindra Kumar, G. V. (1981). Petrological and structural studies of the manganiferous horizons and recrystallized ultramafics around Gundlupet, Karnataka. Geol. Soc. India J., v. 22, 103–111.

Jayaram, S., Venkatasubramanian, V. S., and Radhakrishna, B. P. (1976). Rb-Sr ages of cordierite gneisses of southern Karnataka. Geol. Soc. India J., v. 17, 557–561.

———, ———, and Rajagopalan, P. T. (1984). Geochemistry and petrogenesis of the Peninsular Gneisses, Dharwar craton, India. Geol. Soc. India J., v. 25, 570–584.

Kaila, K. L., Roy Chowdhury, K., Reddy, P. R., Krishna, V. K., Narain, H., Subbotin, S. I., Sollogub, V. B., Chekunov, A. V., Kharetchko, G. E., Lazarenko, M. A., and Ilchenko, T. V. (1979). Crustal structure along the Kavali-Udipi profile in the Indian peninsular shield from deep seismic sounding. Geol. Soc. India J., v. 20, 307–333.

Krishna Murthy, M. (1974). Tectonomagmatic history of the Precambrians in Bellary District, Mysore State. Geol. Soc. India J., v. 15, 37–47.

——— (1978). Geology and mineral resources of Bellary District, Karnataka (Mysore) State. Geol. Surv. India Mem. 108, 98 p.

Krishna Rao, B., Srinivasan, R., Ramachandra, B. L., and Sreenivas, B. L. (1982). Mode of occurrence and origin of manganese ores of Shimoga District, Karnataka. Geol. Soc. India J., v. 23, 226–235.

———, Murthy, P. S. N., and Murthy, D. S. N. (1977). Mineralogy and petrology of the amphibolites and dolerites from Kudremukha, Karnataka State, South India. Geol., Min. Metall. Soc. India Quart. J., v. 49, 75–84.

Kumar, B., Venkatasubramanian, V. S., and Saxena, R. (1983). Carbon and oxygen isotopic composition of the carbonates from greenstone belts of Dharwar craton, India. In Precambrian of South India (ed. S. M. Naqvi and J. J. W. Rogers), Geol. Soc. India Mem. 4, 260–266.

Menon, A. G., Venkatasubramanian, V. S., and Anantha Iyer, G. V. (1981). Sulphur isotope abundance variations in sulphides of the Dharwar craton—Part 1, Kalyadi. Geol. Soc. India J., v. 22, 391–398.

Monrad, J. R. (1983). Evolution of sialic terranes in the vicinity of the Holenarasipur belt, Hassan District, Karnataka, India. In Precambrian of South India (ed. S. M. Naqvi and J. J. W. Rogers), Geol. Soc. India Mem. 4, 343–364.

Mookherjee, A., and Philip, R. (1979). Distribution of copper, cobalt and nickel in ores and host rocks, Ingaldhal, Karnataka, India. Mineralium Deposita, v. 14, 33–55.

Mukhopadhyay, D. (1986). Structural pattern in the Dharwar craton. J. Geol., v. 94, 167–186.

———, and Baral, C. (1985). Structural geometry of the Dharwar rocks near Chitradurga. Geol. Soc. India J., v. 26, 547–566.

———, and Ghosh, D. (1983). Superposed deformation in the Dharwar rocks of the southern part of the Chitradurga schist belt near Dodguni, Karnataka. In Precambrian of South India (ed. S. M. Naqvi and J. J. W. Rogers), Geol. Soc. India Mem. 4, 275–292.

———, Baral, M. C., and Neogi, R. K. (1980). Mineralogy of banded iron formation in the southeastern Bababudan Hills, Karnataka, India. Neues Jahrb. Mineral. Abh., v. 139, 303–327.

———, ———, and Ghosh, D. (1981). A tectono-stratigraphic model of the Chitradurga schist belt, Karnataka, India. Geol. Soc. India J., v. 22, 22–31.

Murali, A. V., Pawaskar, P. B., Reddy, G. R., Subbarao, K. V., Vasudev, V. N., and Sankar Das, M. (1979). Petrogenetic significance of rare earth element patterns of selected samples of Ingaldhal metavolcanics, Karnataka State, India; Consortium studies no. 1. Geol. Soc. India J., v. 20, 334–338.

Murthy, P. S. N., and Krishna Reddy, K. (1984). 2900 m.y. old stromatolite from Sandur greenstone belt of Karnataka craton, India. Geol. Soc. India J., v. 25, 263–266.

Naha, K., and Chatterjee, A. K. (1982). Axial plane folding in the Bababudan hill ranges of Karnataka. Ind. J. Earth Sci., v. 9, 37–43.

———, Srinivasan, R., and Naqvi, S. M. (1986). Structural unity in the early Precambrian Dharwar tectonic province, peninsular India. submitted.

Naqvi, S. M. (1967). The banded pyritiferous chert from Ingaldhal and adjoining areas of Chitaldrug schist belt, Mysore. Natl. Geophys. Res. Inst. Bull. (Hyderabad), v. 5, 173–181.

——— (1972). The petrochemistry and significance of Jogimardi traps, Chitaldrug schist belt, Mysore. Bull. Volcanol., 35, 1069–1093.

——— (1973). Geological structure and aeromagnetic and gravity anomalies in the central part of the Chitaldrug schist belt, Mysore, India. Geol. Soc. Amer. Bull., v. 84, 1721–1732.

——— (1976). Physico-chemical conditions during the Archaean as indicated by Dharwar geochemistry. In The Early History of the Earth (ed. B. F. Windley), John Wiley, New York, 289–298.

——— (1978). Geochemistry of Archaean metasediments; Evidence for prominent anorthosite-norite-troctolite (ANT) in the Archaean basaltic primordial crust. In Archaean Geochemistry; Proc. First Internat. Symp. on Archaean Geochem.: The Origin and Evolution of Archaean Continental Crust (ed. B. F. Windley and S. M. Naqvi), Elsevier, Amsterdam, 343–360.

——— (1981). The oldest supracrustals of the Dharwar craton, India. Geol. Soc. India J., v. 22, 458–469.

——— (1982). Early Archaean evolution of Indian shield with special reference to Dharwar craton. Revista Brasiliera de Geociencias, v. 112, 223–233.

——— (1983). Early Precambrian clastic metasediments of Dharwar greenstone belts; Implications to sial-sima transformation processes. In Precambrian of South India (ed. S. M. Naqvi and J. J. W. Rogers), Geol. Soc. India Mem. 4, 220–236.

———, and Hussain, S. M. (1972). Petrochemistry of some early Precambrian metasediments from the central part of the Chitaldrug schist belt, Mysore, India. Chem. Geol., v 10, 109–135.

———, and ——— (1973a). Geochemistry of Dharwar metavolcanics and the composition of the primeval crust of peninsular India. Geochim. Cosmochim. Acta, v. 37, 159–164.

———, and ——— (1973b). Relations between trace and major element composition of the Chitaldrug metabasalts, Mysore, India, and the Archaean mantle. Chem. Geol., v. 11, 17–30.

———, and ——— (1979). Geochemistry of meta-anorthosites from a greenstone belt in Karnataka, India. Can. J. Earth Sci., v. 16, 1254–1264.

———, Divakara Rao, V., Satyanarayana, K., and Hussain, S. M. (1974). Geochemistry of some post-Dharwar basic dikes and the Precambrian crustal evolution of peninsular India. Geol. Mag., v. 111, 229–236.

———, ———, ———, ——— (1976). Reconnaissance geochemical exploration for copper in the central part of the Chitradurga schist belt, Karnataka, India. Geol. Soc. India J., v. 17, 551–557.

———, Hanumanth Rao, T., Natrajan, R., Satyanarayana, K., Divakara Rao, V., and Hussain, S. M. (1977). Mineralogy, geochemistry and genesis of massive base metal sulfide deposits of Chitradurga (Ingaldhal), Karnataka, India. Precamb. Res., v. 4, 361–386.

———, Divakara Rao, V., Hussain, S. M., Narayana, B. L., Rogers, J. J. W., and Satyanarayana,

K. (1978a). The petrochemistry and geologic implications of conglomerates from Archaean geosynclinal piles of southern India. Can. J. Earth Sci., v. 15, 1085–1100.

———, Viswanathan, S., and Viswanatha, M. N. (1978b). Geology and geochemistry of Holenarasipur schist belt and its place in the evolutionary history of the Indian peninsula. In Archaean Geochemistry; Proc. First Internat. Symp. on Archaean Geochem.: The Origin and Evolution of Archaean Continental Crust (ed. B. F. Windley and S. M. Naqvi), Elsevier, Amsterdam, 109–126.

———, Narayana, B. L., Rama Rao, P., Ahmad, S. M., and Uday Raj, B. (1980). Geology and geochemistry of paragneisses from Javanahalli schist belt, Karnataka, India. Geol. Soc. India J., v. 21, 577–592.

———, Govil, P. K., and Rogers, J. J. W. (1981). Chemical sedimentation in Archaean-early Proterozoic greenschist belts of the Dharwar craton, India. In Second Internat. Symp. on Archaean Geology (ed. J. E. Glover and D. I. Groves), Geol. Soc. Australia Spec. Publ. 7, 245–253.

———, Allen, P., Subba Rao, P., and Ghaneshwar Rao, T. (1983a). Geochemistry of coexisting staurolite and kyanite from an early Archaean greenstone belt of Dharwar craton, India. In Precambrian of South India (ed. S. M. Naqvi and J. J. W. Rogers), Geol. Soc. India Mem. 4, 267–274.

———, Divakara Rao, V., Hussain, S. M., Narayana, B. L., Nirmal Charan, S., Govil, P. K., Bhaskar Rao, Y. J., Jafri, S. H., Rama Rao, P., Balram, V., Ahmad, M., Pantulu, K. P., Ghaneswar Rao, T., and Subba Rao, D. V. (1983b). Geochemistry of gneisses from Hassan District and adjoining areas, Karnataka, India. In Precambrian of South India (ed., S. M. Naqvi and J. J. W. Rogers), Geol. Soc. India Mem. 4, 401–416.

———, Condie, K. C., and Allen, P. (1983c). Geochemistry of some unusual early Archaean sediments from Dharwar craton, India. Precamb. Res., v. 22, 125–147.

Narain, H., and Subrahmanyam, C. (1986). Precambrian tectonics of the South Indian shield inferred from geophysical data. J. Geol., v. 94, 187–198.

Narayana, B. L., and Naqvi, S. M. (1980). Geochemistry of spinifex-textured peridotitic komatiites from Ghatti-Hosahalli, Karnataka, India. Geol. Soc. India J., v. 21, 194–198.

———, Subba Rao, D. V., and Govil, P. K. (1982). High alumina-magnesium sedimentation in the Javanahalli schist belt, Karnataka, India. Geol. Soc. India J., v. 23, 175–182.

———, Naqvi, S. M., Rama Rao, P., Uday Raj, B., and Ahmad, S. M. (1983). Geology and geochemistry of Javanahalli schist belt, Karnataka,

India. In Precambrian of South India (ed. S. M. Naqvi and J. J. W. Rogers), Geol. Soc. India Mem. 4, 143–157.

Narayana Das, G. R. (1970). Arsikere Granite—Its petrography and mode of emplacement. Geol. Soc. India J., v. 11, 253–258.

Nijagunappa, R., and Naganna, C. (1983). Nuggihalli schist belt in the Karnataka craton; An Archean layered complex as interpreted from chromite distribution. Econ. Geol., v. 78, 507–513.

Philip, R. (1977). On the occurrence of welded tuffs around Ingaldhal, Chitradurga, Karnataka, India. Geol. Soc. India J., v. 18, 184–188.

Pichamuthu, C. S. (1976). Some problems pertaining to the Peninsular Gneissic complex. Geol. Soc. India J., v. 17., 1–16.

———, and Srinivasan, R. (1983). A billion-year history of the Dharwar craton (3200 to 2100 m.y. ago). In Precambrian of South India (ed. S. M. Naqvi and J. J. W. Rogers), Geol. Soc. India Mem. 4, 121–142.

———, and ——— (1984). The Dharwar Craton—A Golden Jubilee Publication. Ind. Natl. Sci. Acad., New Delhi, 34 p.

Raase, P., Raith, M., Ackermand, D., Viswanatha, M. N., and Lal, R. K. (1983). Mineralogy of chromiferous quartzites from South India. Geol. Soc. India J., v. 24, 502–521.

———, ———, ———, and Lal, R. K. (1986). Progressive metamorphism of mafic rocks from greenschist to granulite facies in the Dharwar craton of South India. J. Geol., v. 94, 261–282.

Radhakrishna, B. P. (1954). On the nature of certain cordierite-bearing granulites bordering the Closepet granites of Mysore. Mysore Geol Dept. Bull., v. 21, 25 p.

——— (1967). Copper in Mysore State. Mysore Dept. Mines and Geology, Bangalore, 55 p.

——— (1976). Mineralization episodes in the Dharwar craton of peninsular India. Geol. Soc. India J., v. 17, 79–88.

——— (1983). Archaean granite-greenstone terrain of the South Indian shield. In Precambrian of South India (ed. S. M. Naqvi and J. J. W. Rogers), Geol. Soc. India Mem. 4, 1–46.

———, and Naqvi, S. M. (1986). Precambrian continental crust of India and its evolution. J. Geol., v. 94, 145–166.

———, and Sreenivasaiah, C. (1974). Bedded barytes from the Precambrian of Karnataka. Geol. Soc. India J., v. 15, 314–315.

———, and Vasudev, V. N. (1977). The early Precambrian of the southern Indian shield. Geol. Soc. India J., v. 18, 525–541.

———, Pandit, S. A., and Prabhakar, K. T. (1973). Copper mineralization in the ultrabasic complex of Nuggihalli, Hassan District, Mysore State. Geol. Soc. India J., v. 14, 302–312.

Rajagopalan, P. I., Jayaram, S., and Venkatasubramanian, V. S. (1980). Rb-Sr isochron ages of gneisses in the western region of the Dharwar craton. Geol. Soc. India J., v. 21, 54–56.

Ramakrishnan, M. (1981). Nuggihalli and Krishnarajpet belts. In Early Precambrian Supracrustals of Southern Karnataka (ed. J. Swami Nath and M. Ramakrishnan), Geol. Surv. India Mem. 112, 61–70.

———, and Harinadha Babu, P. (1981). Western Ghat belt. In Early Precambrian Supracrustals of Southern Karnataka (ed. J. Swami Nath and M. Ramakrishnan), Geol. Surv. India Mem. 112, 147–161.

———, and Mallikarjuna, C. (1976). Konkanhundi pluton: A new layered basic intrusion from Karnataka. Geol. Soc. India J., v. 17, 207–213.

———, and Viswanatha, M. N. (1981a). Hole Narsipur belt. In Early Precambrian Supracrustals of Southern Karnataka (ed. J. Swami Nath and M. Ramakrishnan), Geol. Surv. India Mem. 112, 115–141.

———, and ——— (1981b). Sigegudda belt. In Early Precambrian Supracrustals of Southern Karnataka (ed. J. Swami Nath and M. Ramakrishnan), Geol. Surv. India Mem. 112, 143–146.

———, and ——— (1983). Crustal evolution in central Karnataka: A review of present data and models. In Precambrian of South India (ed. S. M. Naqvi and J. J. W. Rogers), Geol. Soc. India Mem. 4, 96–101.

———, ———, Chayapathi, N., and Narayanan Kutty, T. R. (1978). Geology and geochemistry of anorthosites of Karnataka craton and their tectonic significance. Geol. Soc. India J., v. 19, 115–134.

———, Moorbath, S., Taylor, P. N., Anantha Iyer, G. V., and Viswanatha, M. N. (1984). Rb-Sr and Pb-Pb whole-rock isochron ages of basement gneisses in Karnataka craton. Geol. Soc. India J., v. 25, 20–34.

Ramiengar, A. S., Chayapathi, N., Raghanandan, K. R., Rao, M. S., and Rama Rao, P. (1978a). Mineralogy and geochemistry of a vanadiferous titano-magnetite deposit and associated copper minerlization in gabbro-anorthosites near Masanikere, Shimoga District, Karnataka, India. In Archaean Geochemistry; Proc. First Internat. Symp. on Archaean Geochem.: The Origin and Evolution of Archaean Continental Crust (ed. B. F. Windley and S. M. Naqvi), Elsevier, Amsterdam, 395–406.

———, Devadu, G. R., Viswanatha, M. N., Chayapathi, N., and Ramakrishnan, M. (1978b). Banded chromite-fuchsite quartzite in the older aupracrustal sequence of Karnataka. Geol. Soc. India J., v. 19, 577–582.

Ranganna, M., Shahidhara, H., and Raghu Prakash, T. R. (1981). Comparison of the Tarikere Valley gneiss and the gneiss pebbles in the Kaldurga conglomerate—A study in the Dharwar

basement problem. Geol. Soc. India J., v. 22, 570–576.

Rogers, J. J. W. (1986). The Dharwar craton and the assembly of peninsular India. J. Geol., v. 94, 129–144.

———, Callahan, E. J., Dennen, K. O., Fullagar, P. D., Stroh, P. T., and Wood, L. F. (1986). Chemical evolution of Peninsular Gneiss in Western Dharwar craton, southern India. J. Geology, v. 94, 233–246.

Roy, A. (1983). Structure and tectonics of the cratonic areas of North Karnataka. Recent Res. Geol., v. 10, 81–97.

———, and Biswas, S. K. (1979). Metamorphic history of the Sandur schist belt, Karnataka. Geol. Soc. India J., v. 20, 179–187.

Satyanarayana, K., Naqvi, S. M., Divakara Rao, V., and Hussain, S. M. (1974). Geochemistry of Archaean amphibolites from Karnataka State, peninsular India. Chem. Geol., v. 14, 305–316.

Sawkar, R. H. (1980). Geology of the manganese ore deposits of North Kanara District, Karnataka State, India. In Manganese Deposits on Continents (ed. I. M. Varentsov et al.), E. Schweizerbart'sche Verlagsbuchhandlung, Stuttgart, v. 2, 279–295.

Seccombe, P. K., Subba Rao, K. V., and Pawar, J. N. (1981). Sulphur isotopic composition of Ingaldhal sulphides, Karnataka State, India. Geol. Soc. India J., v. 22, 326–330.

Seshadri, T., Chaudhuri, A., Harinadha Babu, P., and Chayapathi, N. (1981). Shimoga belt. In Early Precambrian Supracrustals of Southern Karnataka (ed. J. Swami Nath and M. Ramakrishnan), Geol. Surv. India Mem. 112, 163–198.

Sivaprakash, C. (1983). Petrology of quartzo-feldspathic schists/phyllites associated with manganese formation of North Kanara and Kumsi, Karnataka. Geol. Soc. India J., v. 24, 571–587.

Srikantappa, C., Friend, C. R. L., and Janardhan, A. S. (1980). Petrochemical studies on chromites from Sinduvalli, Karnataka, India. Geol. Soc. India J., v. 21, 473–483.

Srikantia, S. V., and Bose, S. S. (1985). Archaean komatiites from Banasandra area of Kibbanahalli arm of Chitradurga supracrustal belt in Karnataka. Geol. Soc. India J., v. 26, 407–417.

Srinivasan, R., and Ojakangas, R. W. (1986). Sedimentology of quartz pebble conglomerates and quartzites of the Archean Bababudan Group: Evidence for early crustal stability. J. Geol., v. 94, 199–214.

Stroh, P. T., Monrad, J. R., Fullagar, P. D., Naqvi, S. M., Hussain, S. M., and Rogers, J. J. W. (1983). 3,000-m.y.-old Halekote Trondhjemite; A record of stabilization of the Dharwar craton. In Precambrian of South India (ed. S. M. Naqvi and J. J. W. Rogers), Geol. Soc. India Mem. 4, 365–376.

Subrahmanyam, C., and Verma, R. K. (1982).

Gravity interpretations of the Dharwar green-stone-gneiss-granite terrain in the southern India shield and its geological implications. Tectonophysics, v. 84, 225–245.

Suresh, R. (1982). Further evidence of biogenic nature of Dodguni microbiota. Geol. Soc. India J., v. 23, 567–574.

Swami Nath, J., and Ramakrishnan, M. (eds.) (1981). Early Precambrian supracrustals of southern Karnataka. Geol. Surv. India Mem. 112, 350 p.

Taylor, P. N., Moorbath, S., Chadwick, B., Ramakrishnan, M., and Viswanatha, M. N. (1984). Petrography, chemistry and isotopic ages of Peninsular Gneiss, Dharwar acid volcanic rocks and the Chitradurga Granite with special reference to the late Archaean evolution of the Karnataka craton, southern India. Precamb. Res., v. 23, 349–375.

Vasudev, V. N. (1983). Geological evolution of Archaean and early Proterozoic sulphide deposits of the Dharwar craton, India. In Precambrian of South India (ed. S. M. Naqvi and J. J. W. Rogers), Geol. Soc. India Mem. 4, 243–259.

——, and Srinivasan, R. (1979). Vanadium bearing titaniferous magnetite deposits of Karnataka, India. Geol. Soc. India J., v. 20, 170–178.

Venkataramana, P. (1982). Chemical remnants of the Archaean protocrust in the Sargur schist belt of Karnataka craton, India. Precamb. Res., v. 19, 51–74.

——, Anantha Iyer, G. V., and Narayanan Kutty, T. R. (1984). Chemical petrology of Archaean ultramafic rocks in the southern parts of Karnataka craton, peninsular India. Ind. J. Earth Sci., v. 11, 9–24.

Venkatasubramanian, V. S., and Narayanaswamy, R. (1974a). The age of some gneissic pebbles in the Kaldurga conglomerate, South India. Geol. Soc. India J., v. 15, 318–319.

——,and —— (1974b). Primary and metamorphic Rb-Sr chronology in some areas of South Mysore. Geol. Soc. India J., v. 15, 200–205.

——, and —— (1974c). Rb-Sr whole-rock isochron studies on granitic rocks from Chitradurga and North Mysore. Geol. Soc. India J., v. 15, 77–81.

——, Menon, A. G., and Anantha Iyer, G. V. (1981). Sulphur isotope abundance variations in sulfides of the Dharwar craton—Part II, Ingaldhal. Geol. Soc. India J., v. 22, 395–398.

——, Jayaram, S., and Subramanian, V. (1982). Lead age measurements on galenas from peninsular India. Geol. Soc. India J., v. 23, 219–225.

Viswanatha, M. N., and Ramakrishnan, M. (1981a). Sargur and allied belts. In Early Precambrian Supracrustals of Southern Karnataka (ed. J. Swami Nath and M. Ramakrishnan), Geol. Surv. India Mem. 112, 41–59.

——, and —— (1981b). Kunigal and related belts. In Early Precambrian Supracrustals of Southern Karnataka (ed. J. Swami Nath and M. Ramakrishnan), Geol. Surv. India Mem. 112, 71–76.

——, and —— (1981c). Bababudan belt. In Early Precambrian Supracrustals of Southern Karnataka (ed. J. Swami Nath and M. Ramakrishnan), Geol. Surv. India Mem. 112, 91–114.

——, ——, and Swami Nath, J. (1982). Angular unconformity between Sargur and Dharwar supracrustals in Sigegudda, Karnataka craton, South India. Geol. Soc. India J., v. 23, 85–89.

Viswanathan, S., Walvekar, A. S., Venkataraman, G., and Patel, B. S. (1981). The Sargurian metaigneous complex in and around Timmappan Beta, Hassan District, Karnataka. Geol., Min. Metall. Soc. India Quart. J., v. 53, 1–13.

Viswanathan, T. V. (1968). Zircons in Peninsular Gneiss, Bangalore, Mysore State. Geol. Surv. India Misc. Publ. 9, 154–157.

Viswanathiah, M. N., Venkatachalapathy, R., and Mahalakshmamma, A. P. (1979). Micro-organic structures from some limestones and quartzites (Sargur schistose rocks) occurring in the northern and southern parts of Sargur and their significance. Geol. Surv. India Misc. Publ. 45, 13–16.

Yellur, D. D., and Nair, R. S. (1978). Assigning a magmatically defined tectonic environment to Chitradurga metabasalts, India, by geochemical methods. Precamb. Res., v. 7, 259–281.

Ziauddin, M., Roy, A., Biswas, S. K., and Gururaja Rao, T. P. (1978). Volcanism in the younger Dharwar rocks near Medur, Dharwar District, Karnataka. Geol. Soc. India J., v. 19, 321–325.

3. EASTERN DHARWAR CRATON

The western margin of the Eastern Dharwar craton is either in the region of the Closepet Granite suite or along a major shear zone somewhat west of the Closepet outcrop (Ch. 2). The southern margin is the gradational transition zone to granulite-facies rocks that extends east-west across southern India (Chs. 1 and 4). The eastern margin of the craton is a thrust fault beneath the granulite facies of the Eastern Ghats (Ch. 5). The fault is between the Cuddapah Basin and the Nellore and related schist belts (Figs. 3.1 and 5.1), where relations between gneiss and granulite are obscured. The contact has been more extensively studied to the north, in the Bastar area (Ch. 6), where it shows charnockitic rocks forming at the expense of amphibolite-facies gneisses. The northern margin of the Eastern Dharwar craton, where it is not overlapped by the Deccan basalts, is the Godavari rift valley (Ch. 10).

The seismic profile of Kaila et al. (1979) shows a number of deep, near-vertical, faults, but most do not have a surface expression. The shear zones proposed by Drury et al. (1984) in southern India do not transect the eastern part of the Dharwar craton. Numerous dike sets have been mapped in various areas (Drury, 1984), and they presumably indicate some development of late fracture patterns.

ROCK SUITES AND COMPARISON WITH WESTERN DHARWAR CRATON

Figure 3.1 shows only four lithologic patterns for Precambrian rocks: schist belts, undifferentiated gneiss, granite, and Late Proterozoic basins (chiefly the Cuddapah basin). Much less work has been done in the eastern part of the Dharwar craton, except in local areas (e.g., the Closepet Granite, Kolar belt, Cuddapah basin), than in the western part or in many other areas of the Indian shield. Only recently has there been a recognition that much of the "gneissic" terrain is occupied by late- or posttectonic granites (Narayanaswamy, 1971; Srinivasan and Sreenivas, 1977; Radhakrishna and Naqvi, 1986).

The schist belts contain both metavolcanic and metasedimentary rocks. All of these belts are generally regarded as correlative with the Bababudan Group of the Western Dharwar craton (Ch. 2). Similarly, the Sakarsanahalli supracrustal en-

claves near the Kolar belt have been correlated with the Sargur enclaves of the western part of the craton.

The granites have been studied in two principal areas, the arcuate Closepet trend, and the apparently more diffuse set of plutons in the vicinity of Hyderabad.

Proterozoic (to early Paleozoic?) undeformed, or slightly deformed, sedimentary suites occur in several areas. The major one is the Cuddapah basin. The Bhima basin occurs in the northwestern part of the craton. A possible equivalent of the Bhima suite is the Kaladgi basin, which is in the Western Dharwar craton but is discussed in this chapter. Some parts of the Pakhal sequence of the Godavari rift (Ch. 10) overlap onto the northeastern part of the craton. The distinction between the Pakhal sequence and the cratonic cover sequences has been based on the fact that Pakhal and Cuddapah outcrops trend at 90° to each other.

Several differences have been proposed between the lithologic assemblages of the Eastern and Western Dharwar cratons. The major ones are as follows:

Ultramafic/anorthositic suites, particularly layered complexes, have not been described in the eastern part of the craton, although they are abundant in the west. Komatiites are preseent in some of the eastern schist belts, but intrusive equivalents are absent or rare.

Mafic volcanic rocks may be more abundant in the eastern schist belts than in those to the west. The set of belts extending north from Kolar contains a dominantly metavolcanic assemblage, but other belts farther east, and the Sandur belt to the west, contain mostly metasedimentary suites. Thus, the significance of this difference is unknown.

The high-Mg, high-Al sediments that characterize some of the older schist belts in the Western Dharwar craton are absent from eastern belts.

Metamorphism of mafic enclaves in the gneisses to cordierite-bearing assemblages is common around the Closepet granite outcrop and in some other areas near the gneiss-granulite transition zone. These suites indicate metamorphism at higher T/p ratios than in the western part of the craton.

The abundance of potassic, Closepet-type granites

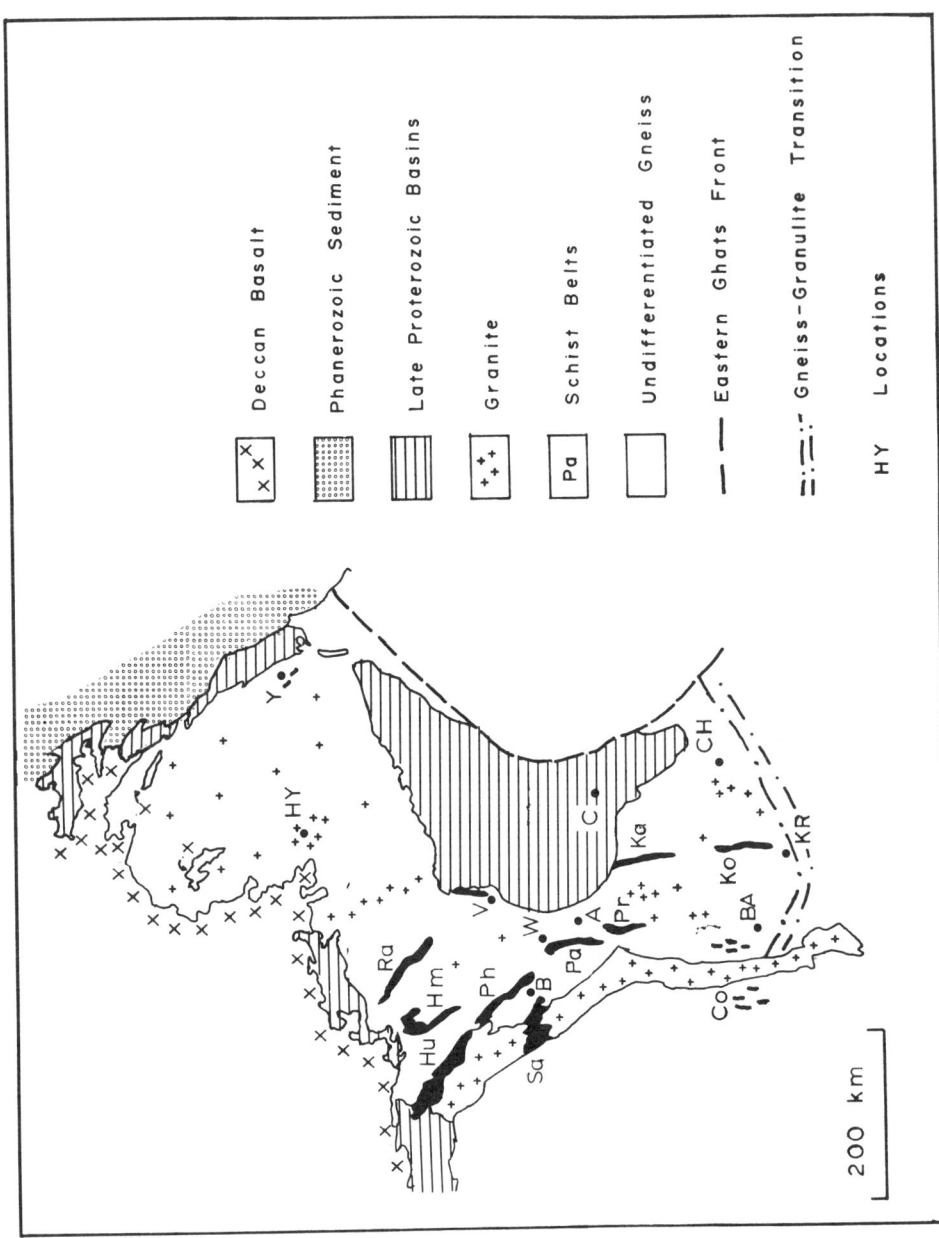

Fig. 3.1. Eastern Dharwar craton. Locations are Y, Yellandlapad; HY, Hyderabad; B, Bellary; V, Veldurti; W, Wajrakarur; A, Anantapur; C, Cuddapah; CH, Chittoor; BA, Bangalore; KR, Krishnagiri. Schist belts include Ra, Raichur; Hm, Hutti-Maski; Hh, Hungund; Pr, Penner-Hagiri. Sa, Sandur; Pa, Pamidi; Pr, Pennukonda-Ramagiri; Ka, Kadiri; Co, cordierite-bearing enclaves; Ko, Kolar.

is much higher in the eastern part of the craton than toward the west.

AGES OF ROCKS AND EVENTS

Most of the ages of rocks and events in the eastern part of the Dharwar craton are inferred from measurements in the Western Dharwar craton. The ages specifically determined within the eastern part of the craton are whole-rock determinations as follows:

3,010 ± 90 m.y. with an initial $^{87}Sr/^{86}Sr$ of 0.701 ± 0.001 on cordierite gneisses on the western and eastern side of the Closepet Granite (Jayaram et al., 1976).

2,950 ± 110 m.y. age with an initial $^{87}Sr/^{86}Sr$ of 0.706 ± 0.004 on gneisses and admixed granites at Bangalore (Venkatasubramanian and Narayanaswamy, 1974); an isochron involving biotite and an aplite dike yielded an age of 2,625 ± 90 m.y. with an initial $^{87}Sr/^{86}Sr$ of 0.703 ± 0.004.

2,886 ± 96 m.y. with an initial $^{87}Sr/^{86}Sr$ of 0.702 ± 0.002 on the Champion Gneiss in the Kolar schist belt (Bhalla et al., 1978).

2,590 ± 40 m.y. with an initial $^{87}Sr/^{86}Sr$ of 0.7016 ± 0.001 on Peninsular Gneiss near Bangalore (Crawford, 1969); this age is lower than is commonly assumed for the gneiss and may represent the effects of the Closepet Granite.

2,380 ± 35 m.y. with an initial $^{87}Sr/^{86}Sr$ of 0.7049 ± 0.0014 on the Closepet Granite (Crawford, 1969).

2,490 ± 115 m.y. with an initial $^{87}Sr/^{86}Sr$ of 0.7045 ± 0.0044 on granites near Hyderabad (Crawford, 1969).

greater than 1,000 m.y. by K/Ar methods on kimberlites at Wajrakarur and Lattavaram (Paul et al., 1975b); a presumably associated dolerite dike, less weathered than the kimberlites, yielded an age of 1,446 ± 60 m.y.

The general chronologic relationships are similar to those in the western part of the craton. Mafic schist belts have been invaded by younger, potassic, granites and perhaps also by suites commonly regarded as part of the Peninsular Gneiss. Relationships between old parts of the gneiss and the schist belts are unknown. Radhakrishna and Naqvi (1986) proposed that the craton was mobilized in the age range of 2,600 to 2,000 m.y. ago, associated with development of potassic granites in the upper crust, where water-rich fluids had been driven by charnockitization in the lower crust. The craton was a stable block by about 2,000 m.y. ago. The Cuddapah and other basins formed during the middle Proterozoic and younger ages, perhaps associated with lithospheric flexure around the craton margins (Radhakrishna and Naqvi, 1986).

METAMORPHISM

Metamorphic conditions have been estimated at various places in the Eastern Dharwar craton.

Most studies (Ch. 4) place metamorphic conditions in the gneiss-granulite transition zone in the range of 6 to 8 kb and 700° to 800°. Rollinson et al. (1981), however, estimated up to 12 kb pressure in rocks along this zone east of the Closepet Granite.

Supracrustal enclaves (Sakarsanahalli association) around the Kolar belt show mineral assemblages that include cordierite-sillimanite-biotite, cordierite-anthophyllite, hornblende-plagioclase-quartz-diopside-epidote, and diopside-plagioclase-hornblende-epidote-clinozoisite/zoisite (Viswanatha and Ramakrishnan, 1981). Narayanan Kutty and Anantha Iyer (1977a, 1977b) estimated metamorphic conditions of about 600° and 3 to 5 kb.

Schistose enclaves around the southern part of the Closepet Granite are characterized by cordierite-bearing assemblages (Radhakrishna, 1954; Jayaram et al., 1976). The western enclaves include the Kunigal and related belts (Ch. 2). Mineral assemblages include cordierite-garnet-sillimanite-quartz-biotite, cordierite-anthophyllite-biotite, and manganiferous calc-silicate rocks. Rollinson et al. (1981) estimated metamorphic pressures of about 6 kb and temperatures about 800°. Harris and Jayaram (1982), however, estimated peak metamorphic conditions of 690° to 730° and 4.5 to 5 kb.

The Sandur schist belt (Ch. 2) occurs mostly in greenschist facies, with amphibolite-facies rocks near the margins (Roy and Biswas, 1979). Some assemblages contain andalusite in presumed metapelites, and Harris and Jayaram (1982) estimated equilibration conditions at about 3 kb and 600°.

STRATIGRAPHY OF MAFIC SCHIST BELTS

Locations of various schist belts are shown in Figure 3.1, based primarily on the maps of Narayanaswamy (1971, 1975), Viswanatha and Ramakrishnan (1975), and Srinivasan and Sreenivas (1977). The belts tend to be complexly deformed, engulfed in at least some phases of the gneisses, and thoroughly metamorphosed. Thus, stratigraphic relationships are difficult to establish.

The majority of the studied belts are just east of the Closepet Granite (Fig. 3.1). The best-studied belt is Kolar, in part because of its economically important gold deposits. Several belts occur in a broad arc north of Kolar, including the Penukonda-Ramagiri belt (Ballal, 1975), and the Pamidi and Hutti-Maski belts (Ballal, 1975; Roy, 1979). Viswanatha and Ramakrishnan (1975) included these belts plus Hungund and Sandur in their suite of high-grade schist belts.

Studies of schist belts farther east in the craton

(Fig. 3.1) are sparse. Some information is available on the Yellandlapad area (Ramamohana Rao and Borreswara Rao, 1972), the Veldurti area (S. K. Bhattacharya, 1975), and Kadiri (Raju et al., 1979; Satyanarayana et al., 1980). The rock suites in these belts are primarily graywackes and phyllites with some volcanic component.

Enclaves of schist belts similar to the Sargurs of the Western Dharwar craton occur in the southern part of the terrain. They include the Sakarsanahalli area, west of the Kolar belt (Narayanan Kutty and Anantha Iyer, 1977a, 1977b) and similar cordierite-bearing enclaves on both sides of the Closepet Granite (Radhakrishna, 1954; Jayaram et al., 1976).

Kolar, Hutti, Maski, and Pamidi belts

Stratigraphic relationships in the Kolar belt are based primarily on the summary by Viswanatha and Ramakrishnan (1981). They regarded the enclaves, such as Sakarsanahalli, around the Kolar belt plus the belt itself as a stratigraphic unit. The Sakarsanahalli association was considered to be older than the Peninsular Gneiss, with the main part of the belt probably younger. Stratigraphic relationships cannot be established in the enclaves because of dispersal of the various rock types throughout the gneisses. The coherent part of the Kolar belt is a multiply refolded synform plunging northward and showing structures similar to those of the Western Dharwar craton (Plate 3.1a). Most of the rocks are amphibolites (Ramachan-

dra Rao, 1937). An approximate stratigraphy is based roughly on a decreasing age westward across the belt (Table 3.1), although Rajamani et al. (1985) considered komatiitic rocks and the Champion Gneiss to be among the oldest members of the suite.

The Hutti, Maski, and Pamidi belts are grouped together, with the Pamidi belt of Ballal (1975) being a southeastern extension of the Hutti belt (Fig. 3.1). Deformation and metamorphism are sufficiently intense that a stratigraphy cannot be well established. More than 90% of the rocks are metabasalts, andesites, and minor silicic volcanic rocks metamorphosed to greenschist to amphibolite facies (Ballal, 1975; Roy, 1979; Anantha Iyer et al., 1980). Talc-chlorite-actinolite schists and serpentine, presumably metaultramafic rocks, are also present; whether these rocks are small intrusive bodies or former komatiites has not been established. Intercalated metasedimentary rocks include banded iron formations, quartzites, quartz-sericite schists, and graywackes-argillites.

A belt to the south of Pamidi, the Penukonda-Ramagiri belt (Ballal, 1975) contains amphibolites, banded iron formations, and phyllites, presumably metasedimentary).

Structural studies in the Hutti-Maski belt yield results very similar to those in the Western Dharwar craton. Roy (1979) described two major episodes of tight folding, the second one with a northwest-southeast orientation and steep plunges, followed by minor and more open folding in a northeast-southwest direction.

TABLE 3.1. *Stratigraphy of Kolar belt (modified from Viswanatha and Ramakrishnan, 1981)*

Upper Unit

Metabasalts consist of hornblende, plagioclase, actinolite, minor epidote, and quartz; contains pillows and amygdules; massive and schistose varieties; intermixed with minor rock types such as pyroxenite and gabbro sills and schistose to phyllitic rocks containing graphite and pyrite

Champion Gneiss consists mainly of fine-grained, micaceous, quartz-feldspar rock with some hornblende, garnet, and accessory minerals; structures have been proposed to be fiamme and flow foliation (e.g., Ziauddin, 1975), and the unit has, therefore, been regarded as metamorphosed extrusive (silicic ignimbrite, rhyolite, agglomerate, etc.); other workers have regarded the Champion Gneiss as intrusive; age is about 2,900 m.y. (Bhalla et al., 1978); suite also contains conglomerates (controversial as to whether they are polymictic sedimentary or autoclastic tectonic); interbedded with ironstones

Amphibolites, which constitute the major part of the belt; consist of hornblende, plagioclase, minor epidote, and accessories; schistose rocks are most abundant, but massive, granular, and minor fibrous varieties are also present (Ramachandra Rao, 1937; Rajamani et al., 1981); some pillow structures

Unconformity above Peninsular Gneiss?; Viswanatha and Ramakrishnan regarded the high degree of deformation in the Peninsular Gneiss, which is not recorded in the amphibolites of the belt, as indicative of an unconformity; no overlapping contact has been found, however, with all margins of the schist belt occupied by younger granodioritic and granitic rocks

Lower Unit

Sakarsanahalli association; enclaves in the gneisses consist of (1) cordierite-sillimanite-muscovite quartzite; (2) fuchsite-quartzite; (3) cordierite-sillimanite-corundum-mica schist (presumably metapelite); (4) cordierite-anthophyllite rocks; and (5) manganiferous marbles and calc-silicates, containing minerals such as manganoan calcite, kutnahorite, manganoan amphiboles, diopside, garnet, epidote, and clinosoisite/zoisite (Narayanan Kutty and Anantha Iyer, 1977a, 1977b)

(a)

(b)

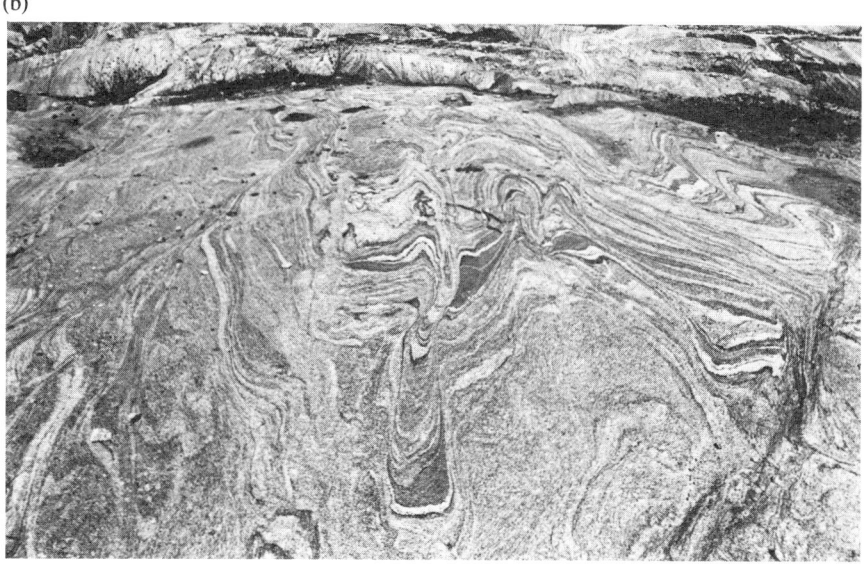

Plate 3.1. (*a*) Minor isoclinal (F_1) and refolded F_2 structures in banded iron formation of Kolar schist belt. The structure to the west of the coin is a minor replica of the Chitradurga schist belt of the Western Dharwar craton. (*b*) Ductile deformation in Peninsular Gneisses near Bangalore. These gneisses are about 3,000 m.y. old and show structural unity with greenstone belts of about 2,600 m.y. age.

Cordierite schists and granulites around Closepet Granite

Conformable mafic bodies in the Peninsular Gneiss were described by Radhakrishna (1954) and Jayaram et al. (1976). Jayaram et al. dated them as about 3,000 m.y. old. Depending on distance from the granulitic rocks in the south of the craton, the schists range from amphibolite to granulite facies. The mineralogy in amphibolite facies is cordierite, sillimanite, quartz, plagioclase ± K-feldspar, ± garnet and accessory minerals. At slightly higher degrees of metamorphism, the mineralogy is cordierite, hypersthene, quartz, biotite, spinel, garnet, and magnetite; some rocks contain anthophyllite.

PETROLOGY OF ROCKS IN SCHIST BELTS

Most rocks in schist belts of the Eastern Dharwar craton are similar to varieties that have been described in Chapter 2 and are not further described here. They include banded iron formations, quartzites, mica schists (metapelites), and some types of calc-silicates. The three rock types discussed here occur in greater abundance in the Eastern Dharwar craton or have been better studied there, including amphibolites, cordierite-bearing schists, and Mn-rich calc-silicates.

Amphibolites

Rajamani et al. (1981, 1985) found two distinct compositional varieties of amphibolites in the Kolar belt: komatiitic rocks with MgO contents > 14%; and low-K tholeiitic rocks with MgO contents < 10% (Table 3.2; Fig. 3.2). The komatiitic rocks are relatively minor in abundance, have a fibrous texture, and contain two different amphiboles (Al-rich hornblende and Al-poor actinolite). The tholeiitic rocks occur as massive, granular, and schistose types, with the schistose amphibolite being the major rock of the belt. The amphibolites contain pillow and other subaqueous flow

TABLE 3.2. *Compositions of amphibolites*

	1	2	3	4	5	6	7
SiO_2	47.1	49.9	50.5	51.8	51.2	49.1	50.5
TiO_2	0.92	0.68	0.78	0.53	0.73	0.62	0.73
Al_2O_3	11.8	14.5	15.7	15.0	14.1	11.1	14.2
Fe_2O_3	13.9	12.3	11.7	9.9	1.9	1.4	12.2
FeO					10.5	10.4	
MnO					0.27	0.16	0.14
MgO	14.9	9.1	8.2	9.0	7.1	14.0	6.0
CaO	11.5	10.5	12.4	12.2	10.6	8.6	11.6
Na_2O	0.97	3.5	1.7	2.3	2.4	2.3	2.3
K_2O	0.21	0.21	0.13	0.13	0.33	0.23	0.30
Rb							
Sr					131	125	106
Y							
Zr					66	50	64
V							
Cr					260	525	110
Ni					165	207	205
Th							
U							

All abundances of major elements are in weight percent of oxides. Where no entry is made in the FeO position, the value at Fe_2O_3 is total iron calculated as ferric oxide. All abundances of trace elements are in parts per million by weight. Values are taken directly from quoted sources and not recalculated to 100%.

Table 3.2. Column References
1. Kolar belt. Fibrous amphibolite (Rajamani et al., 1981).
2. Kolar belt. Granular amphibolite (Rajamani et al., 1981).
3. Kolar belt. Massive amphibolite (Rajamani et al., 1981).
4. Kolar belt. Schistose amphibolite (Rajamani et al., 1981).
5. Kolar belt. Basaltic amphibolite (Anantha Iyer and Vasudev, 1979).
6. Kolar belt. High-Mg basaltic amphibolite (Anantha Iyer and Vasudev, 1979).
7. Hutti belt. Basaltic amphibolite (Anantha Iyer and Vasudev, 1979).

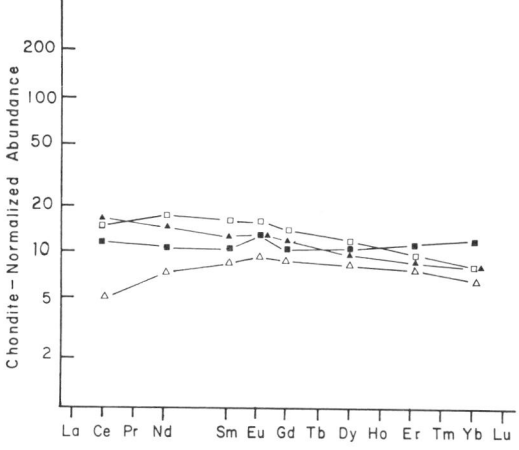

Fig. 3.2. REE patterns in Kolar amphibolites (Rajamani et al., 1985). Open squares and open triangles, high-MgO series from west/central part of belt; closed squares, tholeiitic series from west/central part; closed triangles, amphibolites from central part.

structures. The high-Mg basalts of Anantha Iyer and Vasudev (1979) and the fibrous amphibolites of Rajamani et al. (1981) are similar in whole-rock composition but are not identical. The REE patterns reported by Anantha Iyer and Vasudev are quite different from those shown in Figure 3.2 from Rajamani et al., and the reasons for the difference are not known.

All data show Mg-numbers (ratio of MgO/MgO + FeO) less than 0.6 for Kolar amphibolites. Because a value of > 0.7 is generally regarded as indicative of equilibration of magma with mantle olivine, it is unlikely that even the highest-Mg basalts are direct magmatic fractions from the mantle. Thus, presumably all basalts in the Kolar belt have undergone some type of magmatic differentiation after formation of the initial melts. Variations in REE patterns (Fig. 3.2) cannot be explained by fractionation of different magmas from a source of identical composition, and apparently there were at least two magma source regions, one enriched in light REE and one depleted. Rajamani et al. (1985) concluded that the komatiitic magmas were formed by partial melting at depths greater than 80 km and temperatures higher than 1,575°; tholeiitic magmas were formed by melting above 80 km.

Conclusions concerning basalts based on geochemical data are consistent with other observations concerning the distribution of amphibolites. The geochemical information shows that compositional modification of the mantle must have occurred before the partial melting event that formed the magmas. These magmas apparently formed approximately synchronously along an arc of several hundred kilometers length (from Kolar to Hutti) at the same time as graywacke-argillite sedimentation was occurring elsewhere in the craton. These features are, in very general terms, characteristic of rift valleys and associated volcanic activity.

Calcareous rocks of Sakarsanahalli

The assemblage of calcareous, Mn-rich, and pelitic rocks engulfed in Peninsular Gneiss at Sakarsanahalli (Fig. 3.3) is clearly a metasedimentary suite (Narayanan Kutty and Anantha Iyer, 1977a, 1977b). Metamorphic conditions have been determined by using the temperature dependence of the distribution of Mn and Fe between garnet and pyroxene, the presence or absence of muscovite and/or cordierite in metapelites, and the positions of phase equilibria in the system $CaO-Al_2O_3-SiO_2-H_2O-CO_2$ (Fig. 3.4). The high Mn content in some rocks permits the distribution of Mn to be used for equilibration studies.

At Sakarsanahalli, the presence of cordierite indicates a maximum pressure of about 5 to 6 kb because of breakdown to garnet and sillimanite above those pressures. The presence of muscovite in the pelites indicates a minimum pressure of about 3.5 kb because of the breakdown to K-feldspar below that pressure. The absence of muscovite in surrounding migmatitic gneisses indicates local anatexis, presumably requiring temperatures of at least 600°C in water-rich environments. The measured distribution coefficients of Mn and Fe between garnet and clinopyroxene are consistent with temperatures of about 600°C. The general estimate is for metamorphism at about 600° or higher and 3 to 5 kb.

The partial pressure of CO_2, and the CO_2/H_2O ratio, must have been highly variable from one rock to another in order to permit the coexistence of mineral assemblages stable under conditions of different p_{CO_2} and p_{H_2O}. For example, garnet could only have developed in the calc-silicate rocks at low CO_2 pressure (Fig. 3.4), whereas the coexisting assemblage of tremolite-diopside-calcite-plagioclase required $p_{CO_2}/(p_{CO_2} + p_{H_2O}) > 0.85$. This conclusion is interesting in view of the evidence that charnockitization, which occurred just south

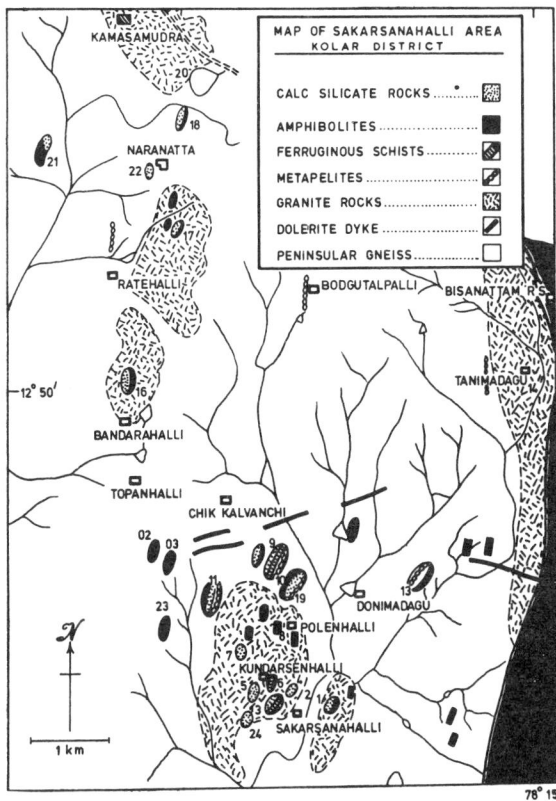

Fig. 3.3. Sakarsanahalli enclaves (Narayanan Kutty and Anantha Iyer, 1977a). Numbers indicate locations of samples used in equilibration studies (Fig. 3.4). Reproduced with permission from the Indian Journal of Earth Sciences, v. 4, p. 142.

of the Sakarsanahalli area, was accomplished partly by CO_2 influx into the rocks (Ch. 4). The Sakarasanahalli area may represent a region that was at a slightly higher elevation than the granulite front during metamorphism, with greater depth in the crust now exposed to the south.

PENINSULAR GNEISS

Data summarized in Chapter 2 led to the conclusion that most of the gneisses in the Western Dharwar craton are metaigneous, with some metasedimentary component and some areas formed by late-tectonic addition of magmatic rocks that have been only slightly deformed after intrusion. The gneisses in both the eastern and western parts of the craton are lithologically similar and presumably formed in similar ways. The gneisses of both parts of the craton have undergone similar patterns of deformation and show evidence of ductile behavior throughout the area (Plate 3.1b).

Information on gneisses in the Eastern Dharwar craton is very limited. The only work in the northern part of the craton is that of Perraju and Natarajan (1977). Ages have been determined only at Bangalore, where the gneisses were estimated to be about 2,600 m.y. old by Crawford (1969) and 3,000 m.y. old by Venkatasubramanian and Narayanaswamy (1974). We discuss two

issues here: the Champion Gneiss of the Kolar schist belt and the geochemical evolution of gneisses in the Krishnagiri-Dharmapuri area.

Champion Gneiss

The Champion Gneiss occurs in the Kolar belt. Because the rock is silicic, foliated, and shows mineral banding, it is properly classified as a gneiss (Rama Rao, 1961). This terminology has led some investigators to group the Champion Gneiss as part of the Peninsular Gneiss suite. The gneiss has also been regarded by some geologists as a stratigraphic unit that could be correlated at various places across the craton (Rama Rao, 1961). Rocks regarded as having a similar lithology were then used as time markers.

The Champion Gneiss is interlayered with amphibolites and undoubted metasediments (Viswanatha and Ramakrishnan, 1981). The formation shows volcanic features such as fiamme structure (Ziauddin, 1975). Included zircons are typically igneous (Srinivas, 1968). Associated conglomeratic rocks are controversial, either autoclastic (tectonic) or oligomictic (sedimentary). Regardless of the origin of the conglomerates, at least part of the Champion Gneiss is volcanic and should be regarded as a member of the Kolar schist belt suite rather than part of the Peninsular

Gneiss. Furthermore, it is clearly impossible to correlate the Champion Gneiss with other gneissic rock suites in the craton.

Evolution of gneisses in the Krishnagiri-Dharmapuri area

The transition between Peninsular Gneiss and charnockite has been studied in detail by Condie et al. (1982) and Allen et al. (1983) in the Dharmapuri-Krishnagiri areas, south of Bangalore (Figs. 3.1 and 4.2; Ch. 4). Briefly, the zone shows a gradational conversion from amphibolite-facies gneiss in the north to charnockite in the south. The three major rock types of the area are tonalitic gneiss, granitic gneiss, and charnockite (Table 3.3).

The principal comparative properties of the three rock suites are as follows. Granitic gneisses have more K_2O and other lithophilic elements (Rb, Ba, Th) than the tonalitic gneisses. Tonalitic gneisses and charnockites are compositionally similar, with the exception of very small, and perhaps statistically insignificant, depletion in lithophilic elements in the charnockites. All three rock types show light REE enrichment, with the tonalitic gneisses containing slightly less total REE than the other two rock types. Most of the rocks have positive Eu anomalies. The size of the Eu anomaly (measured as Eu/Eu*) is inversely correlated with total REE in all rocks. The total REE also decreases with increasing SiO_2 content.

A cogenetic relationship between the tonalitic gneisses, granitic gneisses, and charnockites can be tested using these data. Condie et al. (1982) and Allen et al. (1983) tested five models:

A tonalitic parent rock underwent partial melting to produce the granitic gneiss, leaving charnockite as a residue.

A tonalitic parent magma underwent fractional cystallization, forming a liquid with the composition of the granitic gneiss and leaving a charnockite solid residue.

A tholeiitic parent liquid underwent fractional crystallization to produce a liquid with the composition of the tonalitic gneiss.

An amphibolite source (possibly from the rocks in the schist belt) underwent partial melting to form a magma with the composition of the tonalitic gneiss; development of the charnockite was then caused by a later process.

The compositions of rocks in the area were largely affected by metasomatic processes. One posssibility is metasomatism of the gneisses by water-rich fluids and of the charnockites by CO_2-rich fluids.

The possibility that a tonalitic parent fractionated into granite and charnockite, either by partial melting or fractional crystallization, can be rejected for several reasons. One reason is that REE are preferentially fractionated into melt phases in either process and, therefore, total REE should be inversely correlated with MgO and total Fe con-

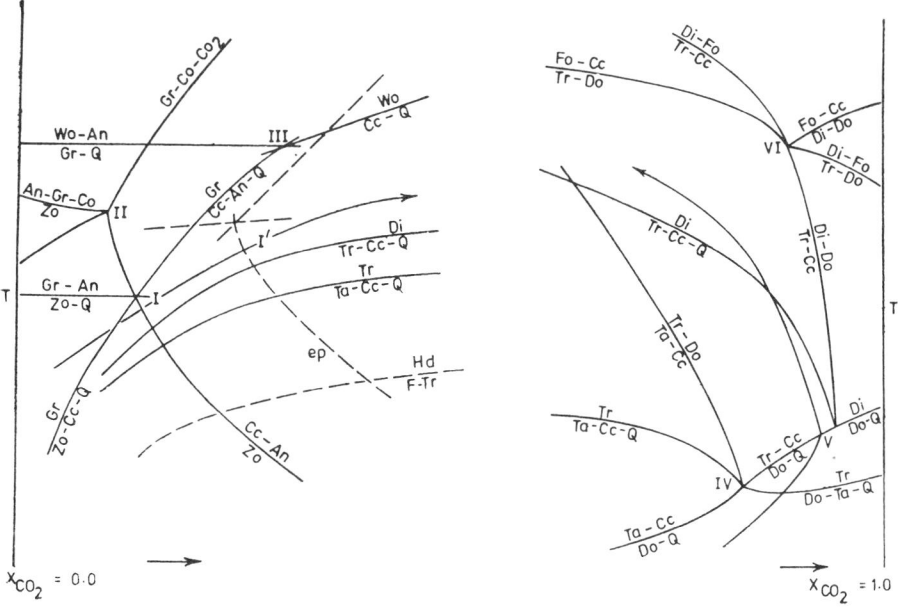

Fig. 3.4. Equilibrium conditions at Sakarsanahalli (Narayanan Kutty and Anantha Iyer, 1977a). Wo, wollastonite; An, anorthite; Gr, grossularite; q, quartz; Co, cordierite; Zo, zoisite; Ca, calcite; Tr, tridymite; Di, diopside; Hd, hedenbergite; F, fluid. Ta, talc; Do, dolomite; Fo, forsterite. Numbers indicate invariant points. Reproduced with permission from the Indian Journal of Earth Sciences, v. 4, p. 155.

TABLE 3.3. *Compositions of silicic rocks in Krishnagiri-Dharmapuri area (from Allen et al., 1983)*

	1	2	3
SiO_2	69.8	68.5	67.2
TiO_2	0.35	0.42	0.45
Al_2O_3	15.6	15.8	16.1
Fe_2O_3	2.7	3.1	3.7
FeO			
MnO			
MgO	0.87	1.1	1.7
CaO	3.3	2.8	4.3
Na_2O	4.9	4.0	5.0
K_2O	1.4	3.3	1.2
Rb	46	87	34
Sr			
Y			
Zr	177	203	180
V			
Cr	12	8	8
Ni			
Th	3.2	13	1.8
U			

All abundances of major elements are in weight percent of oxides. Where no entry is made in the FeO position, the value at Fe_2O_3 is total iron calculated as ferric oxide. All abundances of trace elements are in parts per million by weight. Values are taken directly from quoted sources and not recalculated to 100%.

Table 3.3. Column References
1. Tonalitic gneisses.
2. Granitic gneisses.
3. Charnockite.

tents and positively correlated with SiO_2 and lithophilic elements in all rocks. In fact, the opposite correlations are found (Fig. 3.4). Another problem is the lack of relationship between Eu/Eu* and modal plagioclase content, which should be positively correlated in any process involving equilibration of plagioclase and melt. There are also no negative Eu/Eu* values, which should occur in residual liquids (such as granites) from which plagioclase had been removed.

Fractionation of a tholeiitic parent magma to produce gneisses is attractive because of the necessity to separate continental sialic crust, such as gneiss, from an initial mantle source, possibly through an intermediate basaltic magma. Testing this model with REE elements does produce a negative correlation between total REE and SiO_2 abundance by a two-stage extraction of plagioclase, clinopyroxene, and hornblende, leaving a liquid amounting to 30% of the initial melt and having a SiO_2 content of 70%. Removal of plagioclase, pyroxene, and amphibole, however, cannot create the positive Eu anomalies found in some of the tonalites. Therefore, the initial tholeiite must have had a very large, positive, Eu anomaly at the start of the fractionation process, and basaltic rocks with such an anomaly have not been found anywhere in the world. Another difficulty is that the Co and Cr contents of the tonalites are much higher than would be possible if phases separating from a typical tholeiitic melt caused Rayleigh fractionation of the ferrous metals.

Tests of partial melting of amphibolite are more difficult to evaluate. It is necessary to make assumptions about the compositions of the source rock and its component minerals and the percentages of each mineral melting. The problem with a partial melt origin of the tonalitic gneisses, and more acutely for the granitic gneisses, is the abundance of lithophilic elements. For example, formation of a tonalitic magma with 1.4% K_2O (average for the gneisses being tested) from a tholeiitic amphibolite with about 0.1% K_2O (typical value) permits melting of not more than 7% of the original rock. Even smaller percentages of melting could be calculated for other lithophilic elements.

The problem of the small percentage of the parent rock that could melt to form the tonalite magmas becomes less important if the lithophile element content of the source is higher than is normally observed in amphibolites. Thus, if the tonalites were formed by partial melting of amphibolite, it is likely that the amphibolites had been enriched in lithophilic elements before the melting occurred. This enrichment could have been a metasomatic event.

The difficulties with all of the models involving tonalitic magmas are basically related to the assumption that the gneisses have approximately the same composition as the melts from which they were formed. If, however, their compositions are at least partly controlled by metasomatism, then the problems are reduced. This possibility is particularly attractive because the elements that cause most problems in the models are lithophilic ones (K, Rb, Ba, light REE), which are expected to be most mobile during metasomatism. Condie et al. (1982) and Allen et al. (1983) pointed out that hydrous metasomatism of the gneisses, causing lithophile enrichment, and CO_2 metasomatism of the charnockites, causing mild LIL depletion, could account qualitatively for many of the compositional relationships between the various rock types.

Metasomatism and the evolution of Gneisses

The preceding discussion has shown the near impossibility of evolving the silicic gneisses of a well-studied part of the Dharwar craton by simple magmatic crystallization. Some major elements, such as SiO_2 and Al_2O_3, can be explained by this process, but lithophilic elements and ferrous elements (Co, Cr) are too abundant in the gneisses for such a process. We regard this conclusion as further confirmation of the likelihood that late-stage metasomatism has been a major process in the entire shield. For example, Cr contents are high not only in the gneisses studied by Condie et al. (1982) and Allen et al. (1983) but also in many metasediments, including chert. The abundances of K, possibly Na, and many minor elements probably are quite different now than they were in the original magmas or sediments from which the rocks were formed.

GRANITES

The granitoid rocks of the Eastern Dharwar craton can be separated into two suites. One is bodies of tonalitic to granodioritic composition intermingled with the Peninsular Gneiss. Except on very detailed maps, they are grouped with the Peninsular Gneisses and commonly described as part of that suite. Rocks in this suite probably range from syntectonic to posttectonic.

A second intrusive suite is the granitic rocks discussed in this section. They include some diorites and granodiorites, but the major rock types are rich in K-feldspar (granites and quartz monzonites). Some bodies are diapiric, although gradational margins are present. Flow foliation is locally present near pluton margins, but pervasive foliation is absent. The principal characteristics that distinguish intermediate members of the suite (granodiorites, etc.) from older rocks mapped with the gneisses are their association with potassic granites, which are absent from the general gneissic terrain, and their distinct textural and compositional discordance with the surrounding gneisses. The Closepet and Hyderabad granites have been dated as about 2,400 to 2,500 m.y. old (Crawford, 1969). Dates are not available for other bodies.

The posttectonic granites have not been extensively mapped. Narayanaswamy (1971, 1975) showed general locations of granites in belts trending northwest-southeast. Srinivas and Sreenivasan (1977) located apparent late-tectonic bodies by LANDSAT imagery. The only body that has been extensively studied is the series of plutons that form the Closepet Granite. A number of individual studies have been made of intrusive bodies in the Hyderabad area, but no generalized map is available.

Closepet Granite

The Closepet Granite is a suite of presumably cogenetic bodies aligned along the western edge of the Eastern Dharwar craton (Fig. 3.1). Field characteristics can be summarized as follows (from Radhakrishna, 1956). The granite bodies occur as isolated plutons along the general northwest-southeast trend of the suite. There is no evidence of structural control of the belt, such as faults along the margins. Pink and gray porphyritic granites form hills, and equigranular pink and gray granites, intermixed with gneiss, form lowland areas. Gray phases occupy the outer parts of the porphyritic bodies, with pink phases in the center.

Contacts of all bodies are approximately concordant with gneissic structures. Contacts between porphyritic granites and their wall rocks are gradational from gneiss, veined by pegmatite and aplite, to gneiss and equigranular granite containing metacrysts of K-feldspar, to porphyritic granite characterized by K-feldspar crystals up to 5 cm long. The bodies do not show aureoles of contact metamorphism.

The granites have "normal" mesozonal igneous textures, with plagioclase laths, biotite platelets, and large crystals of K-feldspar. Quartz apparently occurs in two generations, one as rounded inclusions in plagioclase and K-feldspar, and one as late aggregates of sutured grains. The K-feldspar appears to replace plagioclase in some places, which is consistent with its occurrence as xenocrysts in the wall rocks. Biotite and minor hornblende are varietal minerals. Myrmekite is common.

There appears to be no standard or consistent age relationship between porphyritic or equigranular granites that are either gray or pink. All varieties of granite vein each other and contain inclusions of each other.

The Closepet suite shows no change in lithologic characteristics along much of its outcrop area. This observation presumably indicates that it was intruded after any generalized northward tilting of the Dharwar craton. Schistose mafic inclusions are uncommon in the granite. Where found, they tend to be of amphibolite facies in the north and granulite facies toward the south. None of the four phases of the granite (porphyritic or equigranular, gray or pink) is compositionally different from other phases (Divakara Rao and Qureshy, 1969; Divakara Rao et al., 1972a, 1972b).

Figure 3.5 shows compositions of three granites in a q-ab-or diagram. Sinilar data were used by Friend (1984) to investigate the origin of the Closepet Granite. The critical observations are the large scatter of data points around any possible eutectic or cotectic. Simple anatexis to produce the magmas seems impossible. Most samples investigated by Friend plot in a plagioclase-bearing liquidus field rather than an alkali feldspar field. Fractional crystallization must have been important during the evolution of the suite. Fractional crystallization has been proposed for various granitic bodies in and near the southern part of the Closepet outcrop (Allen et al., 1986). Allen et al.

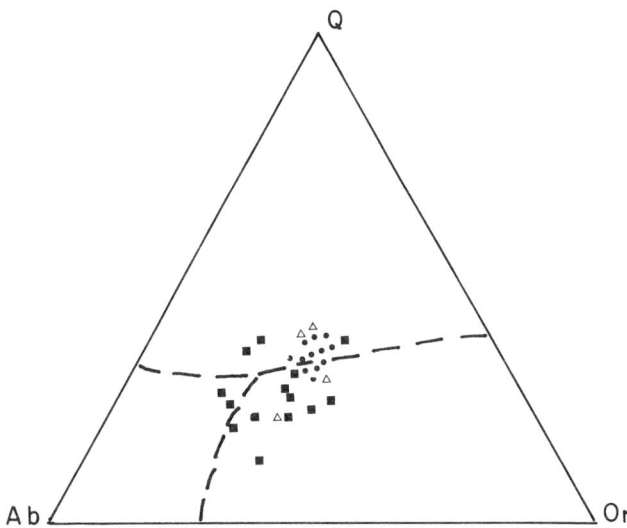

Fig. 3.5. Compositions of potassic granites (from P. T. Stroh). Solid squares are Closepet Granite; dots are Arsikere Granite; open triangles are Chitradurga Granite. Only samples with Ab/An > 3.5 are plotted. Dashed lines are cotectics at 5 kb water pressure.

modeled evolution of a primary magma by partial melting and formation of various granites and granodiorites by crystal settling to develop cumulates and residual melts.

As discussed in Chapter 4, the Closepet Granite is closely associated with the formation of the charnockites south of the gneiss-granulite transition zone. The movement of H_2O and CO_2 in the crust during charnockitization provided a hydrous environment in which the primary Closepet Granites could form. These granite magmas were later modified by fractional crystallization and deuteric processes. Friend (1985) identified specific pathways along which H_2O and CO_2 rose through the crust, promoting the formation of the Closepet Granite at levels above those of granulite metamorphism.

Hyderabad area granites

Granite bodies around Hyderabad occur as a group of apparently separate plutons with a wide range of composition. They are compositionally coherent over distances up to several kilometers or tens of kilometers. Most bodies are nearly massive, but local foliated areas show trends parallel to those in adjacent gneisses (Kanungo et al., 1975). Contacts are gradational over short distances. No consistent sequence of intrusion has been established, and it is likely that emplacement of the different bodies was essentially synchronous. The initial $^{87}Sr/^{86}Sr$ ratio of 0.7045 obtained by Crawford (1969) is not fully diagnostic with regard to source of the melts; possibly they represent some mixture of mantle and crustal material.

Lithologically, the rocks of the Hyderabad suite range from diorites to granites (Sitaramaya, 1971; Janardan Rao and Sitaramaya, 1973). Pyroxene is

a common varietal mineral even in silicic rocks. Textures are typical of calcalkaline magmatic rocks. Late-stage or postemplacement modification is shown by development of unusual features such as sphene coronites (Janardan Rao et al., 1973) and by mobility of trace elements (Janardan Rao and Satyanarayana, 1978). Statistical studies of the granites (B. Satyanarayana, 1977, 1980) indicate that the rocks in this suite have not evolved from each other. If they are cogenetic and magmatic, then the separate rock types must represent separate magma batches.

The Hyderabad igneous rocks appear to be extreme examples of the general tendency for Indian Precambrian rocks to be rich in radioactive elements. Padma Kumari et al. (1977) showed that porphyritic granite suites have average compositions with up to 90 ppm Th and 8 ppm U. Not only are these absolute values high, but also the Th/U ratio greater than 10 is much higher than the normal 4 to 5 for granites.

DIKES AND KIMBERLITES

Dikes are widespread throughout the eastern part of the Dharwar craton (Drury, 1984). East-west trends are dominant, and the dikes do not follow the normal north-south grain of the gneisses and schist belts. The only indication of age is from studies of dikes and sills (along with flow rocks) in the Cuddapah basin, where Crawford and Compston (1973) reported ages of about 1,500 m.y. that potentially have been greatly affected by alteration. With the exception of this limited information, the dikes can only be dated as younger than all major events that affected the gneisses and schist belts.

Mineralogically and texturally, the dikes studied by Drury (1984) are dolerites. Potassium contents are generally low, and some differentiated dikes show an iron-enrichment trend. Thus, the rocks are best classified as tholeiitic. (Table 3.4)

Except for tholeiitic dikes, the only other young igneous activity in the Eastern Dharwar craton is carbonatites/kimberlites found at two locations. One suite is at Wajrakarur and Lattavaram, in the Anantapur area (Fig. 3.1). The other location is near Chelima (west of Giddalur, Fig. 3.6), where dikes cut Cuddapah sediments (Sen and Narasimha Rao, 1971; Bergman and Baker, 1984).

Four kimberlite pipes occur in an area of about 10 km diameter near Wajrakarur and Lattavaram (Fig. 3.1; Satyanarayana Rao and Phadtre, 1966; Karunakaran et al., 1976; S. R. N. Murthy, 1977). The rocks are very weathered, and dates are not reliable. The pipes were presumably the source of diamonds in lower conglomerates in the Cuddapah section, a short distance to the east. Paul et al. (1975b) found minimum K/Ar ages of about 1,000 m.y., with an apparently cogenetic dolerite dike yielding an age of 1,446 m.y. These ages are somewhat lower than the presumed age of initial sedimentation in the Cuddapah basin, but all er-

rors are large, and a mid-Proterozoic age of the kimberlites seems likely.

The pipes have typical kimberlite mineralogy (Akella et al., 1979; Paul, 1981). They contain olivine (fo_{88-92}) that is partly serpentinized. In addition to clayey alteration products, other minerals include spinel, phlogopite, garnet, chromite, and calcite. A few eclogite nodules have been reported. The presence of garnets and eclogites indicates derivation of kimberlitic material from the mantle. The REE patterns are also consistent with mantle derivation, equilibration with garnet residue, and further fractionation (Paul et al., 1975a).

The Chelima dikes were regarded by Bergman and Baker (1984) as being intermediate between kimberlites and carbonatites (Table 3.4). They contain diamonds, and much of the groundmass has been replaced by ferroan dolomite. Bergman and Baker proposed that the dikes are different from the Wajrakarur kimberlites because they were intruded near the margin of the craton, whereas the kimberlite pipes were intruded closer to the center.

TABLE 3.4. *Compositions of dikes*

	1	2	3	4	5
SiO_2	48.1	54.5	51.0	52.6	40.8
TiO_2	1.4	0.61	1.9	0.48	6.3
Al_2O_3	15.5	13.9	13.3	14.5	5.1
Fe_2O_3	1.2	2.8	2.4	4.5	11.4
FeO	9.9	8.7	7.7	8.3	
MnO	0.24	0.08	0.25	0.21	
MgO	9.4	5.3	9.8	10.1	20.6
CaO	10.6	9.2	10.2	5.1	11.4
Na_2O	1.9	2.7	1.9	2.5	0.16
K_2O	0.42	1.3	0.64	0.91	2.9
Rb					160
Sr					1920
Y					47
Zr					1020
V					135
Cr					560
Ni					450
Th					
U					

All abundances of major elements are in weight percent of oxides. Where no entry is made in the FeO position, the value at Fe_2O_3 is total iron calculated as ferric oxide. All abundances of trace elements are in parts per million by weight. Values are taken directly from quoted sources and not recalculated to 100%.

Table 3.4. Column References
1. Bomasamudram (near Chittoor) (Chakrapani Naidu and Jayakumar, 1979).
2. Bomasamudram (near Chittoor) (Chakrapani Naidu and Jayakumar, 1979).
3. Near Kadiri schist belt (K. Satyanarayana et al., 1980).
4. Hyderabad (Divakara Rao et al., 1970).
5. Chelima (west of Giddalur) (Bergman and Baker, 1984).

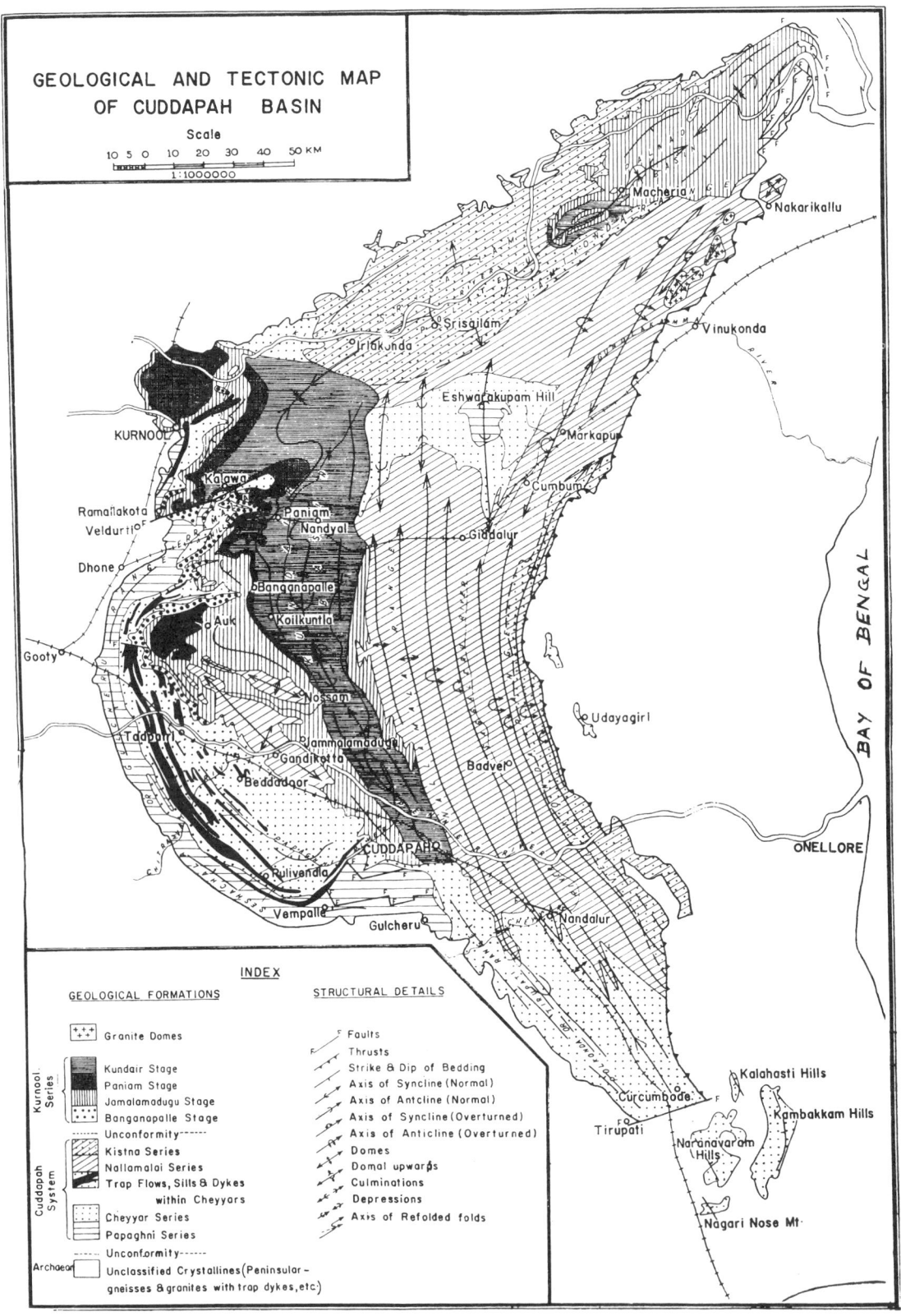

Fig. 3.6. Cuddapah basin (Narayanaswamy, 1966). Reproduced with permission from the Journal of the Geological Society of India, v. 7.

PROTEROZOIC PLATFORMAL SEQUENCES

The deformed, metamorphosed, and intruded sequences of crystalline rock in the Dharwar craton are covered in several places by sedimentary sequences that are mostly undeformed or little deformed. There are three principal basins: the Cuddapah, Bhima, and Kaladgi. (The Kaladgi basin is more properly regarded as a part of the Western Dharwar craton, discussed in Chapter 2, but is considered here for convenience.) As shown in Figure 3.1, the Cuddapah basin is not covered by younger material, and consequently may now show approximately its initial shape. The Bhima and Kaladgi basins, however, are partly covered by Deccan basalts, and their extent under the basalts is unknown.

The Bhima and Kaladgi basins do not show metamorphism or deformation. The Cuddapah basin, however, shows at least mild tectonism throughout its extent. Maps of the basin (Fig. 3.6) show fold axes curving into the northeastern and southeastern arms of the basin (Narayanaswamy,

1966; Geol. Surv. India, 1981). This pattern indicates a refolding of original north-south folds and shows that the present crescentic shape of the basin is at least partly a result of deformation. Folding and metamorphism in the Cuddapah basin are more intense in the east than in the west (Narayanaswamy, 1966; Meijerink et al., 1984). Unfolding of the present basin would yield an approximately circular shape, which the basin may have had initially.

Ages of the sedimentary suites

Some information is available on ages of the sedimentary rocks: Amygdaloidal basalts in the lower part of the Cuddapah sequence (Table 3.5) yielded an age of 1,583 ± 147 m.y. with an initial $^{87}Sr/^{86}Sr$ of 0.7044 ± 0.0014 (Crawford and Compston, 1973). These rocks are altered, but less so than other suites in the basin. Altered dolerite dikes cutting the Vempalle dolomite of the Lower Cuddapah sequence yielded an age of 980 ± 110

TABLE 3.5. *Stratigraphy of Cuddapah basin (Meijerink et al., 1984)*

Kurnool Group		Kundair Fm.	Shales Limestones
		Panem Fm.	Quartzites
		------------------ Paraconformity --------------------	
		Banganapalle Fm.	Conglomerates, sandstones shales, limestones
----------------------------------- Unconformity -----------------------------------			
Nallamalai Group	Cumbum Subgroup	Giddalur Fm.	Quartzites, shales
		Pullampet Fm.	Shales, phyllites
	Bairen- konda Subgroup	Srisailam Fm.	Quartzites sandstones, shales
		Dornala Fm.	
		Iswara Kuppam Fm.	Quartzites, shales
----------------------------------- Unconformity -----------------------------------			
Cuddapah Group	Mogamu- reru Subgroup	Tadpatri Fm.	Shales, volcanics
		Pulivendla Fm.	Grits
		------------------ Paraconformity --------------------	
	Papaghni Subgroup	Vempalle Fm.	Dolomites, shales
		Gulcheru Fm.	Conglomerates
----------------------------------- Unconformity -----------------------------------			
Archean Granites and Gneisses			

Prepared with permission from Precambrian Research, v. 26, p. 159.

m.y. with an initial $^{87}Sr/^{86}Sr$ ratio of 0.7121 ± 0.0015 (Crawford and Compston, 1973). Crawford and Compston regarded the age and high initial ratio as a record of deuteric activity.

The Cuddapah sequence is closely affected by deformation along the Eastern Ghats front. As discussed in Chapter 5, deformational activity in this area is poorly dated but probably about mid-Proterozoic. Thus, at least part of the Cuddapah sequence must be older than about 1,500 m.y., which is consistent with the date on volcanic rocks in the lower Cuddapah beds.

Approximate ages can be based on stromatolites and other microfossils (K. N. Prasad et al., 1979; Venkatachala, 1979). The Cuddapah Supergroup apparently extends through much of the Riphean (lower part of the Upper Proterozoic). Gururaja et al. (1979) proposed an age of 1,900 to 2,000 m.y. for the Upper Cuddapah Cumbum Formation on the basis of faunal content. Fossils in the Kaladgi suite are more restricted in genera and probably occupy an age range somewhere in the middle of the Riphean.

The Kurnool Group (Dutt, 1962) overlies the Cuddapah suite, and the Bhima and Badami sediments overlie the Kaladgi Group. These suites may extend into the Vendian (upper part of the Upper Proterozoic) and possibly into the Paleozoic. Salujha et al. (1973) found microplankton in the Kurnool sequence similar to Early Paleozoic forms. Viswanathiah et al. (1984) dated the Katagiri Formation of the Badami Group as Ordovician, or possibly Silurian, on the basis of acritarchs and other marine microorganisms.

The Kurnool sediments of the Cuddapah basin have undergone only mild folding and metamorphism. They almost certainly formed after the orogeny along the Eastern Ghats front, and they are structurally discordant with the underlying suites. The Bhima and Badami sequences are probably approximate equivalents of the Kurnool Group. The Cuddapah and Kaladgi suites have been correlated with the Vindhyan Supergroup of the Vindhyan basin and the Chattisgarh Supergroup of the Chattisgarh basin by means of faunal similarities (e.g., Raha and Sastry, 1982). The Kurnool-Badami-Bhima suite may correlate with the Bhander Gruop (upper part of the Vindhyan basin).

Geophysics and structure of the Cuddapah basin

The Cuddapah basin contains several major tectonic features (Fig. 3.7; Narayanaswamy, 1966; Balakrishna et al., 1982; Meijerink et al., 1984; Ramanamurthy and Reddy, 1984). From west to east the basin becomes deeper, tilting and open folding change to isoclinal and overturned folds and thrusts, younger sediments occur in the Cuddapah Supergroup, and unconformity between the Cuddapah and overlying Kurnool sediments becomes more pronounced. The eastern part of the basin may be as much as 18,000 feet deep, and geophysical work indicates depression of the basement by as much as 10 km (Bhattacharji and Singh, 1982, 1984).

Faults in the basin are mostly steep, and both normal and reverse types are present. Apparently some faults were active only before deposition of the Kurnool suite, and some were active after Kurnool time. Thrust faults on the eastern edge of the basin have brought old crystalline rocks up over Cuddapah sediments (Kaila and Tewari, 1982). Kaila and Tewari also showed a major fault separating the deep, eastern, Nallamalai basin from the more shallow basin containing older Cuddapah sediments to the west. Depth to mantle increases from normal crustal thickness of about 35 km on the west to nearly 50 km in the deformed Nallamalai area. Total throw on the thrust along the eastern margin is estimated as about 5,000 m. Bhattacharji and Singh (1982, 1984) showed that the downward bowing of the basement generated sufficient extensional force to cause the normal faulting. The Cuddapah basin not only deepens to the east but also plunges to the north into the Palnad basin, and sedimentary units become younger to the north.

The boundary fault on the western side of the Nallamalai Formation is regarded as a major structure by Meijerink et al. (1984). Meijerink et al. used the term "Rudravaram line" for this feature and proposed that it was originally a high-angle reverse fault later converted to a normal fault. Normal fault movement was associated with development of the deep Nallamalai basin to the east, although Meijerink et al. pointed out that Nallamalai sediments overlapped the fault and extended uconformably over the Cuddapah Group to the west.

Kurnool sediments occur in two areas, the western part of the Cuddapah basin and the Palnad basin in the northeast (Vijayam and Reddy, 1976). It is not clear whether these basins formed separately or were once joined and are now separated by erosion. Karunakaran (1982) summarized the likelihood that Kurnool sediments accumulated in local basins that formed as fault-bounded troughs toward the end of the evolution of the Cuddapah basin.

Outliers of the Cuddapah basin sediments occur to the south of the basin, in Tamil Nadu, and also east of the Eastern Ghats front (Narayanaswamy, 1966). If these outliers were once connected to the main basin, then the area of downwarp and sedimentation must have been considerably larger than the present size.

The general evolution of the Cuddapah basin in terms of high-level emplacement of mafic rock, thermal expansion, and then cooling and subsidence was discussed by Bhattacharji and Singh

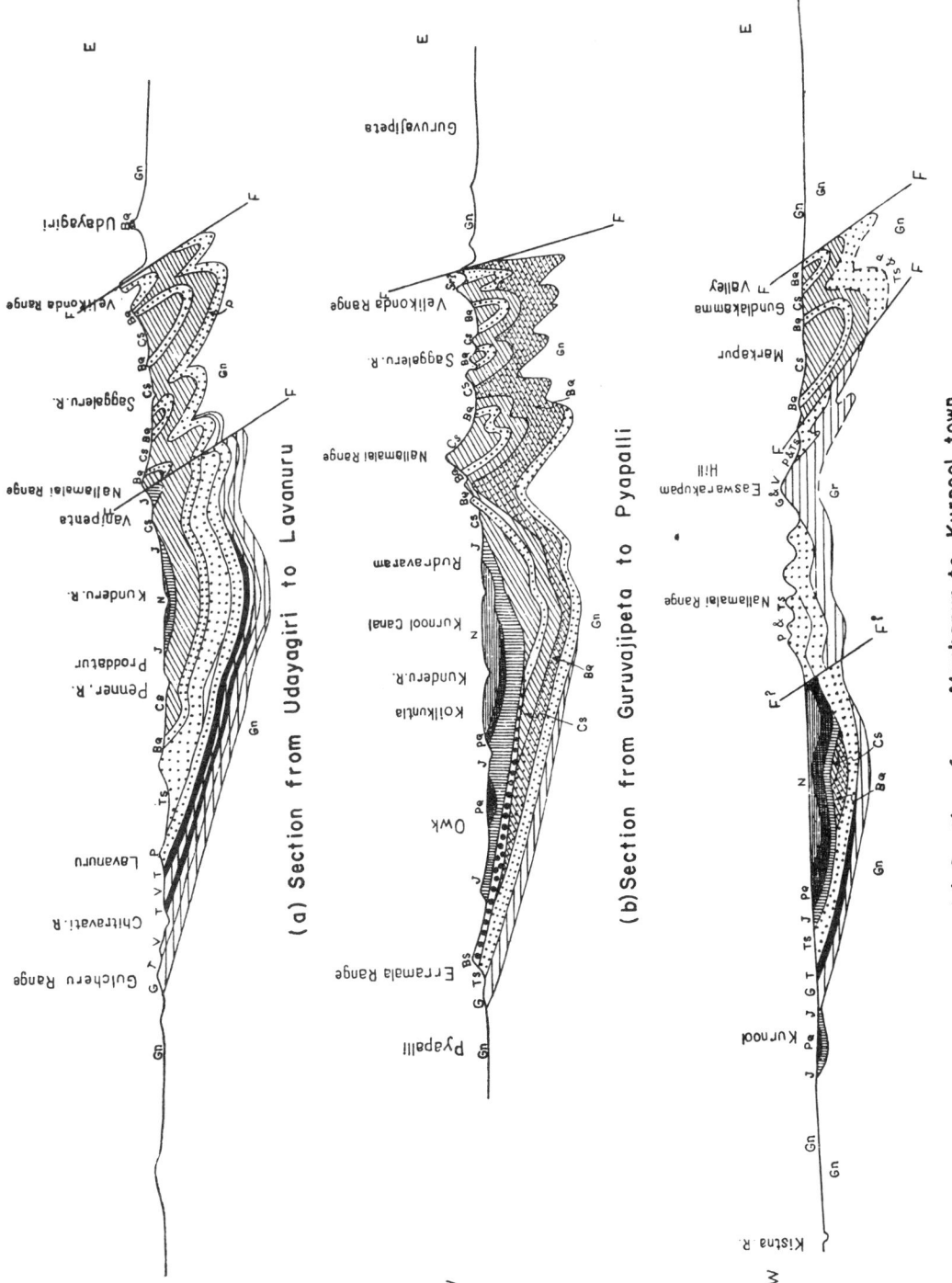

Fig. 3.7. Cross-sections of Cuddapah basin (Narayanaswamy, 1966). Lithologic symbols are the same in Fig. 3.6. Reproduced with permission from the Journal of the Geological Society of India, v. 7.

73

(1982, 1984). Using a total depth to basement of 10 km, it is possible to account for 6 to 7 km of downwarp on the basis of isostatic depression by the sedimentary load. This leaves 3 to 4 km of downwarp to be caused by some type of tectonic process. Bhattacharji and Singh modeled three cycles of upwarp, erosion, and associated igneous activity during the evolution of the basin. Assuming that the basin was active for about 500 million years, the time period for each cycle would have been about 150 m.y., which is approximately normal for mantle thermal cycles.

Based on the preceding calculations, Bhattacharji and Singh (1982, 1984) explained a concurrence of uplift and volcanism/shallow intrusion as a result of diapiric upwelling in the mantle. The uplifts apparently caused crustal thinning, either by erosion or some other process, and the thinned crust subsided to create a deeper basin during the cooling part of each thermal cycle. The three cycles proposed by Bhattacharji and Singh are considerably shorter than the nearly 500-m.y. sedimentational cycles of Meijerink et al. (1984).

An additional factor responsible for downwarping of the Cuddapah basin may have been orogenic activity along the Eastern Ghats front. Radhakrishna and Naqvi (1986) proposed that compression along this zone was responsible for some of the depression of the basin.

Stratigraphy and petrology of the Cuddapah basin

Stratigraphy in the Cuddapah basin was originally proposed by King (1882). Meijerink et al. (1984) correlated three stratigraphic groups (Table 3.5) with tectonic activity in the basin (Table 3.6). The oldest suite is the Cuddapah Group, which crops out in the western and central parts of the basin and probably extends eastward under all other sediments and is only mildly deformed. The Nallamalai Group overlies the Cuddapah Group and occurs primarily in the deep, eastern part of the basin, east of the Rudravaram line. The group overlaps the line, however, and covers Cuddapah rocks with angular unconformity. The Nallamalai Group has been deformed and metamorphosed by orogenic activity along the Eastern Ghats front. The Kurnool Group occurs in the Kurnool and Palnad basins and lies unconformably on mildly deformed Cuddapah rocks. The group does not extend east of the Rudravaram line. The Kurnool sediments have been faulted but only slightly folded.

In general, all sediments in the basin are very mature. Sandstones and conglomerates consist primarily of quartz. Carbonate rocks, including the thick Vempalle sequence (Venkatram and Balasundaram, 1952), are comparatively free of clastic debris. Shales are clay rich, rather than silty, and carbonaceous material is present in some of them. Limestones are micritic, with dolomite in some horizons. It is clear that the sequence is basically a shallow-water assemblage, at least in the western part of the basin. Depositional environments were neritic, intertidal, and probably subaerial.

More specific depositional environments can be inferred from some data. A. K. Bhattacharya et al. (1976) recognized three environments in the Vempalle Formation of the Cuddapah Group, including supratidal, represented by microsparite containing authigenic feldspar, mud cracks, and ripple marks; intertidal, represented by stromatolite-bearing microsparites and microrudites; and subtidal, or shallow open sea, represented by cross-bedded oomicrite and intraclast-bearing microsparites interbedded with shales. Stromatolites, indicative of sea-level deposition, are common in the carbonate rocks of the suite (Schopf and Prasad, 1978; Gururaja et al., 1979). Abundant acritarchs and other microorganisms are also present (Viswanathiah et al., 1982). Ripple marks in clastic sediments are generally indicative of very shallow water (Madhava Rao and Gokhale, 1973).

Sediments in the eastern part of the basin (Nallamalai Group) were apparently deposited in slightly deeper water than in the west. Clastic ratios are higher toward the east. Reducing conditions permitted formation of pyrite in some rocks (Arya and Rao, 1979). Even in the thick sedimentary sequence in the east, however, deposition and downwarp rates seem to have been equivalent throughout much of the development of the Cuddapah suite.

The Vempalle Formation of the Cuddapah Group contains dolerite sills, dikes, and flows. Tuffaceous rocks are also present. These volcanic rocks are associated with, and probably responsible for, base metal (Pb, Zn) mineralization in veins in the Vempalle carbonate rocks. Kimberlitic dikes cut Cuddapah sediments near Chelima (Sen and Narasimha Rao, 1971; Satyanarayana Rao, 1971; and Srikantia, 1984).

The volcanic rocks are associated with two major types of economic mineralization: asbestos and barite. Asbestos occurs in serpentine zones in the dolomite near the mafic rocks (P. B. Murthy, 1950; R. N. Prasad and Prasannan, 1976; Ramam et al., 1979; and Devore, 1982b). The asbestos occurs along stylolitic seams in Vempalle marbles where SiO_2 along the stylolites reacted with Mg in the adjacent dolomite.

Barite occurs at various places in the Vempalle Formation (Neelakantam and Roy, 1979; Neelakantam and Kopresa Rao, 1980; and Devore, 1982a). The barite is in tuffaceous and carbonaceous lenses in the carbonate sequence and also in veins. The lower part of the barite lenses is granular barite, presumed to have been inorganically precipitated. The upper part of the sequence is

TABLE 3.6. *History of Cuddapah basin (Meijerink et al., 1984)*

Age m.y.	Stratigraphy	Events	Diagnostic Features
	Post Kurnool	Eastern thrusting, tear movements along deep faults, drape folding, epeirogenetic movements	Kalava-Gani structure, upthrusts, Vamikonda Siddhavattam tear fault, raised Banganapalle formations, Durgi slip sheet
500?		Folding and metamorphism in the eastern part	Arcuate trend schistosity and fold axes, faults
		------------Denudation------------	Unconformity, bauxite below Nallamalai quartzite
	Kundair and Panem	Subsidence, weak activities along deep faults	Panem paraconformity
	Banganapalle	Transgression, subsidence controlled by deep faults, strong local movement	Banganapalle unconformity, Fault facies
		------------Denudation------------	Unconformity
1090, 980, 870			Dolerite dykes, not cutting through Kurnool rocks
1140			Kimberlite W. of basin
1225			Basic intrusives in folded Nallamalai rocks, Chelima, Vamikonda dyke,
1360 ± 30			metamorphism of lavas
		Folding of Nallamalai Group stopping short at Rudravaram line	Fold directions across the later arcuate trend, Banganapalle unconformity on folded Nallamalai rocks
		Subsidence with onlap of younger formations	Fully developed section around Iswara Kuppam, map evidence
		------------Denudation------------	Unconformity
1450–1490			Nellore mica belt
		Crustal movements resulting in broad folds	
1720		Deep continuous subsidence, extrusion of lavas	Lava dating
1583 ± 147			
1570			
		----------- Interruption------------	Pulivendla unconformity
		Subsidence	Calcareous facies
		Transgression	Basal conglomerates
		------------Denudation------------	Unconformity
			Dating of Closepet Granite dyke swarms
2000		Peninsular Granites and Gneisses	
2440 ± 250			
2520 ± 100			
2580 ± 50			Gneisses NW of basin
2950 ± 120			

Dates are from Crawford and Compston (1973). Prepared with permission from Precambrian Research, v. 26, p. 96.

barite-silica lapilli. The lapilli apparently grew contemporaneously with sedimentation because they depress tuffaceous material underneath them.

Because of their association with volcanic material, the barite deposits are generally regarded as volcanogenic. There are two problems, however, with a simple volcanic origin. One is an excess of SO_3 over BaO in the deposits. The other problem is that $\delta^{34}S$ values in the barite are in the range of ± 40 to 45 parts per thousand (Karunakaran, 1976). This isotopic value is much too large for primary igneous derivation and probably represents bacterial activity converting sulfur in sea water to sulfate. Corresponding $\delta^{34}S$ values for pyrite are near zero, indicating isotopic fraction-

ation of sulfate and sulfide, probably by organic
activity. A consistent explanation is that the Ba
was supplied by volcanic emanation into sea
water, where it reacted with sulfate produced at
least partly bacterially.

Stratigraphy and petrology of
Bhima and Kaladgi basins

The Bhima and Kaladgi basins (Fig. 3.1) are less
well studied than the Cuddapah basin. Because
the Bhima and Kaladgi are partly covered by the
Deccan basalts, their original extent and complete
paleogeography have not been ascertained. The
total thickness of the exposed sections is much
less than in the Cuddapah basin, not more than
2,000 m in the Kaladgi basin and less in the

Bhima sequence (Table 3.7). Although the Bhima
and Kaladgi basins are not joined in outcrop, the
Bhima sequence is regarded as equivalent to the
upper (Badami) part of the section in the Kaladgi
basin. The Bhima and Kaladgi sections are undeformed, and there is no magmatic activity in
either basin.

All of the rock types in the Kaladgi basin are
characteristic of deposition in shallow-water or
subtidal environments. Stromatolites at a variety
of locations indicate continued shallow-water deposition (Chandrasekhara Gowda and Govinda
Rajulu, 1980). There is a mild unconformity between the Kaladgi and Badami Groups, but no
evidence of tectonism. Authigenic or diagenetic
modification of the clastic rocks has been extensive (Viswanathiah and Sreedhara Murthy, 1977).

TABLE 3.7. *Stratigraphy of Kaladgi basin (Viswanathiah, 1977) and Bhima basin (Janardhana Rao et al., 1975).*

Group	Formation	Members	Thickness (m)
		Kaladgi Basin	
Badami	Katageri	Halkurki Argillite	15
		Konkankop Dolomite	35
		Transitional facies contact	
	Ramdurg	Torgal Quartzarenite	70
		Temple Quartzarenite	100
		Murgod Conglomerate	25
	---Unconformity---		
Kaladgi	Mudhol	Machkandi Argillite	70
		Laksanhatti Dolomite	70
		Niralkeri Breccia	25
		Vajramatti Quartzarenite	40
		Transitional facies contact	
	Lokapur	Yadhalli Argillite	20
		Petlur Limestone	100
		Jalikatti Argillite	15
		Chikshellikeri Limestone	260
		Chitrabhanukot Dolomite	330
		Transitional facies contact	
	Bagalkot	Mahakut Breccia	120
		Manoli Argillite	130
		Biligi Quartzarenite	200
		Salagundi Conglomerate	20

---Unconformity---

Metasedimentary and intrusive rock basement, ~ 2000 m.y.

Formation	Lithology
	Bhima Basin
Harwal Shale	Purple shales
Katamadevarhalli Limestone	Well-bedded and flaggy dark gray limestone
Halkal Formation	Greenish yellow to buff colored shales with local quartzitic sandstone and conglomerate
Shahabad Limestone	Limestones of varying physical and chemical properties
Rabanpalli Formation	Greenish yellow and purple shales with thin siltstone and sandstone or conglomerate/grit at the bottom

Kaladgi stratigraphy prepared with permission from Indian Mineralogist, v. 18.

Several stratigraphic sections have been proposed for the Bhima sequence (Dutt, 1975; Janardhana Rao et al., 1975; Mathur, 1977; Table 3.7). The lower part of the Bhima sequence is a thin set of sandstones (about 2 m thick) that grade upward into the carbonate-shale section that constitutes most of the assemblage. Akhtar (1977) regarded the sandstones as probable beach deposits and the overlying shales as tidal-flat sediments. Clastic sediments were dispersed toward the northwest from southern source areas. Salujha et al. (1970) and Viswanathiah et al. (1979) described microfossils consistent with accumulation of the sedimentary sequences in shallow water in Middle Proterozoic time. Suresh (1983) identified the unusual fossil *Chuaria* from shales in the Bhima sequence.

ECONOMIC DEPOSITS

The Eastern Dharwar craton contains several commercially important deposits, with gold and barite being the principal ores. Other economic minerals include asbestos and diamonds (Kurien, 1980).

A variety of deposits occurs in the lower part of the sedimentary sequence in the Cuddapah basin, principally the Vempalle carbonate rocks. The principal asbestos deposit is at Pulivendla, which is well known for its long-fiber chrysotile (P. B. Murthy, 1950; R. N. Prasad and Prasannan, 1976; Ramam et al., 1979). Major barite occurrences are in the Vempalle and other formations (P. B. Murthy, 1950; R. N. Prasad and Prasannan, 1976; Neelakantam and Roy, 1979). Total reserves of barite amount to approximately 25% of the world's supply. Other economic materials in the Cuddapah basin are in conglomerates in the lower part of the section, including uranium (Suryanarayana Rao and Rao, 1977) and diamonds (Rajaraman and Deshpande, 1978). The diamonds were derived from kimberlitic rocks at Wajrakarur and Lattavaram.

Gold occurs in both the Hutti and Kolar belts, with the latter being the most important deposit. The deposits at Hutti (Vasudev and Naganna, 1973) consist of gold-quartz veins containing pyrite, arsenopyrite and sphalerite in shear zones in metamorphosed mafic rocks. Isotopic studies of the sulfide minerals indicate a δ^{34}S consistent with derivation of the ore fluids from a primitive mantle (Menon et al., 1981).

The gold deposits at Kolar have been investigated for many years (Narayanaswamy et al., 1960; Anantha Iyer and Vasudeva Murthy, 1976). The gold occurs mostly in vein deposits in shear zones, with some ore dispersed in amphibolites. Safonov et al. (1984) described six stages of mineralization: early quartz deposition; quartz-feldspar (pegmatite) formation; deposition of scheelite; early-stage deposition of sulfides and gold;

later-stage deposition of sulfides, gold, tellurides, and quartz; and late deposition of quartz and carbonate. Studies of fluid inclusions indicate mineralization conditions in the range of 250° to 300° and 1.8 to 3.5 kb (Safonov et al., 1984). Sulfur isotopic ratios show derivation of mineralizing fluids from the mantle (Safonov et al., 1984).

REFERENCES

Akella, J., Rao, P. S., McAllister, R. H., Boyd, F. R., and Meyer, H. C. A. (1979). Mineralogical studies on the diamondiferous kimberlite of the Wajrakarur area, southern India. In Kimberlites, Diatremes, and Diamonds: Their Geology, Petrology, and Geochemistry (ed. F. R. Boyd and H. C. A. Meyer), Proc. Internat. Kimberlite Conf. 2, Amer. Geophys. Union, Washington, v. 1, 172–177.

Akhtar, K. (1977). Depositional environment, dispersal pattern and palaeogeography of the clastic sequence in the Bhima Basin. Ind. Mineralogist, v. 18, 65–72.

Allen, P., Condie, K. C., and Narayana, B. L. (1983). The Archaean low- to high-grade transition in the Krishnagiri-Dharmapuri area, Tamil Nadu, southern India. In Precambrian of South India (ed. S. M. Naqvi and J. J. W. Rogers), Geol. Soc. India Mem. 4, 450–461.

——, ——, and Bowling, G. P. (1986). Geochemical characteristics and possible origins of the southern Closepet batholith, South India. J. Geol., v. 94, 283–299.

Anantha Iyer, G. V., and Vasudev, V. N. (1979). Geochemistry of the Archaean metavolcanic rocks of Kolar and Hutti gold fields, Karnataka, India. Geol. Soc. India J., v. 20, 419–432.

——, and Vasudeva Murthy, A. R. (1976). Metamorphism, geochemistry and mineralization in the Precambrian of Kolar. Geol. Surv. India Misc. Publ. 23, Part 2, 596–614.

——, Vasudev, V. N., and Jayaram, S. (1980). Rare earth element geochemistry of metabasalts from Kolar and Hutti gold-bearing volcanic belts, Karnataka craton, India. Geol. Soc. India J., v. 21, 603–608.

Arya, B. C., and Rao, C. N. (1979). Bioturbation structures from the middle Proterozoic Narji Formation, Kurnool Group, Andhra Pradesh, India. Sediment. Geol., v. 22, 127–134.

Balakrishna, S., Venkatanarayana, B., and Venkateswara Rao, T. (1982). Geological and geophysical investigations along a few selected east-west traverses across the south western margin of Cuddapah Basin. In Evolution of the Intracratonic Cuddapah Basin (ed. S. Bhattacharji and S. Balakrishna), Inst. Ind. Peninsular Geology, Hyderabad, 19–46.

Ballal, N. R. R. (1975). The geology of Pamidi and Penukonda-Ramagiri schist belts and the surrounding Archaean gneisses and granites,

Anantapur District, Andhra Pradesh. Geol. Surv. India Misc. Publ. 23, Part 1, 97–104.

Bergman, S. C., and Baker, N. R. (1984). A new look at the Proterozoic dikes from Chelima, Andhra Pradesh, India: Diamondiferous lamproites? (Abstract). Geol. Soc. Amer. Abstracts with Programs, v. 16, 444.

Bhalla, N. S., Gupta, J. M., Chabris, T., and Vasudeva, S. G. (1978). Rb-Sr geochronology of rocks from the Kolar schist belt, South India. In Archaean Geochemistry; Proc. First Internat. Symp. on Archaean Geochem.: The Origin and Evolution of Archaean Continental Crust (ed. B. F. Windley and S. M. Naqvi), Elsevier, Amsterdam, 79–84.

Bhattacharji, S., and Singh, R. N. (1982). Mantle perturbations, thermal episodes and origin of intracratonic Proterozoic Cuddapah Basin on South Indian shield. In Evolution of the Intracratonic Cuddapah Basin (ed. S. Bhattacharji and S. Balakrishna), Inst. Ind. Peninsular Geology, Hyderabad, 107–134.

——, and —— (1984). Thermo-mechanical structure of the southern part of the Indian shield and its relevance to Precambrian basin evolution. Tectonophysics, v. 105, 103–120.

Bhattacharya, A. K., Rao, C. N., and Kaul, I. K. (1976). Thermoluminescence characteristics, correlation and depositional environment of the Vempalle Dolomite (Algonkian) in parts of Andhra Pradesh, India. Modern Geol., v. 5, 237–254.

Bhattacharya, S. K. (1975). Stratigraphic and structural investigation of the Dharwar schist belt in the Veldurti-Kurnool-Gadwal section in Kurnool and Mahbubnagar Districts, Andhra Pradesh. Geol. Surv. India Misc. Publ. 23, Part 1, 105–113.

Chakrapani Naidu, M. G., and Jayakumar, D. (1979). Dyke swarms in the Bommasumadram area, Chittoor District, Andhra Pradesh. Ind. Mineralogist, v. 20, 6–12.

Chandrasekhara Gowda, M. J., and Govinda Rajulu, B. V. (1980). Stromatolites of the Kaladgi basin and their significance in paleoenvironmental studies. Geol. Surv. India Misc. Publ. 44, 220–239.

Condie, K. C., Allen, P., and Narayana, B. L. (1982). Geochemistry of the Archean low- to high-grade transition zone, southern India. Contrib. Mineral. Petrol., v. 81, 157–167.

Crawford, A. R. (1969). Reconnaissance Rb-Sr dating of the Precambrian of southern peninsular India. Geol. Soc. India J., v. 10, 117–166.

——, and Compston, W. (1973). The age of the Cuddapah and Kurnool systems, southern India. Geol. Soc. Australia J., v. 19, 453–464.

Devore, G. W. (1982a). The Cuddapah Basin baryte deposits: A model for their origin. In Evolution of the Intracratonic Cuddapah Basin (ed.

S. Bhattacharji and S. Balakrishna), Inst. Ind. Peninsular Geology, Hyderabad, 63–70.

—— (1982b). Certain aspects of origin of chrysotile seams with special reference to the Cuddapah Basin deposits. In Evolution of the Intracratonic Cuddapah Basin (ed. S. Bhattacharji and S. Balakrishna), Inst. Ind. Peninsular Geology, Hyderabad, 71–80.

Divakara Rao, V., and Qureshy, M. N. (1969). Major element geochemistry of parts of the Closepet Granite pluton, Mysore State, India. Geophys. Res. Bull. (Hyderabad), v. 7, 145–157.

——, Hasnain, I., and Hussain, S. M. (1970). Petrochemical studies on a dyke near Hyderabad. Geophys. Res. Bull. (Hyderabad), v. 8, 27–37.

——, Aswathanarayana, U., and Qureshy, M. N. (1972a). Petrochemical studies in parts of the Closepet Granite pluton, Mysore State. Geol. Soc. India J., v. 13, 1–12.

——, ——, and —— (1972b). Trace-element geochemistry of parts of the Closepet Granite, Mysore State, India. Mineral. Mag., v. 38, 678–686.

Drury, S. A. (1984). A Proterozoic intra-cratonic basin, dike swarms and thermal evolution in South India. Geol. Soc. India J., v. 25, 437–449.

——, Harris, N. B. W., Holt, R. W., Reeves-Smith, G. J., and Wightman, R. T. (1984). Precambrian tectonics and crustal evolution in South India. J. Geol., v. 92, 3–20.

Dutt, N. V. B. S. (1962). Geology of the Kurnool system of rocks in Cuddapah and the southern part of Kurnool Districts, Andhra Pradesh. Geol. Surv. India Records, v. 87, Part 3, 549–604.

—— (1975). The Bhima Group of Hyderabad District, Andhra Pradesh, Geol. Surv. India Misc. Publ. 23, Part 1, 185–189.

Friend, C. R. L. (1984). The origins of the Closepet Granites and the implications for the crustal evolution of southern Karnataka. Geol. Soc. India J., v. 25, 73–84.

—— (1985). Evidence for fluid pathways through Archaean crust and the formation of the Closepet Granite, Karnataka, India. Precamb. Res., v. 27, 239–250.

Geol. Surv. India (1981). Geologic Map of Cuddapah Basin.

Gururaja, M. N., Jagannatha Rao, B. R., and Bhaskara Rao, B. (1979). Stromatolitic microbiota from black chert of Cumbum Formation, Upper Cuddapah, Andhra Pradesh. Geol. Soc. India J., v. 20, 138–142.

Harris, N. B. W., and Jayaram, S. (1982). Metamorphism of cordierite gneisses from the Bangalore region of the Indian Archaean. Lithos, v. 15, 89–98.

Janardan Rao, Y., and Satyanarayana, B. (1978). Studies on trace elemental distribution in the granites of Hyderabad. Ind. Acad. Geosci. J., v. 21, 1–8.

——, and Sitaramaya, S. (1973). Petrochemistry of the granitic rocks of the Ghatkesar area, Hyderabad, A. P. Ind. Mineralogist, v. 12, 40–50.

——, Murthy, I. S. N., and Ramulu, C. S. (1973). Sphene coronites from Hyderabad granites, Andhra Pradesh. Contrib. Mineral. Petrol., v. 41, 57–60.

Janardhana Rao, L. H., Srinivasa Rao, C., and Ramakrishna, T. L. (1975). Re-classification of the rocks of Bhima basin, Gulbarga District, Mysore State. Geol. Surv. India Misc. Publ. 23, Part 1, 177–184.

Jayaram, S., Venkatasubramanian, V. S., and Radhakrishna, B. P. (1976). Rb-Sr ages of cordierite-gneisses of southern Karnataka. Geol. Soc. India J., v. 17, 557–561.

Kaila, K. L., and Tewari, H. C. (1982). Structure and tectonics of the Cuddapah Basin in the light of DSS studies. In Evolution of the Intracratonic Cuddapah Basin (ed. S. Bhattacharji and S. Balakrishna), Inst. Ind. Peninsular Geology, Hyderabad, 53–62.

——, Roy Chowdhury, K., Reddy, P. R., Krishna, V. K., Narain, H., Subbotin, S. I., Sollogub, V. B., Chekunov, A. V., Kharetchko, G. E., Lazarenko, M. A., and Ilchenko, T. V. (1979). Crustal structure along the Kavali-Udipi profile in the Indian peninsular shield from deep seismic sounding. Geol. Soc. India J., v. 20, 307–333.

Kanungo, D. N., Rama Rao, P., Murthy, D. S. N., and Ramana Rao, A. V. (1975). Structural features of granites around Hyderabad. Geophys. Res. Bull. (Hyderabad), v. 13, 337–358.

Karunakaran, C. (1976). Sulphur isotopic composition of barytes and pyrites from Mangampeta, Cuddapah District, Andhra Pradesh. Geol. Soc. India J., v. 17, 181–185.

—— (1982). The Cuddapah basin, India: Its significance in geology and human affairs. In Evolution of the Intracratonic Cuddapah Basin (ed. S. Bhattacharji and S. Balakrsihna), Inst. Ind. Peninsular Geol., Hyderabad, 1–18.

——, Murthy, S. R. N., and Das Gupta, S. P. (1976). Kimberlites of Wajrakarur and Lattavaram, Andhra Pradesh. Geol. Surv. India Misc. Publ. 23, Part 2, 538–547.

King, W. (1882). The Kadapah and Karnul Formations in the Madras Presidency. Geol. Surv. India Mem. 8, Part 1, 346 p.

Kurien, T. K. (1980). Geology and mineral resources of Andhra Pradesh. Geol. Surv. India Bull., Ser. A, v. 44, 242 p.

Madhava Rao, D., and Gokhale, K. V. G. K. (1973). Ripple marks in quartzites from Kurnool Supergroup, Cuddapah Basin, India. J. Sed. Petrol., v. 43, 1122–1124.

Mathur, S. M. (1977). Some aspects of the stratig-

raphy and limestone resources of the Bhima Group. Ind. Mineralogist, v. 18, 59–64.

Meijerink, A. M. J., Rao, D. P., and Rupke, J. (1984). Stratigraphic and structural development of the Precambrian Cuddapah basin, S.E. India. Precamb. Res., v. 26, 57–104.

Menon, A. G., Venkatasubramanian, V. S., Vasudev, V. N., and Anantha Iyer, G. V. (1981). Sulphur isotope abundance variations in sulphides of the Dharwar craton—Part III, Hutti. Geol. Soc. India J., v. 22, 448–450.

Murthy, P. B. (1950). Genesis of asbestos and barite, Cuddapah District, Rayalasema, South India. Econ. Geol., v. 45, 681–696.

Murthy, S. R. N. (1977). Petrochemistry and origin of the kimberlites of Wajrakarur and Lattavaram, Andhra Pradesh. Geol. Surv. India Records, v. 109, Part 2, 148–160.

Narayanan Kutty, T. R., and Anantha Iyer, G. V. (1977a). Chemical petrology of calc-silicate rocks and associated metamorphics around Sakarsanahalli, Kolar. Ind. J. Earth Sci., v. 4, 141–159.

——, and —— (1977b). Mineralogy of coexisting zoisite-clinozoisite and epidote from Sakarsanahalli, Kolar, Karnataka. Geol. Soc. India J., v. 18, 78–89.

Narayanaswamy, S. (1966). Tectonics of the Cuddapah basin. Geol. Soc. India J., v. 7, 33–50.

—— (1971). Tectonic setting and manifestation of the upper mantle in the Precambrian rocks of South India. Proc. Second Symp. on Upper Mantle, National Geophys. Res. Inst., Hyderabad, 377–403.

—— (1975). Proposal for charnockite-khondalite system in the Archaean shield of peninsular India. Geol. Surv. India Misc. Publ. 23, Part 1, 1–16.

——, Ziauddin, M., and Ramachandra Rao, A. V. (1960). Structural control and localization of gold-bearing lodes, Kolar gold field, India. Econ. Geol., v. 55, 1429–1459.

Neelakantam, S., and Kopresa Rao, K. S. (1980). Structure and chemistry of the Mangampeta barytes deposits and their bearing on its genesis. Geol. Surv. India Spec. Publ. 1, 645–662.

——, and Roy, S. (1979). Barytes deposits of Cuddapah basin. Geol. Surv. India Records, v. 112, Part 5, 51–64.

Padma Kumari, V. M., Venkat Rao, N., Sitaramayya, S., and Bhimasankaram, V. L. S. (1977). Variation of uranium, thorium and potassium contents in the granitic rocks of Hyderabad, Andhra Pradesh—A preliminary study. Ind. J. Earth Sci., v. 4, 192–196.

Paul, D. K. (1981). Indian diamonds and kimberlites. In Kimberlites and Diamonds (ed. J. E. Glover), Geol. Dept. and Extension Service, Univ. Western Australia, Publ. 5, 15–31.

——, Potts, P. J., Gibson, I. L., and Harris, P.

G. (1975a). Rare earth abundances in Indian kimberlites. Earth Planet. Sci. Lett., v. 25, 151–158.

————, Rex, D. C., and Harris, P. G. (1975b). Chemical characteristics and K-Ar ages of Indian kimberlites. Geol. Soc. Amer. Bull., v. 86, 364–366.

Perraju, P., and Natarajan, V. (1977). "Peninsular Gneiss" in the northern parts of Andhra Pradesh. Geol. Soc. India J., v. 18, 224–232.

Prasad, K. N., Jagannatha Rao, B. R., and Gururaja, M. N. (1979). Observations on the stromatolites from the Precambrian formations of South India. Geol. Surv. India Misc. Publ. 45, 23–29.

Prasad, R. N., and Prasannan, E. B. (1976). Asbestos-barytes-steatite mineralisation in the Lower Cuddapah of Andhra Pradesh. Geol. Surv. India Misc. Publ. 23, Part 2, 560–568.

Radhakrishna, B. P. (1954). On the nature of certain cordierite-bearing granulites bordering the Closepet Granites of Mysore. Mysore Geol. Dept. Bull., v. 21, 25 p.

———— (1956). The Closepet Granite of Mysore State, India. Mysore Geol. Assoc. (Bangalore) Spec. Publ. 3, 110 p.

————, and Naqvi, S. M. (1986). Precambrian continental crust of India and its evolution. J. Geol., v. 94, 145–166.

Raha, P. K., and Sastry, M. V. A. (1982). Stromatolites and Precambrian stratigraphy in India. Precamb. Res., v. 18, 293–318.

Rajamani, V., Shivkumar, K., Hanson, G. N., and Granath, J. W. (1981). Petrogenesis of amphibolites in the Kolar schist belt, India—A preliminary report. Geol. Soc. India J., 22, 470–487.

————, ————, ————, and Shirey, S. B. (1985). Geochemistry and petrogenesis of amphibolites, Kolar schist belt, South India: Evidence for komatiitic magma derived by low percentages of melting of the mantle. J. Petrol., v. 26, 92–123.

Rajaraman, S., and Deshpande, M. L. (1978). Banganapalle diamondiferous conglomerates in Kurnool District, Andhra Pradesh. Ind. Minerals, v. 32, no. 3, 33–43.

Raju, K. C. C., Kareemuddin, M., and Prabhakara Rao, P. (1979). Operation Anantapur. Geol. Surv. India Misc. Publ. 47, 1–57.

Rama Rao, B. (1961). The "Champion Gneiss" in the Archaean complex of Mysore, southern India—A review. Geol. Soc. India J., v. 2, 31–38.

Ramachandra Rao, M. B. (1937). The petrology of the hornblende rocks of the Kolar schist belt. Mysore Geol. Dept. Bull., v. 16, 103 p.

Ramam, P. K., Anantha Iyer, G. V., Narayanan Kutty, T. R., and Murthy, M. V. N. (1979). Stylolite origin of chrysotile asbestos near Puliven-dla, Cuddapah district, Andhra Pradesh. Geol. Soc. India J., v. 20, 467–480.

Ramamohana Rao, I., and Borreswara Rao, C. (1972). Trace elements in minerals and rocks of the Precambrian group of the Yellandlapad area, Andhra Pradesh. Geol. Soc. India J., v. 13, 165–171.

Ramanamurthy, M. V., and Reddy, D. S. N. (1984). Evaluation of aeromagnetic data and regional geological interpretation for a part of the Cuddapah basin. Geol. Soc. India J., v. 25, 666–676.

Rollinson, H. R., Windley, B. F., and Ramakrishnan, M. (1981). Contrasting high and intermediate pressures of metamorphism in the Archaean Sargur schists of southern India. Contrib. Mineral. Petrol., v. 76, 420–429.

Roy, A. (1979). Polyphase folding deformation in the Hutti-Maski schist belt, Karnataka. Geol. Soc. India J., v. 20, 598–607.

————, and Biswas, S. K. (1979). Metamorphic history of the Sandur schist belt, Karnataka. Geol. Soc. India J., v. 20, 179–187.

Safonov, Yu. G., Genkin, A. D., Vasudev, V. N., Krishna Rao, B., and Anantha Iyer, G. V. (1984). Genetic features of gold ore deposit at Kolar, Dharwar craton, India. Geol. Soc. India J., v. 25, 145–154.

Salujha, S. K., Rehman, K., and Arora, C. M. (1970). Microplankton from the Bhimas. Paleontol. Soc. India J., v. 15, 10–16.

————, ————, and ———— (1973). Early Paleozoic microplankton from the Kurnools, Andhra Pradesh. J. Palynol. (Palynol. Soc. India), v. 8, 125–131.

Satyanarayana, B. (1977). Quantitative petrological studies of the granites of Padhashareef area, Hyderabad District, Andhra Pradesh. Ind. Acad. Geosci. J., v. 20, 37–46.

———— (1980). Areal composition trends in the granites of Palmakole area, Hyderabad District, Andhra Pradesh. Geoviews, v. 7, 79–89.

Satyanarayana, K., Siddilingam, J., and Surya Prakash Rao, K. (1980). Geochemistry of basic dykes from Kadiri schist belt, Anantapur District, Andhra Pradesh. Geophys. Res. Bull. (Hyderabad), v. 18, 153–164.

Satyanarayana Rao, P. (1971). Volcanic and other igneous rocks in the Cuddapah formations of Gani-Veldurti area, Kurnool District, Andhra Pradesh. Geol. Surv. India Records, v. 96, Part 2, 205–213.

————, and Phadtre, P. N. (1966). Kimberlite pipe-rocks of Wajrakarur area, Anantapur District, Andhra Pradesh. Geol. Soc. India J., v. 7, 118–123.

Schopf, J. W., and Prasad, K. N. (1978). Microfossils in Collenia-like stromatolites from the Proterozoic Vempalle Formation of the Cuddapah basin, India. Precamb. Res., v. 6, 347–366.

Sen, S. M., and Narasimha Rao, Ch. (1971). Chelima dykes. Proc. Second Symp. on Upper Mantle Project, Natl. Geophys. Res. Inst, Hyderabad, 435–439.

Sitaramaya, S. (1971). The pyroxene-bearing granodiorites and granites of Hyderabad area (the Osmania Granite). Geol., Min. Metall. Soc. India Quart. J., v. 43, 117–129.

Srikantia, S. V. (1984). Kappalapalle volcanics—A distinct Upper Papaghni volcanic activity in the Cuddapah basin. Geol. Soc. India J., v. 25, 775–779.

Srinivas, G. (1968). Study of zircons from the Champion Gneisses, Kolar gold fields, Mysore State. Geol. Soc. India Bull., v. 5, 112–116.

Srinivasan, R., and Sreenivas, B. (1977). Some new geological features from the LANDSAT imagery of Karnataka. Geol. Soc. India J., v. 18, 589–597.

Suresh, R. (1983). Problematic Chuaria from the Bhima Basin, South India. Precamb. Res., v. 23, 79–85.

Suryanarayana Rao, C., and Rao, T. N. R. (1977). Uranium mineralization in the conglomerate horizons of Cuddapah and Kaladgi basins of peninsular India. Proc. Seminar of Kaladgi, Badami, Bhima and Cuddapah Sediments (ed. A. S. Janardhan), Ind. Mineralogist, v. 18, 7–17.

Vasudev, V. N., and Naganna, C. (1973). Mineragraphy of gold-quartz-sulphide reefs of Hutti gold mines of Raichur District, Mysore State. Geol. Soc. India J., v. 14, 378–383.

Venkatachala, B. S. (1979). Precambrian palynology—A review. Geol. Surv. India Misc. Publ. 45, 1–11.

Venkatasubramanian, V. S., and Narayanaswamy, R. (1974). Studies in Rb-Sr geochronology and trace element geochemistry in granitoids of Mysore craton, India. Indian Inst. Sci. J. (Bangalore), v. 56, 19–42.

Venkatram, M. S., and Balasundaram, M. S. (1952). Geology of the Vempalle Limestone belt. Geol. Surv. India Records, v. 78, Part 4, 419–480.

Vijayam, B. E., and Reddy, P. H. (1976). Tectonic framework of sedimentation in the western part of the Palnad basin, Andhra Pradesh. Geol. Soc. India J., v. 17, 439–448.

Viswanatha, M. N., and Ramakrishnan, M. (1975). The pre-Dharwar supracrustal rocks of Sargur schist complex and their tectono-metamorphic significance. Ind. Mineralogist, v. 16, 48–65.

——, and —— (1981). Kolar belt. In Early Precambrian Supracrustals of Southern Karnataka (ed. J. Swami Nath and M. Ramakrishnan), Geol. Surv. India Mem. 112, 221–245.

Viswanathiah, M. N. (1977). Lithostratigraphy of the Kaladgi and the Badami Groups, Karnataka. Proc. of Seminar of Kaladgi, Badami, Bhima and Cuddapah Sediments (ed. A. S. Janardhan), Ind. Mineralogist, v. 18, 122–132.

——, and Sreedhara Murthy, T. R. (1977). Potash feldspathization of conglomerate and subfeldsarenites of Kaladgi Group, Biligi, Bijapur District, Karnataka. Proc. of Seminar of Kaladgi, Badami, Bhima and Cuddapah Sediments (ed. A. S. Janardhan), Ind. Mineralogist, v. 18, 85–93.

——, Venkatachalapathy, V., and Doddiah, D. (1979). Acritarchs from the Bhima Basin, Karnataka, South India. Geol. Surv. India Misc. Publ. 45, 17–22.

——, ——, and Shekhar, R. G. C. (1982). Acritarchs and related microbiota from Pulivendla, Tadapatri and Gandikota formations of the Chitravati Group, Cuddapah Super Group. In Evolution of the Intracratonic Cuddapah Basin (ed. S. Bhattacharji and S. Balakrishna), Inst. Ind. Peninsular Geology, Hyderabad, 91–106.

——, ——, and Shankara, M. (1984). Acritarchs and associated microplankton from the Katagiri Formation of the Badami Group, South India. Rev. Paleobotany and Palynology, v. 41, 13–30.

Ziauddin, M. (1975). The acid volcanic and pyroclastic rocks (Champion Gneiss and autoclastic conglomerate) of the Kolar schist belt (Karnataka and Andhra Pradesh). In Studies in Precambrians (ed. C. Naganna), Bangalore Univ. Press, Bangalore, 142–162.

4. GRANULITE TERRAIN

The southernmost part of the Indian peninsula is a complex zone particularly characterized by rocks of granulite facies (Fig. 4.1). We have separated the Granulite terrain from an area of similar geology in the Eastern Ghats (Chapter 5; Fig. 1.1) because the Eastern Ghats is a recognizable province delineated on its western margin by a steep gravity gradient. This gradient is absent at the northern end of the Granulite terrain (Subrahmanyam, 1983), which correlates with the gradational nature of the contact zone. The Granulite terrain is not composed solely of rocks of granulite facies. Amphibolite-facies gneisses and "supracrustal" rocks are abundant, and some of these lower-grade rocks may represent areas of retrogression from granulites.

Major structural features in the Granulite terrain include highland areas and shear zones (Fig. 4.1). The principal highland areas consist primarily of intensely metamorphosed rocks of granulite facies. The most characteristic rock type in the massifs is charnockite highly depleted in lithophilic elements. The margins of the massifs show a gradation into less metamorphosed rocks, either in a lower granulite facies or amphibolite facies.

Figure 4.1 shows a number of shear zones based on the terminology of Harris et al. (1982) and Drury et al. (1984). The shear zones are generally areas of amphibolite-facies rocks or they mark the border between highland granulites and lower-grade lithologies. The area between the Noyil-Cauvery and the Moyar-Bhavani-Attur shear zones is the site of the Sathyamangalam Group of supracrustal rocks, generally in high-amphibolite facies (Gopalakrishnan et al., 1975).

The Achankovil shear zone separates charnockitic rocks of the Kodaikanal massif from the NW-SE-trending khondalite belt of southern Kerala (Fig. 4.1). The Moyar lineament has been proposed to offset the Billigirirangan Hills, on the north, from the Nilgiri Hills, on the south. Many of the lineaments contain recrystallized cataclastic rocks, foliations, and lineations parallel to the shears, and diminution of strain away from the zones. Drury et al. (1984) referred to the shear zones as probably Late Proterozoic, but definite ages and magnitude of displacements must await further work.

Shear zones along the edges of charnockite massifs are not responsible for all of the juxtaposition of terrains of different metamorphic grade. For example, Janardhan et al. (1982, 1983) and Hansen et al. (1985a) demonstrated a smooth gradation from amphibolite-facies terrain southward through lowland charnockites and into the high-pressure granulites of the Nilgiri massif (Fig. 4.1). If this transition exists, then the Moyar shear zone on the north side of the Nilgiris must be a zone of very small vertical displacement or must have existed prior to charnockitization, which seems unlikely. Most of the contacts between granulite- and amphibolite-facies rocks have been described as gradational (e.g., S. Narayanaswamy and Lakshmi, 1967, near Tinnevelly; Saravanan and Ramanathan, 1982, near Salem; Holt and Wightman, 1983, on the eastern side of the Kodaikanal massif).

SIGNIFICANCE OF GRAVITY FIELD

The free-air (Fig. 1.6) and Bouguer gravity maps of India yield perplexing information with regard to the Granulite terrain. On a regional scale, the granulite area shows only minimal free-air anomalies, indicating isostatic balance and "normal" thickness of continental crust. Locally, highland areas show some positive free-air anomalies relative to the lowland terrain. Because the highland massifs contain less-hydrous, higher-density rocks than the lowland areas, these local gravity anomalies appear to be explainable by uplift of the massifs by normal faulting around their margins (Radhakrishna, 1968). Another possibility is that the highland massifs lie above a sole thrust that placed them as detached slices on the lower-grade terrain. Such a thrust could have formed during a collisional event that accompanied, perhaps was responsible for, the granulite metamorphism.

The difficulty with the preceding explanations is that fault contacts have not been located around most massifs. As discussed previously, all current evidence indicates a gradation between amphibolite- and lower granulite-facies rocks in the lowlands and upper granulite-facies rocks in the highlands. Gradational contacts accompanied only by small, local gravity anomalies require that the denser crust under the massifs be thicker than the less-dense crust under the lowlands. Furthermore, because the highland massifs expose formerly deeper crustal levels, they must have un-

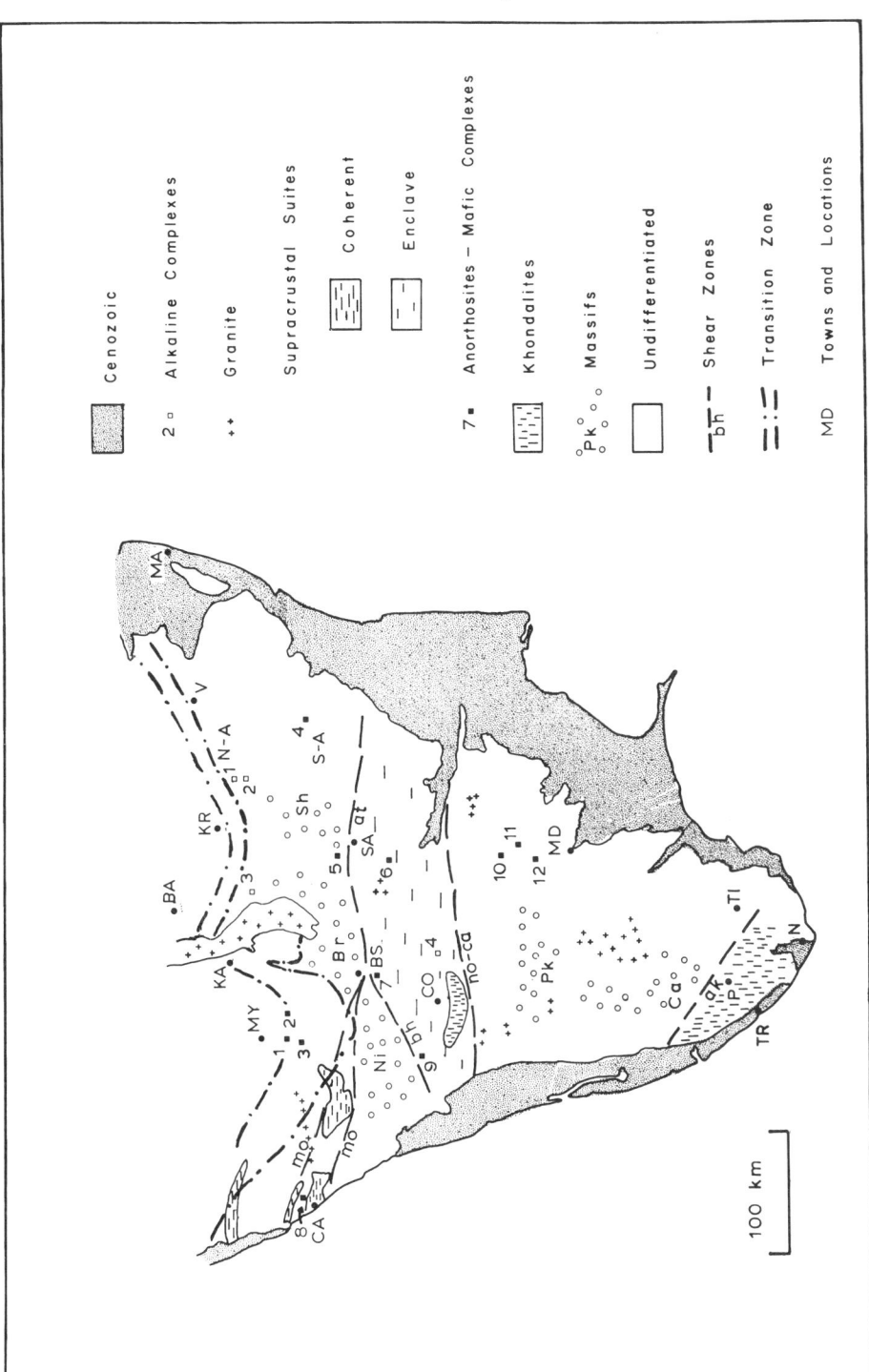

Fig. 4.1. Granulite terrain. Locations include MA, Madras; BA, Bangalore; KR, Krishnagiri; KA, Kabbaldurga; MY, Mysore; CA, Cannanore; BH, Bhavanisagar; SA, Salem; N-A, North Arcot; S-A, South Arcot; CO, Coimbatore; MD, Madurai; TI, Tinnevelly; TR, Trivandrum; N, Nagercoil. Massifs include Sh, Shevaroy; Br, Billigirirangan; Ni, Nilgiri; Pk, Palni-Kodaikanal; Ca, Cardamom.

Shear zones include mo, Moyar; bh, Bhavani; at, Attur; no-ca, Noyil-Cauvery.

Anorthosites and mafic complexes include 1, Sindhuvalli; 2, Konkanhundi; 3, Kurihundi; 4, Mamandur; 5, Salem (Chalk Hills); 6, Sittampundi; 7, Bhavani; 8, Ezhimala; 9, Agali; 10, Chinnadharampuram; 11, Kadavur; 12, Oddanchattram.

Alkaline complexes include 1, Simalpatti (Tirupattur); 2, Sevathur; 3, Hogenekal; 4, Sivamalai.

dergone more erosion than the surrounding lowlands. Thus, crustal thickness differences between massifs and the surrounding terrains must have been even larger before erosion than they are now, or there must have been selective later crustal thickening under the massifs.

ROCK SUITES MAPPED

The choice of rock suites to show in Figure 4.1 is very difficult. Most of the rocks in the Granulite terrain are sialic, but the charnockites, amphibolite-facies gneisses, and khondalites all contain conformable enclosures of more mafic rock types. These mafic (metasedimentary and metaigneous) rocks are generally not mappable at the scale of Figure 4.1 and probably do not constitute more than 10% of the area. Furthermore, it is not clear whether the minor mafic rocks in the different sialic suites are similar to each other. In this book, we follow the terminology of T. V. Viswanathan (1969) and S. Narayanaswamy (1975).

Charnockite. An assemblage of orthopyroxene (generally hypersthene or ferrohyperthene) ± quartz ± K-feldspar ± plagioclase ± garnet ± other mafic minerals.
Enderbite. An assemblage of orthopyroxene ± quartz ± plagioclase (commonly antiperthite) ± K-feldspar ± garnet ± other mafic minerals.
Mafic granulite. A pyroxene-plagioclase rock with a metamorphic texture, as distinguished from gabbroic or diabasic textures. These rocks commonly contain both ortho- and clinopyroxenes and small amounts of hornblende. Another common mineral is garnet. Mafic enclaves in charnockite tend to be lenticular and amphibole bearing. Enclaves in khondalites tend to be sheet like and contain more pyroxene and less amphibole.
Khondalite. An assemblage of quartz ± sillimanite ± garnet ± K-feldspar ± plagioclase ± graphite. Orthopyroxene and clinopyroxene are absent, but minor amounts of other mafic minerals may be present.
Leptynite. An assemblage of quartz ± garnet ± K-feldspar ± plagioclase ± amphibole ± biotite ± sillimanite. Pyroxene is absent. The garnets are abundant and large, and the amphibole and biotite are present in minor amounts. Rocks referred to as leptynite occur: in migmatized gneiss terrain, where the leptynite is a garnet-bearing rock of the leucosome; on the margins between khondalites and charnockites; and as sillimanite-bearing, garnet-rich, layers in the khondalite sequences.
Gneiss. An assemblage of quartz ± K-feldspar ± plagioclase ± amphibole ± biotite and having a gneissic foliation. Garnet may be present, but pyroxene is absent. With decrease in the amount of amphibole and biotite, gneiss can have the same mineral assemblage as leptynite.

The preceding rock types cannot be shown on Figure 4.1 because of the small size of most units, and we have chosen the following suites:

Massifs. These areas are highlands consisting largely of charnockites depleted in lithophilic elements. Mafic granulites are present as screens and lenses.
Khondalites. These areas consist primarily of khondalites and associated rock types, including calc-silicates and minor mafic granulites. Some areas mapped as khondalite also contain gneiss and incipient charnockite.
Undifferentiated (sialic terrain). This suite consists either of areas that have not been mapped adequately or in which the admixture of charnockites, khondalites, mafic granulites, and gneisses is too complex to depict. Jacob (1962) indicated that about 80% of southernmost India consists of intermingled gneiss, granite, and schist.
Supracrustal suites. The area of the Sathyamangalam series of Gopalakrishnan et al. (1975) is shown as enclaves. This suite consists of amphibolite-facies gneisses with intercalations of amphibolites, quartzites (± fuchsite, ± sillimanite, ± iron ores), mica schists, and calc-silicate rocks. All rocks are intruded by minor mafic and ultramafic bodies that occur in small lenses. Small coherent belts, such as the Wynad (Vengad) suite of northern Kerala, are mapped in the northwestern part of the terrain.
Anorthosite-mafic complexes. These assemblages appear to represent layered complexes (generally fragments of such complexes) intruded into other rock types. Many, but not all, suites have been metamorphosed to granulite facies.
Alkaline complexes. These suites are posttectonic intrusions occurring as plugs, ring dikes, etc.
Late- or posttectonic granites. Most bodies are too small to show on Figure 4.1. In contrast to the large plutons of the northern part of the Dharwar craton, plutons in the Granulite terrain commonly have gradational boundaries into surrounding rocks, and many may represent extreme development of potassic neosomes.

AGES OF ROCK TYPES

Information on the absolute ages of events and rocks in the Granulite terrain is shown in Table 4.1. Based on this table, plus field observations, it is clear that major granulite-facies metamorphism occurred about 2,600 m.y. ago. This process reset Rb/Sr clocks throughout the region. The event coincided approximately with the emplacement of the Closepet Granite and possibly of other gra-

TABLE 4.1. *Radiometric ages*

Rock Type and Location (Reference)	Method	Age (in m.y.)	Initial $^{87}Sr/^{86}Sr$ or $^{238}U/^{204}Pb$ (μ)	Notes
Charnockites				
At Pallavaram, near Madras (1)	Rb/Sr	2,580 ± 95	0.7059 ± 0.0042	
At Pallavaram (2)	Rb/Sr	1,980 ± 124	0.7037 ± 0.007	
At Pallavaram (3)	K/Ar	2,085 ± 43		Biotite age
In Nilgiri hills (1)	Rb/Sr	2,615 ± 80	0.7023 ± 0.0012	
Near Salem (1)	Rb/Sr	2,476 ± 115	0.7042 ± 0.0002	
Charnockites and associated rocks in transition zone				
mixed rocks (2)	Rb/Sr	2,618 ± 46	0.7039 ± 0.0005	
gneisses (4)	Rb/Sr	2,670 ± 60	0.704	
gneisses (5)	U/Pb	2,515 to 2,535		Zircons
pegmatite (6)	U/Pb	2,507 to 2,520		Allanite
Gneisses				
(1)	Rb/Sr	3,065 ± 65	0.70002 ± 0.0117	"Oldest Gneisses"
(7)	Rb/Sr	550	0.737	Retrograde Gneisses
Syenite				
at Sivamalai (1)	Rb/Sr	1,020 ± 670	0.7008 ± 0.0046	
Anorthosite-gabbro-granite suite				
Ezhimala complex (8)	Rb/Sr	670		

Table 4.1. Sources
1. Crawford (1969).
2. Spooner and Fairbairn (1970).
3. Balasubrahmanyam (1975).
4. Venkatasubramanian and Narayanaswamy (1974).
5. Buhl et al. (1983).
6. Grew and Manton (1984).
7. Hansen et al. (1985b).
8. M. M. Nair and Vidyadharan (1982).

nitic suites in the Eastern and Western Dharwar cratons (Chs. 2 and 3). The 2,600-m.y. age is much greater than the mid-Proterozoic age of granulite metamorphism in the Eastern Ghats, confirming that these two areas are not genetically related.

The parent material of the charnockites, probably largely an orthogneiss, is clearly older than 2,600 m.y., but the age of about 3,000 m.y. obtained by Crawford (1969) is constructed from a variety of rock types and needs further confirmation. Because the khondalite suite and other supracrustal rocks (including the Sathyamangalam suite) are so closely admixed with the charnockites, they also can be presumed to be older than 2,600 m.y., but no direct age data are available. Some anorthosite magmatism occurred prior to the granulite event, as shown by anorthosites metamorphosed to granulite facies, but some anorthosite suites are unmetamorphosed and must be younger. Alkaline suites are clearly young but not adequately dated. One suite of gneisses, apparently formed by retrogression from charnock-

ites, has been dated at 550 m.y., the age of Pan-African activity (Hansen et al., 1985b); the extent of this resetting in southern India is unknown.

PROGRADE AND RETROGRADE CHARNOCKITES AND GNEISSES

It is possible that large areas of the Granulite terrain mapped as gneiss and, by implication, correlated with the Peninsular Gneiss of the Western and Eastern Dharwar cratons is retrogressive from higher-grade rocks rather than a primary igneous or metasedimentary material. Distinction of retrograde and prograde processes is clearly important.

One criterion for distinguishing prograde and retrograde charnockite-gneiss relationships is the paragenesis of minerals in the rocks. Janardhan et al. (1982), for example, recognized prograde conversion of amphibole to orthopyroxene at Kabbaldurga and retrograde replacement of orthopyroxene and clinopyroxene by amphibole and biotite at Bhavanisagar (Fig. 4.1). Retrogression is commonly accompanied by bleaching of the rocks, presumably by destruction of the chlorite plates that give a greenish color to high-grade charnockites.

Some compositional criteria exist for distinction of prograde and retrograde processes. The best one is probably the fluorine (and chlorine) content of amphibole and coexisting biotite and apatite. During prograde metamorphism, halogens attain an equilibrium distribution between all three minerals, commonly forming amphiboles with F contents between 0.5% and 1.0% and biotites with slightly higher F abundances. This equilibrium is apparently not maintained during retrogression. Fluorine contents near zero are a characteristic feature of amphiboles (and probably also biotite) in retrograde rocks (Janardhan et al., 1982). There is also the possibility that retrograde gneisses can inherit the generally low abundance of lithophile elements that characterizes high-grade charnockites, thus distinguishing them from the lithophile-rich prograde gneisses, but this discrimination needs further study.

GNEISS-CHARNOCKITE TRANSITION AT THE NORTHERN EDGE OF THE GRANULITE TERRAIN

The amphibolite-facies gneisses and supracrustal enclaves in southern Karnataka are bordered on the south by the Granulite terrain. The contact between these suites is gradational (Hansen et al., 1985a). Metasomatic conversion of gneiss into charnockite was first shown by Pichamuthu (1960, 1961, 1965).

Major investigations in the transition zone have taken place in two principal areas. One is the famous quarry at Kabbaldurga and neighboring areas (Fig. 4.2; Pichamuthu, 1960; Devaraju and Sadashivaiah, 1969, 1971; Janardhan and Srikantappa, 1974; Janardhan and Ramachandra, 1977; Janardhan et al., 1979; Hansen et al., 1985a). The area is in the complex border between the southern end of the Closepet Granite suite (Ch. 3) and the northern edge of the Billigirirangan Hills, occupied primarily by massif charnockites. The other principal area is in the vicinity of Krishnagiri and Dharmapuri (Fig. 4.1), east of the Closepet Granite, south of the Kolar schist belt (Ch. 3), and north of the highland charnockite massif of the Shevaroy Hills (Condie et al., 1982; Allen et al., 1983). Similar relationships to those at Krishnagiri-Dharmapuri occur within the Granulite terrain in the North and South Arcot Districts, which show extensive intermingling of rocks of granulite and amphibolite facies (Saravanan and Ramanathan, 1973, 1982; S. Viswanathan and Nagendra Kumar, 1982; Narayanan Kutty et al., 1983).

Mineralogical changes occurred in all rock types along the transition zone. Charnockite

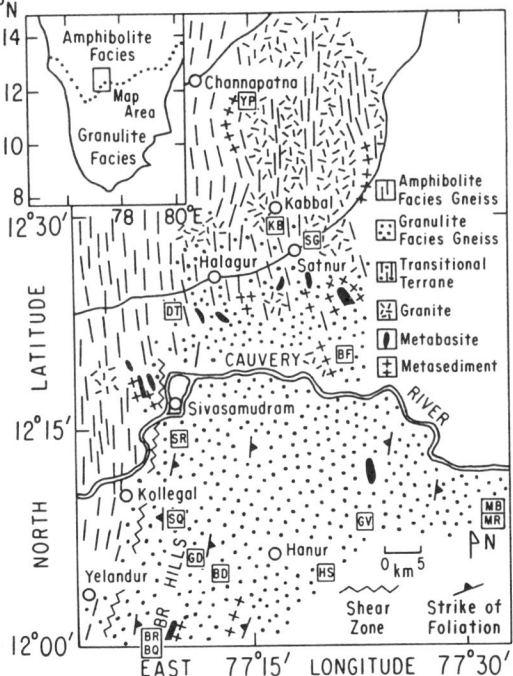

Fig. 4.2. Kabbaldurga area (Hansen et al., 1984). Reproduced with permission from the Journal of Metamorphic Geology, v. 2, p. 253.

formed in silicic rocks by development of orthopyroxene from biotite and hornblende, although some rocks have all three mafic minerals. Mafic rocks with amphibole, biotite, and plagioclase north of the transition zone have been converted to orthopyroxene, clinopyroxene, ± amphibole, ± biotite, ± garnet, ± plagioclase in the granulite area. There is abundant pegmatization, K-feldspar metasomatism, and other evidence of neosome development. Because some of this material is now charnockite, the migmatizing fluids preceded (if only briefly) the granulite metamorphism (Friend, 1985).

A patchy, irregular, development of charnockite and other granulite-facies rocks occurs throughout the transition zone (Fig. 4.3). Characteristic features include stringers of charnockite crossing gneissic structures, obliteration of foliation by irregularly shaped bodies of charnockite (with dimensions measured in centimeters and meters), and zones of granulite along preexisting rock contacts. The charnockite patches are megascopically shown by darkening caused largely by chlorite veinlets and specks, particularly in plagioclase, which probably formed by retrogression of the characteristic mafic minerals.

Metamorphic conditions are indicated by mineralogical and compositional information. Biotites and amphiboles in the transition rocks generally have high F and Cl contents, indicating a

Fig. 4.3. Incipient charnockite at Kabbaldurga. Dark streaks are the result of conversion of light-colored gneisses to charnockite by the drying action of CO_2 streaming along preferred pathways in the gneissic terrain. (Courtesy R. C. Newton)

primary (high-temperature) origin. In contrast, halogen contents are low in the biotites and amphiboles of the massif charnockites, suggesting a retrogressive origin (Janardhan et al., 1982). Fluid inclusions in minerals commonly have high CO_2 contents, confirming the activity of CO_2 in causing the granulite metamorphism (Newton et al., 1980; Hansen et al., 1984; Gopalakrishna et al., 1986). Compositional differences between charnockites and gneisses are very small, even for lithophilic elements (Table 4.2), and the conversion was isochemical on a local scale. Typical metamorphic conditions in the transition zone were about 750° and 6 kb (discussed later).

The general model that has emerged from these studies along the transition zone is as follows. At some time, probably about 2,600 m.y. ago, gneisses and associated rocks were converted into granulite along a front that, in its present orientation, moved from south to north. At least part of the movement was upward, as shown by higher equilibration pressures in the Granulite terrain than farther north, and the present exposure is at least partly explained by northward tilting of the shield. The charnockitization was probably also, in large part, the result of drying out of the terrain by the passage of CO_2 upward from depth, which served to reduce P_{H2O} in the fluid phase. With the

exception of water, abundances of components did not change very much along the front. The CO_2 was, however, preceded by a wave of water that metasomatized, pegmatized, and may have caused partial melting of the overlying crust.

CONFORMABLE MAFIC SUITES

Mafic granulites with various mineral assemblages and amphibolite suites with similar compositions occur as thin, conformable bands throughout all areas of the Granulite terrain. In areas of low-grade metamorphism, mafic suites of various origins commonly can be distinguished by their lithologic properties. In granulite terrains, however, bands and lenses of dark rock are more difficult to identify. Aluminous metapelites should have a mineralogy of plagioclase plus either cordierite or garnet-sillimanite, without pyroxene. Conversely, mafic igneous rocks generally do not contain excess aluminum, and their mineralogy should consist of plagioclase, ortho- and clinopyroxenes, ± amphibole. The problem of discriminating mafic suites arises from the fact that some sediments may not be highly aluminous, and their mineral suite could be similar to that of metamorphosed igneous rocks.

Realizing the overlap of the mafic granulites

TABLE 4.2. *Comparison of compositions of charnockites and precursor gneisses*

	1	2	3	4
SiO_2	68.4	71.4	69.8	67.2
TiO_2	0.62	0.49	0.35	0.45
Al_2O_3	13.6	13.7	15.6	16.1
Fe_2O_3	4.4	3.4	2.7	3.7
FeO				
MnO				
MgO	0.77	0.79	0.9	1.7
CaO	2.5	2.0	3.3	4.3
Na_2O	3.7	3.8	4.9	5.0
K_2O	3.3	3.7	1.4	1.2
Rb	70	60	46	34
Sr	130	110		
Y	310	230	177	180
Zr				
V				
Cr				
Ni				
Th	15	15		
U	1.4	1.0		

All abundances of major elements are in weight percent of oxides. Where no entry is made in the FeO position, the value at Fe_2O_3 is total iron calculated as ferric oxide. All abundances of trace elements are in parts per million by weight. Values are taken directly from quoted sources and not recalculated to 100%.

Table 4.2, Column References
1. Kabbaldurga gneiss (Janardhan et al., 1982).
2. Kabbaldurga charnockite (Janardhan et al., 1982).
3. Krishnagiri-Dharmapuri tonalitic gneiss (Allen et al., 1983).
4. Krishnagiri-Dharmapuri charnockite (Allen et al., 1983).

with charnockite and khondalite suites, we have chosen to describe all of the mafic granulites, mafic conformable rocks of amphibolite facies, and associated ironstones and other unquestionable metasediments in this section. Most of the suites are conformable with layering in the sialic rocks, but some show cross-cutting relationships. To the extent that they are recognizable, the mafic rocks associated with anorthositic layered intrusions are considered in a later section. The categories discussed here are mafic granulites in charnockite terrains, mafic suites occurring as apparent schist belts and as small bands (possibly engulfed relics) in gneissic terrains, and minor rock types of special significance.

Mafic granulites with charnockites

Mafic granulites were originally thought to have been comagmatic with "acid" charnockites (quartz-K-feldspar-plagioclase-orthopyroxene rocks) (Howie, 1955). Recent evidence against

such an origin is based on several observations. One is that the mafic granulites and silicic charnockites generally form a bimodal suite, with "intermediate" members being hybrids of the reaction between the original mafic rocks and some medium that was responsible for development of the charnockites (Devaraju and Sadashivaiah, 1969; Sen, 1974; Weaver et al., 1978; Weaver, 1980).

A second observation is that the mineralogy and composition of the mafic granulites and charnockites show relationships that are incompatible with magmatic fractionation. The orthopyroxenes in charnockites and associated mafic granulites have substantially the same composition (enstatite content), which could not occur if the silicic melt was a fractionation product (A. P. Subramaniam, 1959). Mafic granulites near Madras show high contents of REE elements and large, negative Eu anomalies (Weaver et al., 1978; Weaver, 1980). If associated charnockites had formed by magmatic fractionation, they would be expected to have even more REE and larger Eu anomalies, and this set of properties is not found.

The typical mafic granulite consists of a texturally equilibrium assemblage of orthopyroxene, clinopyroxene, plagioclase, hornblende, $\pm$ biotite $\pm$ garnet $\pm$ accessory minerals. Almost all of the rocks contain some hornblende, which has equilibrated at the same temperature as the other minerals (as shown by fractionation coefficients for Fe and Mg; Sen and Chakraborty, 1968; Ray and Sen, 1970; Sen and Ray, 1971; Sen, 1973). Garnet is generally associated with high-pressure suites and can form in low-pressure rocks only where aluminum contents are high. The prograde reaction to form garnet is presumably an adjustment of plagioclase, orthopyroxene, and clinopyroxene to crystallize garnet and either make the plagioclase richer in albite or the clinopyroxene richer in jadeite. On textural bases, both primary and secondary (retrogressive) biotite is found in different mafic granulites. The retrogressive biotite is recognized as a replacement of other ferromagnesian minerals.

Compositions of mafic granulites are shown in Table 4.3. The suites from Madras (Pallavaram) are from the type charnockite area and were originally regarded as igneous. Weaver et al. (1978) and Chakraborty and Sen (1983) both subdivided the Madras mafic granulites into two suites on the basis of geochemical parameters. Group I (Weaver et al.) and Group A (Chakraborty and Sen) are iron-rich residual liquids. Group II was regarded by Weaver et al. as a cumulate, and Group B was regarded by Chakraborty and Sen as more primitive and possibly komatiitic. The two suites near Agali were regarded as the mafic paleosome for amphibolite-facies migmatites (Sinha Roy and Radhakrishna, 1982).

The mafic granulites of the Madras area prob-

TABLE 4.3. *Compositions of mafic granulites*

	1	2	3	4	5	6	7	8
SiO_2	50.1	49.4	50.0	48.9	50.3	51.4	55.3	50.2
TiO_2	1.8	0.93	1.8	1.1	0.32	0.85	0.18	0.41
Al_2O_3	13.2	13.1	12.9	14.2	14.5	14.1	2.9	4.1
Fe_2O_3	18.5	12.7	19.7	13.6	16.0	12.1	11.4	10.5
FeO								
MnO								
MgO	5.1	8.4	4.9	7.8	5.9	6.9	23.4	14.2
CaO	7.4	11.8	7.8	11.6	10.0	9.3	4.5	17.2
Na_2O	2.4	3.2	2.9	2.2	3.3	2.6	0.93	1.6
K_2O	0.87	0.41	0.86	0.43	0.45	1.9	0.19	0.40
Rb	17	4	16	4				
Sr	150	113	143	126				
Y	41	20						
Zr	112	56	111	59				
V								
Cr								
Ni	33	152	34	137				
Th								
U								

All abundances of major elements are in weight percent of oxides. Where no entry is made in the FeO position, the value at Fe_2O_3 is total iron calculated as ferric oxide. All abundances of trace elements are in parts per million by weight. Values are taken directly from quoted sources and not recalculated to 100%.

Table 4.3. Column References
1. Pallavaram. Group I of Weaver et al. (1978).
2. Pallavaram. Group II of Weaver et al. (1978).
3. Pallavaram. Group A of Chakraborty and Sen (1983).
4. Pallavaram. Group B of Chakraborty and Sen (1983).
5. Pachaimalai (near Madras) (Leelananda Rao, 1955a).
6. Vellore (Narayana et al., 1979).
7. Near Agali. Orthopyroxene-dominated pyroxenite (Sinha Roy and Radhakrishna, 1982).
8. Near Agali. Clinopyroxene-dominated pyroxenite (Sinha Roy and Radhakrishna, 1982).

ably are metaigneous rocks. Their present mineralogy is consistent with an original gabbroic mineralogy of plagioclase, ortho- and clinopyroxene, amphibole, ± biotite. Their compositional variation trends are consistent with such an interpretation if later metamorphism was isochemical except for lithophilic elements (Weaver et al., 1978; Chakaborty and Sen, 1983). The very mafic samples from the Agali area are also probably metaigneous, and the sample from Vellore could easily represent a magmatic differentiate. No suites representative of aluminous metapelites are shown in Table 4.3.

The Granulite terrain was clearly a site of mafic magmatism at one or more times prior to the major granulite event. These magmas were not associated with anorthositic layered intrusions, which are very localized and presumably represent another tectono/magmatic event. Some basaltic magmas may have formed synchronously with the (silicic) precursors to the charnockites in a bimodal igneous assemblage dominated by the

acid member, but other suites appear to represent solely mafic magmatism.

Mafic schist belts and possible fragments

This section describes a suite of rocks characterized by assemblages of mafic granulites or amphibolites, depending on the grade of metamorphism, associated with banded iron formations, quartzites (some fuchsitic), calc-silicate rocks formed from former carbonates, and minor rocks of uncertain parentage containing quartz ± sillimanite ± kyanite ± mica. These suites occur in a few coherent belts but mostly in isolated small outcrops of individual rock types engulfed comformably in silicic gneisses or their charnockitic equivalents.

Coherent units of mafic schists have been described in two parts of the Granulite terrain. The southern part of the Kolar schist belt, in the gneiss-granulite transition zone, has been referred to as the Bargur series by Gopalakrishnan et al.

(1975). It contains hornblende schist, amphibolite, banded iron formation, and quartz-muscovite schist, none of which has been raised completely to granulite grade. The other suite is the Vengad (or Wynad) Group and nearby belts in Kerala (Fig. 4.1). The Vengad Group contains conglomerate, probably basal, overlain by quartz-biotite-muscovite schist (M. M. Nair et al., 1975, 1980; P. K. R. Nair et al., 1981). The Vengad rocks show such sedimentary features as relic current bedding, are highly quartzose, and do not contain volcanic rocks. They are apparently not equivalent to any of the meta-sedimentary suites in the Dharwar area to the north, largely because of their more platformal characteristics.

The principal area of mafic, partly metasedimentary, rocks engulfed in gneisses is the Sathyamangalam Group in the area between the Moyar-Bhavani-Attur and the Noyil-Cauvery lineaments (Fig. 4.1). Here the non-gneissic components occur in screens conformable to the generally E-W structures in the gneisses and range upward in width from a few meters and in length from a few tens of meters. Rock types include amphibolites, quartzites ± fuchsite ± kyanite or sillimanite, banded iron formations, and calc-silicate rocks (Gopalakrishnan et al., 1975; Sinha Roy and Radhakrishna, 1982).

The Sathyamangalam suite is clearly in amphibolite facies. It is partly bordered on the north by the highland massif of the Billigirirangan hills, whose southern contact is an area regarded by Janardhan et al. (1979) as a retrogressive zone, in which charnockite has been converted back to gneiss. It is possible that the entire area of amphibolite-facies rocks between the Moyar-Bhavani and Noyil-Cauvery lineaments is retrogressive, perhaps related to Proterozoic shearing.

Iron formations

Banded iron formations or other rocks consisting largely of quartz and iron oxides are significant components of the mafic schist terrains just described. Their mineralogy differs with the grade of metamorphism. In amphibolite facies, common minerals are quartz and magnetite; grunerite and cummingtonite may also be present. In granulite facies, the same minerals can be present, but the higher degree of metamorphism is demonstrated by the presence of orthopyroxene, hedenbergitic clinopyroxene, and garnet.

Prasad et al. (1982) and Subba Reddy and Prasad (1982) described iron formations in the Granulite terrain and the Sargur schist area just north of the transition zone. The iron formations are generally thin lenses (< 20 m thick) associated with quartzites (± fuchsite), calc-silicate rocks, amphibolites or mafic granulites (former igneous rocks), and fragments of anorthositic layered complexes. Prasad et al. contrasted this assemblage with other areas of iron formations (Table 4.4) and concluded that neither the Keewatin (volcanic) environment nor the Algoman (graywacke) environment were tectonic analogues of the southern Indian iron formations.

The most likely environment of deposition of the iron formations was a platformal area of quartzite, carbonate, and shale. Thus, these iron formations represent a suite of metasediments that has no correlative in any of the Dharwar schist belts described in Chapter 2. As further evidence for this conclusion, Subba Reddy and Prasad (1982) demonstrated that mafic granulites associated with the iron formations must be metaigneous, rather than metasedimentary, on the basis of much higher contents of Cr, Ni, Mn, V, and Ti in the magnetites of the mafic granulites than in the clearly sedimentary magnetites of the iron formations.

Calc-silicate rocks

Because of their importance as indices of metamorphic grade, numerous studies have been made of former carbonate or carbonate-sand-shale rocks (Jacob, 1962; P. L. Narayanaswami, 1962; Krishnanath, 1981; and Devaraju and Coolen, 1983). The calc-silicate rocks are associated with fragments of mafic schist belts (including some types of mafic granulites, banded iron formations, etc.). They are also interbedded with khondalites and have been mapped as members of the "khondalite suite" where that term has been used for a map unit.

The more common mineral assemblages include the following:

Amphibolite facies
1. Salitic to hedenbergitic clinopyroxene; garnet (rich in grossularite and, in some cases, uvarovite); quartz.
2. Diopside; wollastonite; scapolite; calcite.
3. Clinopyroxene; scapolite; ± calcite; ± hornblende; ± plagioclase.

Granulite facies
1. Hedenbergite; anorthite; scapolite; grossularite.
2. Humite; chondrodite; spinel; magnesian calcite.

One calc-silicate assemblage in the gneiss-granulite zone contains an equilibrium assemblage of hypersthene, salite (with 6% jadeite), garnet (alm_{50} pyr_{32} $gross_{17}$), edenitic hornblende, scapolite (me_{70}), and plagioclase (an_{39}) (Devaraju and Coolen, 1983). The scapolite contains high concentrations of both S and C, which indicates high-grade metamorphism. This assemblage is consistent with equilibration at 9 to 10 kb and 810 ± 20°, as estimated by Devaraju and Coolen using a variety of other methods.

TABLE 4.4. *Characteristics of iron formations (Prasad et al., (1982)*

Character	Tamil Nadu Type in Archaean Granulite-Gneiss Belts	Algoma Type in Archaean Greenstone Belts	Superior Type in Proterozoic Sequences
1. Occurrence	Bands less than 20 m thick and a few km long	Lensoid bodies of up to more than 100 m thick and a few km long	Usually more than 100 m thick and persistent up to hundreds or more than a thousand km
2. Associated lithology	Predominantly tonalitic gneiss. Anorthosite complexes locally common. Thin basaltic lavas. Supracrustal rocks make up C.5% of the belts.	Predominantly various volcanics together with fine grained clastic sediments. Supracrustal rocks make up C.95% of belts.	Coarse clastics, quartzites, conglomerates, dolomites and black shales. Supracrustal rocks make up C.95% or more.
3. Sedimentary facies	Quartzite-pelite-carbonate association	Graywacke-flysch-conglomerate-shale association	Black shale, chert, quartzite, dolomite association
4. BIF facies	Only oxide	Oxide, silicate, carbonate, sulphide	Oxide, silicate, carbonate, sulphide
5. Associated minerals	Cummingtonite/grunerite $\pm$ opx $\pm$ cpx $\pm$ hedenbergite $\pm$ garnet	Hornblende, garnet, biotite, chlorite, grunerite, ankerite, siderite	Greenalite, stilpnomelane, minnesotaite, ankerite, siderite, calcite
6. Oolitic texture	Absent	Present	Present
7. Silica	Quartz	Chert, jasper, or quartz	Chert or jasper
8. Iron oxide	Magnetite	Hematite or magnetite	Predominantly hematite
9. Percentage of Fe	< 40	> 50	> 50

Prepared with permission from the Journal of the Geological Society of India, v. 23, p. 113.

Sapphirine-bearing rocks

Sapphirine and the related mineral kornerupine have been reported at a number of locations in southern India (Grew, 1982; Subba Reddy and Thonthiappan, 1982). Sapphirine and kornerupine are rare minerals with the formulas:

Sapphirine (Mg, Fe^{+2}) Al_4SiO_7
Kornerupine $Mg_3Al_6(Si, Al, B)_5$ $O_{21}(OH)$

The boron is apparently an essential component of kornerupine.

Clearly, both minerals are the product of metamorphism of rocks rich in Al and Mg, presumably mafic metasediments. The only sapphirine locality in southern India that is not metasedimentary is in the anorthositic Sittampundi complex, described later. Sapphirine commonly occurs in quartz-bearing rocks but is rarely in direct contact with (equilibrated with) the quartz. Sapphirine may occur in the rare association of sillimanite and orthopyroxene.

Grew (1982) used the assemblage sillimanite-orthopyroxene-garnet-quartz-cordierite-spinel to fix an invariant point in the system $MgO-FeO-Al_2O_3-SiO_2$ (Fig. 4.4). The two problems with this system are higher p_{H2O} stabilizes cordierite to higher temperatures, thus preventing the forma-

tion of sapphirine, and Fe^{+3} stabilizes spinel to higher temperatures and sapphirine to lower temperatures. In general, sapphirine and kornerupine form in the temperature range of 800° to 850° and through a broad pressure range of 6 to 10 kb.

Spinel-bearing rocks

Harris (1981) described a spinel-bearing mafic assemblage in amphibolite-facies terrain south of the highland charnockite massif of Kodaikanal (Fig. 4.1). The assemblage occurs in migmatites and is regarded as the solid residue of anatexis. For given p_{H2O} and MgO contents, the five-phase assemblage cordierite-garnet-sillimanite-spinel-corundum is invariant. The composition of the assemblage, even after anatexis, is probably still indicative of a metasedimentary parent rock. Harris used the assemblage to estimate equilibration pressures of 4.8 $\pm$ 0.5 kb and temperatures of 740° $\pm$ 20°, assuming $p_{H2O} < 0.2p_T$.

KHONDALITES

The only large area of southern India consistently mapped as khondalite is in southern Kerala, south of the Achankovil shear zone (Fig. 4.1). It is not known whether the shear zone juxtaposes

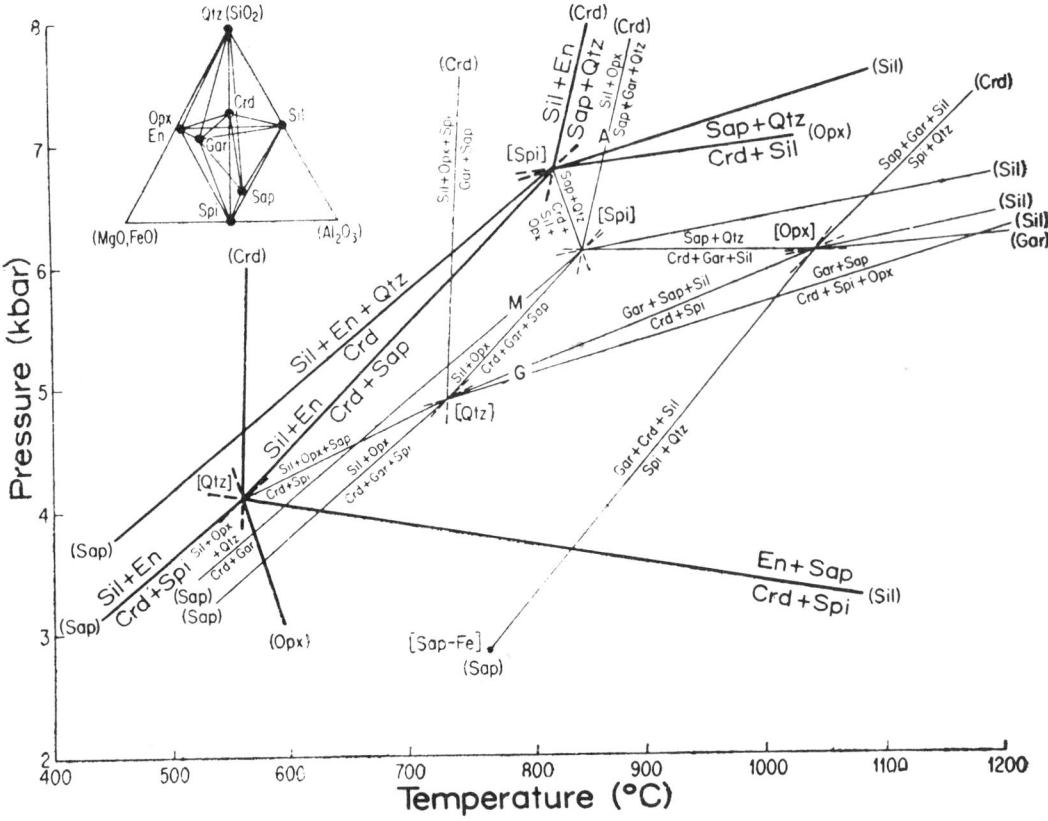

Fig. 4.4. Equilibrium in sapphirine-bearing rocks (Grew, 1982). Petrogenic grid for the system, MgO-FeO-Al$_2$O$_3$-SiO$_2$ and the mineral phases (abbreviations in parentheses) quartz (Qtz), sillimanite (Sil), enstatite (En), orthopyroxene (Opx), cordierite (Crd), sapphirine (Sap), spinel (Spi), and garnet (Gar). Compositions of the minerals are illustrated in the triangular diagram. Invariant points are identified by the absent phase in square brackets and univariant lines by the second absent phase in round brackets. Positions of the univariant curves in the iron-bearing system relative to those in the iron-free system are calculated from compositions of coexisting minerals. Heavy lines—MgO-Al$_2$O$_3$-SiO$_2$ system. Light lines—FeO-MgO-Al$_2$O$_3$-SiO$_2$ system. Temperature-pressure conditions are indicated for Ganguvarpatti (G), Madras (M), and Tula Mountains, Enderby Land, Antarctica (A). Reproduced with permission from the Journal of the Geological Society of India, v. 23, p. 491.

rocks that would otherwise be unrelated or whether it is a small displacement along the typical gradational contact between khondalite and charnockite (Sinha Roy, 1983). In this book, khondalite is an assemblage of quartz, garnet, sillimanite ± feldspar ± graphite ± minor minerals (Table 4.5). Associated rock types are nongarnetiferous quartzites, some calc-silicate rocks, banded iron formations, and minor mafic granulites.

The khondalites are clearly metasediments (Leelananda Rao, 1955a, 1956; Jacob, 1962; S. Narayanaswamy, 1975; Weaver, 1980). The high content of aluminum (shown by the sillimanite) and quartz, plus the presence of graphite in many varieties, is inconsistent with any kind of igneous parent.

Statements concerning khondalites are generally inconsistent, largely because of different definitions of khondalites and the "khondalite suite."

The khondalite suite, as used on many maps, may include mafic granulites, fragments of mafic schists and iron formations, calc-silicates, amphibolite-facies gneisses, "pelitic gneisses," etc. For example, Jacob (1962) estimated that about 10% of southern India is underlain by charnockite, 10% by khondalite, and 80% by granite, gneiss, and schist (in amphibolite facies). Conversely, Gopalakrishnan et al. (1975) estimated that 75% of Tamil Nadu was underlain by a "supergroup" of khondalite and charnockite. Some authors include mafic granulites in the khondalite suite, and others do not.

The general history deduced from the southern Indian khondalites is one of platformal sedimentation of quartzose, pelitic sediments, plus some calcareous and ferruginous rocks, followed by deformation and metamorphism. This history implies the presence of a stable continental crust in southern India early enough for platformal sedi-

TABLE 4.5. *Compositions of rocks in Granulite terrain*

	1	2	3	4	5	6	7	8	9	10	11	12	13	14
SiO$_2$	76.2	70.1	67.6	77.9	75.2	70.9	67.8	56.9	61.6	58.4	68.2	64.7	67.3	70.2
TiO$_2$	0.15	0.15	0.72	0.31	0.15	0.29	0.97	1.1	0.53	1.1	0.60	0.58	0.42	0.68
Al$_2$O$_3$	14.8	13.8	16.9	10.7	13.7	16.2	14.8	17.9	15.7	16.2	14.4	15.2	15.5	15.4
Fe$_2$O$_3$	1.2	3.0	6.1	0.99	0.53	2.4	7.0	10.0	2.9	8.2	5.5	1.6	0.21	3.9
FeO	2.8	8.1		2.5	1.6				3.2			3.2	3.1	
MnO	0.03		0.06	0.04		0.03	0.08	0.22		0.12	0.10	0.19	0.23	0.07
MgO	1.0	2.3	1.4	0.18	0.58	0.48	1.0	5.2	3.5	4.2	2.4	3.5	3.7	0.66
CaO	0.54	0.42	2.2	0.40	3.3	2.7	2.8	3.3	5.7	5.8	2.4	4.9	3.8	2.3
Na$_2$O	0.78	0.38	3.1	2.2	3.6	4.6	2.3	3.1	5.4	3.7	3.0	4.1	4.1	2.8
K$_2$O	2.2	0.85	3.7	4.5	0.92	2.4	4.8	2.7	1.4	2.5	3.3	1.5	1.5	5.6
Rb			185			59	221	56		115			56	150
Sr			257			438	204	359		433			185	366
Y			43			3	50	49		29				32
Zr			260			144	303	194		157				465
V			63			9	59	191		56				74
Cr			217			110	168	220		197				82
Ni			43			8	12	64		26				13
Th														
U														

All abundances of major elements are in weight percent of oxides. Where no entry is made in the FeO position, the value of Fe$_2$O$_3$ is total iron calculated as ferric oxide. All abundances of trace elements are in parts per million by weight. Values are taken directly from quoted sources and not recalculated to 100%.

Table 4.5. Column References
1. Pachaimalai (near Madras). Khondalite (Leelananda Rao, 1955a).
2. Pallavaram. Khondalite (Leelananda Rao, 1956).
3. Kerala. Khondalite (Chacko, personal communication).
4. Madras. Leptynite (Howie and Subramaniam, 1957).
5. Pachaimalai (near Madras). Enderbite (Leelananda Rao, 1955a).
6. Kerala. Sodic gneiss (Chacko, personal communication).
7. Kerala. Potassic gneiss (Chacko, personal communication).
8. Kerala. Cordierite gneiss (Chacko, personal communication).
9. Shevaroy Hills. "Intermediate charnockite" (Ramanathan, 1956b).
10. Kerala. Intermediate charnockite (Chacko, personal communication).
11. Pallavaram. Charnockite (Sen, 1974).
12. Vellore. Charnockite (Narayana et al., 1979).
13. North Arcot. Charnockite (Narayanan Kutty et al., 1983).
14. Kerala. Felsic massif charnockite (Chacko, personal communication).

ments to accumulate and then be metamorphosed about 2,600 m.y. ago. Comparison of the khondalites with the metasedimentary suite of the Western Dharwar craton described in Chapter 2 shows very little similarity. Certainly there is no relationship between khondalites and the graywacke sequences of the younger greenschist belts or the thick metapelites of the older greenstone belts.

CHARNOCKITES (AND ENDERBITES, GNEISSES, AND LEPTYNITES)

Although charnockites are the characteristic rock of the Dharwar Granulite terrain, they actually constitute less than 25% of the outcrop area. Early discussions of charnockites followed their definition by Holland (1900) at Pallavaram, near Madras. The major questions, until recently, centered around the problems of whether charnockites are igneous or metamorphic, whether they form an igneous differentiation sequence from mafic to acidic, and what the relationship was to associated rocks such as enderbites and leptynites (Howie, 1955; Leelananda Rao, 1956; Ramanathan, 1956a, 1956b).

The first major proposals for metasomatic transformation of gneiss to charnockite were made by Pichamuthu (1960, 1961). The concept of transformation of amphibolite-facies rocks into charnockite by lowering p_{H2O} by means of CO_2 streaming through the rock has been described earlier. Some of the contacts between these different facies may be retrogressive. The tranformation process was not restricted to the transition zone at the northern border of the terrain, and evidence for incipient charnockitization and development of charnockite bodies is widespread throughout the entire Granulite terrain (Ravindra Kumar et al., 1984).

Field characteristics of charnockites were summarized by A. P. Subramaniam (1967) and T. V. Viswanathan (1969). Charnockites are younger than khondalites, as shown by cross-cutting relationships in detail. Early workers described the cross-cutting as caused by injection of charnockite magmas. Khondalites and charnockites tend to be conformable and have clearly been deformed together in at least one tectonic episode. Gradational contacts are common between charnockite and biotite-amphibole gneisses. Mafic granulite is common as lenses and screens conformable in the charnockite. Metasomatism and hybridization of the mafic rocks has been caused by the development of the charnockites. Both massive and foliated varieties of charnockite occur. "Igneous-looking" (massive) rocks may, in part, have been formed by anatexis, but it is also possible that purely metamorphic reconstitution led to the development of massive bodies.

Charnockites in complex gneiss-khondalite-granulite terrains appear to be most abundant where there has been prior development of a quartzofeldspathic neosome, as shown by blurring of earlier textures, pegmatite formation, and migmatization (Janardhan and Ramachandra, 1977; Weaver, 1980; Condie and Allen, 1984; Friend, 1985). Narayanan Kutty et al (1983) described "pink granulites" in the North Arcot District that are chemically distinct from associated charnockites by having much higher K_2O content and showing a negative Eu anomaly.

The mineralogy of charnockites is comparatively simple (Pichamuthu, 1979). Major minerals are quartz, K-feldspar, sodic plagioclase, orthopyroxene, $\pm$ garnet and varietal minerals. Most charnockites show a granulitic fabric in which all minerals appear to have grown at the same time in equilibrium. Disequilibrium is most commonly represented by development of amphibole or other retrogressive minerals. Compositional variation among charnockites is limited, particularly for major elements (Table 4.5). There is a complete gradation from charnockites (in which the alkali feldspar is perthitic) to enderbites (in which the alkali feldspar is antiperthitic), but the compositional differences between the two rock types is small. The highland, or massif, charnockites are depleted in lithophilic elements relative to lowland charnockites.

Most of the charnockitization can be explained by a single prograde metamorphic event and local retrogression. It is also possible that more than one granulite event occurred. For example, Pichamuthu (1959) showed that plagioclase in dikes that cut charnockites is clouded, apparently by the same process that caused clouding of plagioclase in the charnockites. This clouding may imply a second granulite metamorphism.

A common occurrence along the contact between charnockite and khondalite is the development of highly garnetiferous rocks (leptynites). They apparently are the result of interaction between the charnockitizing medium and the adjacent khondalite. T. V. Viswanathan and Murthy (1968) showed that zircons in charnockites range from rounded to euhedral; those in khondalites are all rounded, confirming a metasedimentary origin; and zircons in leptynites show extensive overgrowths, presumably by reaction of a fluid with the khondalite zircons.

The likelihood that much, if not all, of the charnockite in the granulite terrain was formed by reconstitution of preexisting material raises the question of the nature of that parent rock. Along the transition zone, various types of Peninsular Gneiss have been converted to charnockites. Formation of paracharnockites by metamorphism of khondalites has probably occurred also. There is no evidence, however, that rock types of greatly different lithology (such as mafic granulite) have been converted.

CONDITIONS OF METAMORPHISM

Various geothermometric and geobarometric techniques have been used to determine metamorphic conditions in southern India (summaries by Raith et al., 1983a, 1983b). Most estimates place the peak of granulite-facies metamorphism at about 750° and a fluid composition dominated by CO_2 (i.e., $p_{H2O} < 0.25\ p_T$). Estimates of total pressure throughout the Granulite terrain, however, vary from about 5 to 12 kb depending on location and rock type. Low pressures are indi-

cated at a number of places by cordierite-bearing assemblages (Sinha Roy et al., 1984). Estimates of metamorphic conditions are shown in Table 4.6.

The data in Table 4.6 must be evaluated in connection with two other observations. One is that the cores of grains commonly yield temperature estimates nearly 100°C higher than grain margins, presumably because of retrogression after the metamorphic peak (Raith et al., 1983a, 1983b). A second observation is the limited amount of anatexis. Raith et al. concluded that temperatures could not have exceeded about 800°C in most

TABLE 4.6. *Metamorphic conditions*

Rock Type and Location (Reference)	Pressure (in kb)	Temperature (in °C)	Notes
Transition zone			
Sargur area in Western Dharwar craton (1)	12	800	Recalculated to lower pressures by Drury et al. (1984)
Sargur area in Western Dharwar craton (2)	8 to 9	740 to 880	
Sargur area in Western Dharwar craton (3)	9 to 10	800	
Near Kabbaldurga (4, 5)	5 to 7	700 to 800	
Near Kabbaldurga (6, 7)	6	750	
West of Kabbaldurga (8)		800	
North side of Nilgiri massif (9)	8	750	Between highland charnockite and gneisses
North side of Nilgiri massif (6, 7)	9.3	700	
Pallavaram (near Madras)			
(10)	9 to 10	780 ± 60	Recalculated to 6.2 to 7.2 kb and 750° by Harris et al. (1982)
Salem			
(6, 7)	8 to 9	700	
Nilgiri massif			
(6, 7)	9	730	North side
(11)	8.3 ± 1	760 ± 40	North side
(11)	6.4 ± 1	735 ± 40	Central part
Kollaimalai massif			
(6, 7)	8.6	750	
South side of Cauvery lineament			
(6, 7)	4 to 6		
Nagercoil			
(11)	5 ± 1		
In Achankovil shear zone			
(12)	7	720	May be retrogressive

Table 4.6. Sources
1. Rollinson et al. (1981)
2. Janardhan and Gopalakrishna (1983)
3. Devaraju and Coolen (1983)
4. Janardhan et al. (1982)
5. Hansen et al. (1984)
6. Raith et al. (1983a)
7. Raith et al. (1983b)
8. Mahabaleswar and Naganna (1981)
9. Harris et al. (1982)
10. Weaver et al. (1978)
11. Harris et al. (1982)

areas because anatexis should have occurred broadly even in dry silicic rocks at that temperature and pressures in the range of 6 to 8 kb. This conclusion, however, is in conflict with the conclusion reached by Grew (1982) based on his study of the distribution of sapphirine and kornerupine in southern India. Grew's data indicated temperatures in the range of 800° to 850° and a broad range of pressures from 6 to 10 kb.

The best generalization that we can make concerning metamorphic conditions in the Granulite terrain is that temperatures through much of the area now exposed at the surface were around 750° to 800°C. The area is essentially an exposed isotherm. Pressures, however, varied enormously. The implication is that temperatures showed only a limited variation through a very broad depth range (an advective gradient; Drury et al., 1984), and this condition could have been established by thickening of the crust by thrusting.

Prior to the use of geothermometers and geobarometers in southern India, Katz (1978) attempted to use more qualitative data on metamorphic facies to plot tectonic belts through southern India and the rest of Gondwanaland. In an attempt to recognize Precambrian analogs to more recent paired metamorphic belts (blueschists and greenschists), Katz recognized two lithologic suites: low pressure, high temperature, characterized by cordierite or wollastonite in rocks of appropriate composition; and intermediate pressure, high temperature, characterized by typical khondalites and wollastonite-free calcsilicates.

ANORTHOSITES AND LAYERED MAFIC COMPLEXES

Anorthosites and associated rocks occur at a variety of localities in the Granulite terrain and the transition zone. Windley and Selvan (1975) listed the major ones as Sittampundi, Kadavur, Chinnadharampuram, Mamandur, and Oddanchatram (Fig. 4.1). Another complex occurs at Bhavani (Gopalakrishnan et al., 1975; Janardhan and Gopalakrishna, 1983). The major suites in the transition zone are at Kurihundi, Konkanhundi, and Sindhuvalli (Fig. 2.2; Ramakrishnan et al., 1978). The Ezhimala complex of Kerala contains mafic rocks and rapakivi granites, is only about 700 m.y. old, and probably is not related to other mafic complexes (with anorthosite) in the region (M. M. Nair and Vidyadharan, 1982). Anorthosite-rich mafic complexes do not occur in southern India outside of the Granulite terrain and transition zone, the possible layered complexes in older schist belts of the Western Dharwar craton having a much higher ratio of mafic rock to anorthosite (Ch. 2).

The amount of deformation and metamorphism in the mafic complexes varies greatly. Most have been subjected to granulite-facies metamor-

phism and intense folding, prime examples being Sittampundi (described shortly) and Kadavur (A. P. Subramaniam, 1956b). At least one complex, Konkanhundi, in the transition zone is virtually undeformed and unmetamorphosed (Ch. 2). The Oddanchatram complex shows undeformed igneous textures but a granulite facies mineralogy, including garnet. Janardhan and Wiebe (1985) proposed that it was intruded after major orogeny but while high temperatures still prevailed. Thus, it is likely that the suite of anorthositic/mafic complexes was emplaced roughly synchronously with granulite metamorphism.

The best studied complex is at Sittampundi (Nehru, 1955a; A. P. Subramaniam, 1956a; Naidu, 1963a; Chappell and White, 1970; Janardhan and Leake, 1975; Ramadurai et al., 1975). The body is a cross-folded remnant of an isoclimally folded layered intrusion. After correction for structural complexity, the body is only about 1,000 m thick. Stratigraphy in the complex is given by Ramadurai et al. (1975) as follows:

Upper layer: clinozoisite-bearing anorthosite
Intermediate: chromite- and hornblende-bearing anorthosite; shows internal layering
Lower layer: gabbro with inclusions of pyroxenite

Metamorphism of the initial complex converted the original anorthositic rocks into an assemblage of calcic plagioclase, edenitic amphibole, corundum, clinozoisite, garnet, and anthophyllite. The gabbroic rocks at Sittampundi are now an assemblage of garnet, clinopyroxene, orthopyroxene, plagioclase, and amphibole. Much of the garnet has been converted on the margins into a symplectic intergrowth of plagioclase and hornblende, presumably during metamorphic retrogression.

There appears to be no genetic relationship between the isolated layered complexes and the mafic granulites that are such a common part of charnockite terrains. Conversely, some of the mafic suites regarded as fragments of schist belts could be associated with, or parts of, layered complexes (e.g., the mafic rocks described by Sinha Roy and Radhakrishna, 1982). Similarly, isolated ultramafic bodies such as those described by Narayanan (1955) and Ramanathan (1956c) could either be associated with the intrusion of layered complexes or with the posttectonic, alkalic magmatism described later.

LATE- OR POSTTECTONIC GRANITES

Numerous small bodies of granite are depicted on the generalized Geology and Mineral Deposits maps prepared by the Geological Survey of India for Kerala and Tamil Nadu, and some have been desribed briefly (Nehru, 1955b; Narasimha Rao, 1969). Many of the mapped granites may actually be extensive areas of potassic neosome development.

Four small granite bodies with similar features have been studied by N. G. K. Nair and co-workers (N. G. K. Nair et al., 1982, 1983; Santosh et al., 1983; N. G. K. Nair and Santosh, 1984). These bodies have elongated, irregular, partly branching outcrop patterns (Fig. 4.5), and they are very rich in alkali elements and silica. Trace elements, particularly lithophilic ones, are extremely abundant. These properties have been proposed to represent diapiric magmatism in rift zones or triple junctions. The bodies are different from those in the Eastern and Western Dharwar cratons on the basis of their unusual composition and irregular outcrop patterns.

Condie et al. (1985) surveyed the properties of granites in the Granulite terrain, referring to high-grade granites as bodies emplaced at depths greater than 20 km. Relative to low-grade granites, occurring in the amphibolite-facies terrain of the Dharwar craton, the high-grade granites have similar major-element compositions but are greatly depleted in Cs, Rb, Nb, Y, and total REE. Most of the high-grade granites have positive Eu anomalies, the opposite of granites emplaced at higher levels. Condie et al. proposed that the low- and high-grade granites were both produced by partial melting of tonalitic rocks in the lower crust, with the low-grade granites being a residual liquid and the high-grade granites being a crystal cumulate from this initial liquid.

POSTTECTONIC ALKALIC AND RELATED SUITES

Posttectonic magmatic suites in the Granulite terrain include alkalic rocks, carbonatites, kimberlites, and mafic/lamprophyric dikes (Fig. 4.1; Sugavanam et al., 1976; Krishnamurthy, 1977;

Narasimban and Sundaram, 1977; Srinivasan, 1977; Suryanarayana Rao et al., 1978). The alkaline suites and related carbonatites occur only in the granulite areas of southern India and the Eastern Ghats.

Detailed studies have been made of several of the alkalic/carbonatite complexes. The Sivamalai complex (Bose, 1971) is about 400 m in diameter and consists of syenite and nepheline syenite with inclusions of mafic alkaline rocks. The differentiation sequence is from hornblende syenite ($\pm$ nepheline) to a leuconepheline syenite. The oldest rocks (in inclusions) are ferrosyenites with minor olivine and salitic clinopyroxene. The suite is silica undersaturated but alumina oversaturated, and the agpaicity index, $(Na_2O \pm K_2O)/Al_2O_3$, remains near 1.0 throughout the fractionation sequence. Biotite and hastingsite are varietal minerals in the syenites.

The Simalpatti complex consists of numerous isolated outcrops (V. Subramaniam et al., 1978). The major body is roughly circular, with a diameter of 5 km. The fractionation sequence is from dunite to pyroxenite to syenite to carbonatite. Fenitisation has occurred in the wall rocks.

In addition to the alkalic suites, a number of mafic to ultramafic rocks occur as dikes and other small bodies throughout the Granulite terrain. These bodies include kimberlites (S. R. N. Murthy, 1979) and dunites or alkalic ultramafic rocks (Naidu, 1963b). Alkalic and lamprophyric dikes are also common (Leelananda Rao, 1955b; Ramanathan, 1954, 1955; D. S. N. Murthy, 1972). Alkalic (nepheline-bearing) dikes are apparently much more abundant in the Granulite terrain than in the lower-grade rocks of the Dharwar cra-

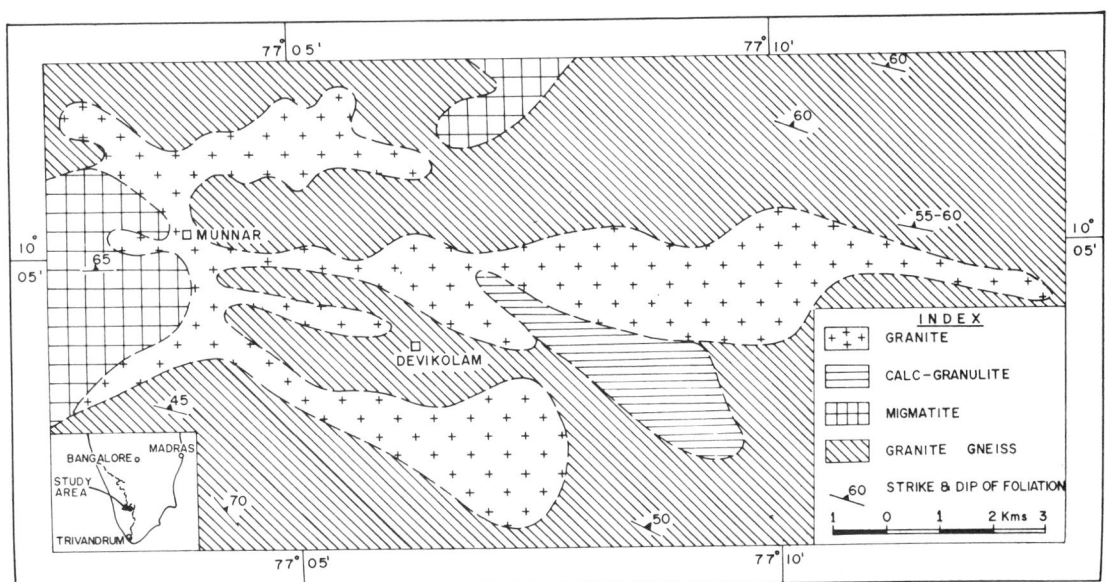

Fig. 4.5. Munnar Granite, Kerala (N. G. K. Nair et al., 1983). Reproduced with permission from the Neues Jahrbuch fur Mineralogie, Abhndlungen, v. 148, p. 224.

ton, where most of the dikes are tholeiitic (non-nepheline bearing) (Chs. 2 and 3). In areas of overlap, such as the gneiss-granulite transition zone, alkalic dikes appear to be younger than tholeiites. Differences in thermal gradients apparently continued under the various terrains long after their metamorphic characteristics were imposed on them. Lower gradients under present granulite areas permitted formation of alkalic magmas by melting at greater depth than those at which the tholeiitic magmas evolved in areas now in amphibolite facies.

REFERENCES

Allen, P., Condie, K. C., and Narayana, B. L. (1983). The Archaean low- to high-grade transition in the Krishnagiri-Dharmapuri area, Tamil Nadu, southern India. In Precambrian of South India (ed. S. M. Naqvi and J. J. W. Rogers), Geol. Soc. India Mem. 4, 450–461.

Balasubrahmanyam, M. N. (1975). Significance of K-Ar age of biotite from charnockite type area, Madras. Geol. Surv. India Misc. Publ. 23, Part 1, 233–235.

Bose, M. K. (1971). Petrology of the alkalic suite of Sivamalai, Coimbatore, Tamil Nadu. Geol. Soc. India J., v. 12, 241–261.

Buhl, D., Grauert, B., and Raith, M. (1983). U-Pb zircon dating of Archaean rocks from the South Indian craton; Results from the amphibolite to granulite facies transition zone at Kabbal Quarry, southern Karnataka. Fortschritte Mineralogie, v. 61, Beiheft 1, 43–45.

Chakraborty, K. R., and Sen, S. K. (1983). A comparative analysis of geochemical trends in basic granulites, greenstones, and recent basalts based on three basic granulite suites from India. In Precambrian of South India (ed. S. M. Naqvi and J. J. W. Rogers), Geol. Soc. India Mem. 4, 462–487.

Chappell, B. W., and White, A. J. R. (1970). Further data on an "eclogite" from the Sittampundi complex, India. Mineral. Mag., v. 37, 555–560.

Condie, K. C., and Allen, P. (1984). Origin of Archean charnockites from southern India. In Archaean Geochemistry (ed. A. Kroner, G. N. Hanson and A. M. Goodwin), Springer Verlag, New York, 183–203.

———, ———, and Narayana, B. L. (1982). Geochemistry of the Archean low- to high-grade transition zone, southern India. Contrib. Mineral. Petrol., v. 81, 157–167.

———, Bowling, G. P., and Allen, P. (1985). Missing Eu anomaly and Archean high-grade granites. Geology, v. 13, 633–636.

Crawford, A. R. (1969). Reconnaissance Rb-Sr dating of the Precambrian rocks of southern peninsular India. Geol. Soc India J., v. 10, 117–166.

Devaraju, T. C., and Coolen, J. J. M. M. M. (1983). Mineral chemistry and P-T conditions of formation of a basic scapolite-garnet-pyroxene granulite from Doddakanya, Mysore District. Geol. Soc. India J., v. 24, 404–411.

Devaraju, T. C., and Sadashivaiah, M. S. (1969). The charnockites of Satnur-Halaguru area, Mysore State. Ind. Mineralogist, v. 10, 67–88.

———, ——— and (1971). Some orthopyroxene-bearing rocks constituting an integral part of high-grade metapelites of Satnur-Halaguru area, Mysore State. Geol. Soc. India J., v. 12, 1–13.

Drury, S. A., Harris, N. B. W., Holt, R. W., Reeves-Smith, G. J., and Wightman, R. T. (1984). Precambrian tectonics and crustal evolution in South India. J. Geol., v. 92, 3–20.

Friend, C. R. L. (1985). Evidence for fluid pathways through Archaean crust and the formation of the Closepet Granite, Karnataka, India. Precamb. Res., v. 27, 239–250.

Gopalakrishna, D., Hansen, E. C., Janardhan, A. S., and Newton, R. C. (1986). The southern high-grade margin of the Dharwar craton. J. Geol., v. 94, 247–260.

Gopalakrishnan, K., Sugavanam, E. B., and Venkata Rao, V. (1975). Are there rocks older than Dharwars? A reference to rocks in Tamil Nadu. Ind. Mineralogist, v. 16, 26–34.

Grew, E. S. (1982). Sapphirine, kornerupine, and sillimanite ± orthopyroxene in the charnockite region of South India. Geol. Soc. India J., v. 23, 469–505.

———, and Manton, W. I. (1984). Age of allanite from Kabbaldurga Quarry, Karnataka. Geol. Soc. India J., v. 25, 193–195.

Hansen, E. C., Newton, R. C., and Janardhan, A. S. (1984). Fluid inclusions in rocks from the amphibolite-facies gneiss to charnockite progression in southern Karnataka, India: Direct evidence concerning the fluids of granulite metamorphism. J. Metam. Geol., v. 2, 249–264.

———, ———, and ——— (1985a). Geochemistry, geobarometry, and fluid inclusions of a continuous prograde amphibolite-facies gneiss to charnockite succession in southern Karnataka. In Archaean Geochemistry (ed. A. Kroner, G. N. Hanson and A. M. Goodwin), Springer Verlag, New York, 161–181.

———, Hickman, N. H., Grant, N. K., and Newton, R. C. (1985b). Pan-African age of "Peninsular Gneiss" near Madurai, S. India (abstract). EOS (Trans. Amer. Geophys. Union), v. 66, 419–420.

Harris, N. B. W. (1981). The application of spinel-bearing metapelites to P/T determinations; An example from South India. Contrib. Mineral. Petrol., v. 76, 229–233.

———, Holt, R. W., and Drury, S. A. (1982). Geobarometry, geothermometry, and late Archaean geotherms from the granulite facies terrain of South India. J. Geol., v. 90, 509–527.

Holland, T. H. (1900). The charnockite series; A group of Archaean hypersthenic rocks in peninsular India. Geol. Surv. India Mem. 28, Part 2, 192–249.

Holt, R. W., and Wightman, R. T. (1983). The role of fluids in the development of a granulite facies transition zone in S. India. Geol. Soc. London J., v. 140, 651–656.

Howie, R. A. (1955). The geochemistry of the charnockite series of Madras, India. Trans Roy. Soc. Edinburgh, v. 62, 725–768.

——, and Subramaniam, A. P. (1957). The paragenesis of garnet in charnockites, enderbite and related granulites. Mineral. Mag., v. 31, 565–586.

Jacob, S. C. (1962). Geology of the western Ghats around Nagercoil, Kanniyakumari District, Madras State. Ind. Mineralogist, v. 3, 1–45.

Janardhan, A. S., and Gopalakrishna, D. (1983). Pressure-temperature estimates of the basic granulites and conditions of metamorphism in Sargur terrain, southern Karnataka and adjoining areas. Geol. Soc. India J., v. 24, 219–228.

——, and Leake, B. E. (1975). The origin of the meta-anorthositic gabbros and garnetiferous granulites of the Sittampundi complex, Madras, India. Geol. Soc. India J., v. 16, 391–408.

——, and Ramachandra, H. M. (1977). Charnockites from Hullahalli, Mysore District, Karnataka State. Geol. Soc. India J., v. 18, 331–337.

——, and Srikantappa, C. (1974). Petrology of metasedimentary bands of Doddakanaya, Karnataka State. Indian Acad. Geosci. J., v. 17, 15–26.

——, and Wiebe, R. A. (1985). Petrology and geochemistry of the Oddanchatram anorthosite and associated basic granulites, Tamil Nadu, South India. Geol. Soc. India J., v. 26, 163–176.

——, Newton, R. C., and Smith, J. V. (1979). Ancient crustal metamorphism at low p_{H2O}: Charnockite formation at Kabbaldurga, South India. Nature, v. 278, 511–514.

——, ——, and Hansen, E. C. (1982). The transformation of amphibolite facies gneiss to charnockite in southern Karnataka and northern Tamil Nadu, India. Contrib. Mineral. Petrol., v. 79, 130–149.

——, ——, and —— (1983). Transformation of Peninsular Gneiss to charnockite in southern Karnataka. In Precambrian of South India (ed. S. M. Naqvi and J. J. W. Rogers), Geol. Soc. India Memoir 4, 417–435.

Katz, M. B. (1978). Tectonic evolution of the Archaean granulite facies belt of Sri Lanka-South India. Geol. Soc. India J., v. 19, 185–205.

Krishnamurthy, P. (1977). On some geochemical aspects of the Sevattur carbonatite complex, North Arcot District, Tamil Nadu. Geol. Soc. India J., v. 18, 265–274.

Krishnanath, R. (1981). Coexisting humite-chondrodite-spinel-magnesian calcite assemblage from the calc-silicate rocks of Ambasamudram, Tamil Nadu, India. Geol. Soc. India J., v. 22, 235–242.

Leelananda Rao, N. (1955a). The geology and petrology of Pachaimali. Madras Univ. J., v. 25B, no. 1, 55–81.

—— (1955b). Further studies on dyke rocks of Pallavaram. Madras Univ. J., v. 25B, no. 3, 323–339.

—— (1956). Charnockites and associated rock types of the type area of Sir Thomas Holland; Part I. Madras Univ. J., v. 26B, no. 1, 93–115.

Mahabaleswar, B., and Naganna, C. (1981). Geothermometry of Karnataka charnockites. Bull. Mineralogie, v. 104, 848–855.

Murthy, D. S. N. (1972). Petrography of dolerites of Namakkal Taluk, Tamil Nadu. Ind. Acad. Geosci. J., v. 15, 47–56.

Murthy, S. R. N. (1979). Petrology of the ultramafic rocks of Chalk Hills, Salem, Tamil Nadu. Geol. Surv. India Records, Part 5, Southern Region, v. 112, 15–35.

Naidu, P. R. J. (1963a). A layered complex in Sittampundi, Madras State, India. Mineral. Soc. Amer. Spec. Paper 1, 116–123.

—— (1963b). Crystallization of leucite-nepheline-sanidine in basic differentiates from a peridotite-dunite mass in Salem, Madras State, India. Mineral. Soc. Amer. Spec. Paper 1, 251–257.

—— (1963c). Hypersthene-bearing rocks of Madras State, India. Proc. 50th Ind. Sci. Cong., Part II, 1–15.

Nair, M. M., and Vidyadharan, K. T. (1982). Rapakivi granite of Ezhimala complex and its significance. Geol. Soc. India J., v. 23, 46–51.

——, ——, Powar, S. D., Sukumaran, P. V., and Murthy, Y. G. K. (1975). The structural and stratigraphic relationship of the schistose rocks and associated igneous rocks of the Tellicherry-Manantoddy area, Cannanore District, Kerala. Ind. Mineralogist, v. 16, 89–100.

——, Powar, S. D., Vidyadharan, K. T., and Senthiappan, M. (1980). Vengad Group of rocks of Kerala—An equivalent of the Dharwars of South India. Geol. Surv. India Spec. Publ. 5, 9–14.

Nair, N. G. K., and Santosh, M. (1984). Petrochemistry and tectonic significance of the Peralimala alkali granite, Cannanore District, Kerala. Geol. Soc. India J., v. 25, 35–44.

——, ——, Thampi, P. K., and Balasubramanian, G. (1982). Petrochemistry of the Ambalawayal granite, Wynad District, Kerala. Geol., Min. Metall. Soc. India Quart. J., v 54, nos. 3/4, 28–42.

——, ——, and —— (1983). Geochemistry and petrogenesis of the alkali granite of Munnar, Kerala (India) and its bearing on rift tectonics. Neues Jahrb. Mineral. Abh., v. 148, 223–232.

Nair, P. K. R., Prasannakumar, V., and Mathai,

PRECAMBRIAN GEOLOGY OF INDIA

T. (1981). Structure of the western termination of the Bhavani lineament. Geol. Soc. India J., v. 22, 285–291.

Narasimban, A. S., and Sundaram, D. (1977). Geology of the carbonatite complex, Sevattur, Tiruppattur Taluk, North Arcot District, Tamil Nadu. Geol. Surv. India Records, v. 109, Part 2, 1–6.

Narasimha Rao, P. (1969). Studies on some granitic rocks of the Palni-Dindigul area, Tamil Nadu. Geol. Soc. India J., v. 10, 65–76.

Narayana, B. L., Rama Rao, P., Ahmad, S. M., Hussain, S. M., Jafri, S. H., and Satyanarayana, K. (1979). Geochemistry of granulites from Vellore and Pallavaram in relation with Dharwar craton, India. Geophys. Res. Bull. (Hyderabad), v. 17, 11–26.

Narayanan, K. (1955). Granulitization of charnockites in the Chalk Hills region of Salem District. Madras Univ. J., v. 25B, no. 1, 7–18.

Narayanan Kutty, T. R., Anantha Iyer, G. V., and DePoli, E. (1983). REE geochemistry of pink granulites from North Arcot District, Tamil Nadu. Geol. Soc. India J., v. 24, 113–133.

Narayanaswami, P. L. (1962). Wollastonite bearing calc-silicate rocks of Tinnevelly District. Geol. Soc. India J., v. 3, 147–156.

Narayanaswamy, S. (1975). Proposal for charnockite-khondalite system in the Archaean shield of peninsular India. Geol. Surv. India Misc. Publ. 23, Part 1, 1–16.

——, and Lakshmi, P. (1967). Charnockitic rocks of Tinnevelly District, Madras. Geol. Soc. India Jour., v. 8, 38–50.

Nehru, C. E. (1955a). Geology and petrochemistry of the anorthite gneiss and associated rocks of Sittampundi, Salem Dt. Madras Univ. J., v. 25B, no. 2, 173–188.

—— (1955b). Tectonics of the granites and gneisses of Tiruchengode. Madras Univ. J., v. 25B, no. 3, 227–233.

Newton, R. C., Smith, J. V., and Windley, B. F. (1980). Carbonic metamorphism, granulites and crustal growth. Nature, v. 288, 45–50.

Pichamuthu, C. S. (1959). The significance of clouded plagioclase in the basic dykes of Mysore State, India. Geol. Soc. India J., v. 1, 68–79.

—— (1960). Charnockite in the making. Nature, v. 188, 135–136.

—— (1961). Transformation of Peninsular Gneiss into charnockite in Mysore State, India. Geol. Soc. India J., v. 2, 46–49.

—— (1965). Regional metamorphism and charnockitization in Mysore State, India. Ind. Mineralogist, v. 6, 119–126.

—— (1979). Mineralogy of Indian charnockites. Geol. Soc. India J., v. 20, 257–276.

Prasad, C. V. R. K., Subba Reddy, N., and Windley, B. F. (1982). Iron formations in Archaean granulite-gneiss belts with special reference to southern India. Geol. Soc. India J., v. 23, 112–122.

Radhakrishna, B. P. (1968). Geomorphological approach to the charnockite problem. Geol. Soc. India J., v. 9, 67–74.

Raith, M., Raase, P., Ackermand, D., and Lal, R. K. (1983a). Metamorphic conditions in charnockite-khondalite zone of South India; Geothermobarometry on garnet-pyroxene-plagioclase rocks. In Precambrian of South India (ed. S. M. Naqvi and J. J. W. Rogers), Geol. Soc. India Mem. 4, 436–459.

——, ——, ——, and —— (1983b). Regional geothermobarometry in the granulite-facies terrane of South India. Roy. Soc. Edinburgh Trans., Earth Sci., v. 73, 221–244.

Ramadurai, S., Sankaran, M., Selvan, T. A., and Windley, B. F. (1975). The stratigraphy and structure of the Sittampundi complex, Tamil Nadu, India. Geol. Soc. India J., v. 16, 409–414.

Ramakrishnan, M., Viswanatha, M. N., Chayapathi, N., and Narayanan Kutty, T. R. (1978). Geology and geochemistry of anorthosites of Karnataka craton and their tectonic significance. Geol. Soc. India J., v. 19, 115–134.

Ramanathan, S. (1954). Shonkinites from the ultrabasic areas of Salem. Madras Univ. J., v. 24B, no. 3, 315–333.

—— (1955). Vogesites and noritic olivine dolerites from Salem and Dodkanaya. Madras Univ. J., v.25B, no.1, 29–54.

—— (1956a). On the hypersthene-bearing rocks of Salem, I. Madras Univ. J., v. 26B, no. 1, 117–159.

—— (1956b). On the hypersthene-bearing rocks of Salem, II. Madras Univ. J., v. 26B, no. 2, 161–187.

—— (1956c). Petrofabric analysis of the rocks of the charnockite-dunite area of Salem. Madras Univ. Jour., v. 26B, no. 2, 209–228.

Ravindra Kumar, D. C., Srikantappa, C., and Hansen, E. C. (1984). Charnockite formation at Ponmudi in southern India. Nature, v. 313, 207–209.

Ray, S., and Sen, S. K. (1970). Partitioning of major exchangeable cations among orthopyroxene, calcic pyroxene and hornblende in basic granulite from Madras. Neues Jahrb. Mineral. Abh., v. 114, 61–88.

Rollinson, H. R., Windley, B. F., and Ramakrishnan, M. (1981). Contrasting high and intermediate pressures of metamorphism in the Archaean Sargur schists of southern India. Contrib. Mineral. Petrol., v. 76, 420–429.

Santosh, M., Rajan, P. K., and Nair, N. G. K. (1983). The Pariyaram Granite, Trichur District, Kerala—Its petrochemistry. Ind. Geol. Assoc. Bull., v. 16, 33–44.

Saravanan, S., and Ramanathan, S. (1973). Calc-granulites around Manalur, South and North Arcot Districts, Tamil Nadu, India. Geol. Soc. India J., v. 14, 60–70.

———, and ——— (1982). A transitional metamorphic zone around Manalur, South and North Arcot Districts, Tamil Nadu. Geol., Min. Metall. Soc. India Quart. J., v. 54, 36–44.

Sen, S. K. (1973). Compositional relations among hornblende and pyroxenes in basic granulites and an application to the origin of garnets. Contrib. Mineral. Petrol., v. 38, 299–306.

——— (1974). A review of some geochemical characters of the type area (Pallavaram, India) charnockites. Geol. Soc. India J., v. 15, 413–420.

———, and Chakraborty, K. R. (1968). Magnesium-iron exchange equilibrium in garnet-biotite and metamorphic grade. Neues Jahrb. Mineral. Abh., v. 108, 181–207.

———, and Ray, S. (1971). Hornblende-pyroxene granulites versus pyroxene granulites; A study from the type charnockite area. Neues Jahrb. Mineral. Abh., v. 115, 291–314.

Sinha Roy, S. (1983). Structural evolution of the Precambrian crystalline rocks of South Kerala. Recent Res. Geol., v. 10, 127–143.

———, and Radhakrishna, T. (1982). Geochemistry of the ultramafic and mafic complex of Agali, Kerala, and its implications for Archaean greenstones of South India. Neues Jahrb. Mineral. Abh., v. 143, 309–330.

———, Mathai, J., and Narayanaswamy (1984). Structural and metamorphic characteristics of cordierite-bearing gneisses of South Kerala. Geol. Soc. India J., v. 25, 231–244.

Spooner, C. M., and Fairbairn, H. W. (1970). Strontium 87/strontium 86 initial ratios in pyroxene granulite terranes. J. Geophys. Res., v. 75, 6706–6713.

Srinivasan, V. (1977). The carbonatite of Hogenakal, Tamil Nadu, South India. Geol. Soc. India J., v. 18, 598–604.

Subba Reddy, N, and Prasad, C. V. R. K. (1982). Trace element studies and origin of magnetite quartzites of Tamil Nadu, India. Geol. Soc. India J., v. 23, 80–84.

———, and Thonthiappan, S. (1982). Kornerupine from Kavuthimalai, Tiruvannamalai area, Tamil Nadu. Geoviews, v. 10, 489–492.

Subrahmanyam, C. (1983). An overview of gravity anomalies, Precambrian metamorphic terrains and their boundary relationships in the southern Indian shield. In Precambrian of South India (ed. S. M. Naqvi and J. J. W. Rogers), Geol. Soc. India Mem. 4, 553–566.

Subramaniam, A. P. (1956a). Mineralogy and petrology of the Sittampundi complex, Salem District, Madras State, India. Geol. Soc. Amer. Bull., v. 67, 317–390.

——— (1956b). Petrology of the anorthosite-gabbro mass at Kadavur, Madras, India. Geol. Mag., v. 93, 287–300.

——— (1959). Charnockites of the type area near Madras—A reinterpretation. Amer. J. Sci., v. 257, 321–353.

——— (1967). Charnockites and granulites of southern India: A review. Dansk Geol. Foren. Medd., v. 17, 473–493.

Subramaniam, V., Viladkar, S. G., and Upendran, R. (1978). Carbonatite alkali complex of Samalpatti, Dharmapuri District, Tamil Nadu. Geol. Soc. India J., v. 19, 206–216.

Sugavanam, E. B., Balasubramanian, G., and Vemban, N. A. (1976). A note on the association of alkali syenite and quartz-barytes veins in North Arcot District, Tamil Nadu. Geol. Surv. India Misc. Publ. 23, Part 2, 385–394.

Suryanarayana Rao, C., Narayanan Das, G. R., Krishnaiah Setty, B., and Perumal, N. V. S. (1978). Radioactive carbonatites of Pakkanadu and Mulakaddu, Salem District, Tamil Nadu. Geol. Soc. India J., v. 19, 53–63.

Venkatasubramanian, V. S., and Narayanaswamy, R. (1974). Studies in the Rb-Sr geochronology and trace element geochemistry in granitoids of Mysore craton, India. Ind. Inst. Sci. J. (Bangalore), v. 56, 19–42.

Viswanathan, S., and Nagendra Kumar, P. (1982). Charnockites and associated ultramafic rocks around Shevaroy Hills, Salem District, Tamil Nadu. Geol., Min. Metall. Soc. India Quart. J., v. 54, 36–44.

Viswanathan, T. V. (1969). The granulitic rocks of the Indian Precambrian shield. Geol. Surv. India Mem. 100, 37–66.

———, and Murthy, M. V. N. (1968). Zircon studies in interpreting the origin of charnockites and associated rocks from Pallavaram, Madras, and Puri District, Orissa, India. 22nd Internat. Geol. Cong. Rept., Delhi, Sect. 13, 97–120.

Weaver, B. L. (1980). Rare earth element geochemistry of Madras granulites. Contrib. Mineral. Petrol., v. 71, 271–279.

———, Tarney, J., Windley, B. F., Sugavanam, E. B., and Venkata Rao, V. (1978). Madras granulites: Geochemistry and P-T conditions of crystallization. In Archaean Geochemistry; Proc. First Internat. Symp. on Archaean Geochem: The Origin and Evolution of Archaean Continental Crust (ed. B. F. Windley and S. M. Naqvi), Elsevier, Amsterdam, 177–204.

Windley, V. F., and Selvan, T. A. (1975). Anorthosites and associated rocks of Tamil Nadu, southern India. Geol. Soc. India J., v. 16, 209–215.

5. EASTERN GHATS

The Eastern Ghats province is identified on Figure 1.1 and shown in more detail in Figure 5.1. The province consists largely of high-grade igneous and metamorphic rocks whose principal structural trends are parallel to the north-northeast to south-southwest elongation of the Eastern Ghats (M. V. N. Murthy et al., 1971; Narayanaswamy, 1975). The western boundary is a zone of abrupt variation in Bouguer gravity, and a profile across this boundary (Fig. 5.2) is almost identical to a profile across the Grenville front of the Canadian shield. The high Bouguer gravity values over the Eastern Ghats belt are clearly the result of high density of the rocks that constitute the belt because they were calculated by assuming a constant crustal density of 2.67 across the entire area, thereby making the mass correction too small within the belt.

A thrust along the western margin of the Eastern Ghats has been confirmed by deep seismic sounding (Kaila and Bhatia, 1981). The northeastern termination of the belt is the thrust zone on the northern side of the Mahanadi rift valley, which is at least partly referred to as the Sukinda thrust. The western margin of the belt intersects the coastal plain north of Madras. This distinction separates the Eastern Ghats belt from the Granulite terrain of southernmost India (Ch. 4) and places the type charnockite area at Pallavaram, near Madras, outside of the Eastern Ghats. Another interpretation of the Eastern Ghats belt by Radhakrishna and Naqvi (1986) extends the belt southward to the southern tip of India, including Pallavaram.

The Godavari rift (Fig. 1.5; Ch. 10) intersects the Eastern Ghats belt and disrupts its gravity profile. This intersection means that at least some of the movement on the rift occurred after development of the Eastern Ghats suite. Age relations have not been determined, however, between the early stages of rifting and the possible collisional event that juxtaposed the Eastern Ghats terrain against the remainder of the Indian crust.

MAJOR ROCK SUITES

All rocks within the Eastern Ghats belt are either igneous or are sediments that have been metamorphosed to a very high rank. The principal suites are listed below, and nomenclature is discussed in Chapter 4.

Mafic schists include biotite-, muscovite-, and amphibole-bearing schists. The principal occurrences are in the Nellore schist belt and the Khammam area (Fig. 5.1).

Charnockites are silica-rich, quartzofeldspathic assemblages containing hypersthene.

Khondalites and related rocks, including calc-silicates, are metasediments. In addition to sillimanite and garnet, other minerals indicative of metamorphic conditions are cordierite and sapphirine.

Mafic granulites commonly occur as conformable layers in the charnockites and khondalites and consist of plagioclase, clinopyroxene, ± orthopyroxene, and minor amphibole.

Anorthosites/layered complexes consist of a sequence of anorthosites, gabbros, norites, and minor more differentiated rocks. Dunites and serpentinites, commonly with chromite, form small, generally conformable bodies.

Alkaline rocks occur as a number of posttectonic bodies throughout the Eastern Ghats.

GENERAL STRUCTURE AND STRATIGRAPHY

A general distribution of rock types in the Eastern Ghats is impossible to show on a small scale such as Figure 5.1. The detailed map pattern of the principal lithologies (charnockite, khondalite, leptynite, and gneiss) is very complex, characteristic of a zone of intense deformation. In general, charnockites occupy the western, marginal part of the belt, and khondalites and other pyroxene-free rocks occur predominantly to the east (Viswanathan, 1969). Viswanathan regarded this succession and the general southeasterly dip of the suites as a stratigrahic succession from metavolcanic (now charnockitic) rocks at the base to metasedimentary (now khondalitic rocks) at the top.

Contacts between the charnockites, khondalites, and gneisses are mostly gradational, with interlayering on a scale of meters. Some gneisses, however, show clear cross-cutting relationships, indicating that they are igneous intrusions younger than at least the protolith. Whether the

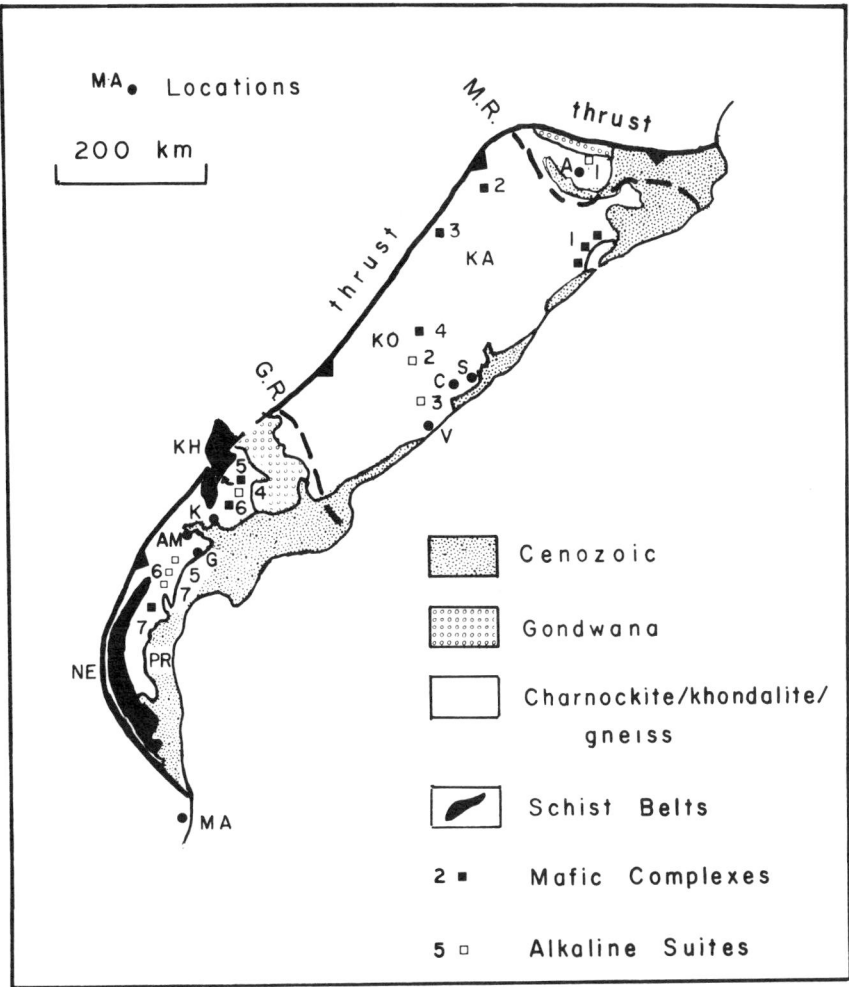

Fig. 5.1. Eastern Ghats. Locations include A, Angul; KA, Kalahandi; KO, Koraput; S, Srikakulam; C, Chipurupalle; V, Visakhapatnam; KH, Khammam; K, Kondapalle; AM, Amaravathi; G. Guntur; PR, Prakasam; NE, Nellore; MA, Madras. M.R. is the Mahanadi River, and G.R. is the Godavari River.

Anorthosite/mafic complexes include 1, Chilka Lake; 2, Bolangir; 3, Kalahandi; 4, Koraput; 5, Kondapalle; 6, Chimalpahad; 7, Chimakurti. Alkaline complexes include 1, Baradangua; 2, Koraput; 3, Borra; 4, Kunavaram; 5, Elchuru; 6, Purimetla; 7, Uppalapadu.

gneisses also postdate the khondalite metamorphism is not clear, and most workers assume that deformation, metamorphism, and intrusion are approximately synchronous. Silicic granulites (charnockites) and mafic granulites are generally conformable and interlayered.

Most of the anorthosites/layered complexes are broadly conformable and concordant to the surrounding charnockite and khondalite terrain, and many appear to be sheets. Sufficient cross-cutting relationships occur locally, however, to show that the mafic intrusive suite is younger than most of the other rocks in the Eastern Ghats. Alkaline bodies, including carbonatites, clearly postdate all tectonic events.

Each rock suite in the Eastern Ghats has been affected by at least four events, although some may have been synchronous, and the available evidence yields little information on the sequence of these events. The four events are the time of formation of the sediment or igneous rock, the time or times of metamorphism, the time or times of deformation, and the time that the Eastern Ghats was accreted to the Dharwar, Bhandara and Singhbhum cratons, assuming that the gravity gradient represents a tectonic suture.

At present, only a few definitive statements can be made about the sequence of events. The mafic and khondalitic schists are probably among the oldest rocks, possibly sharing that age with some

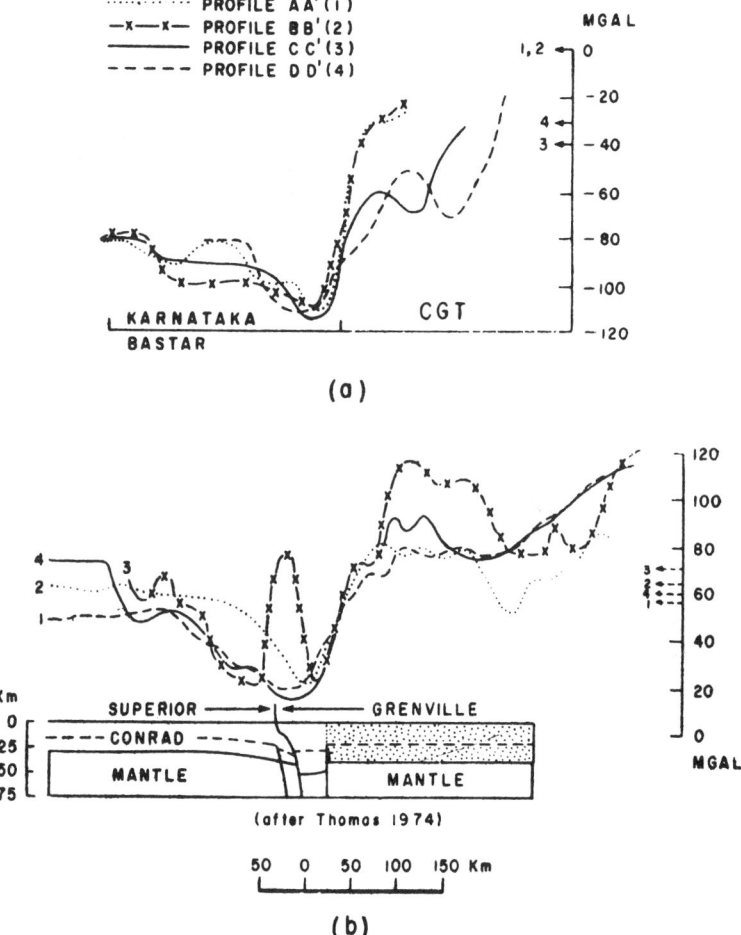

Fig. 5.2. Bouguer gravity profile across western margin of Eastern Ghats compared with profile across Grenville front (Subrahmanyam, 1983). (*a*) Profile from Karnataka (Eastern Dharwar) and Bastar (Bhandara) cratons into Coastal Granulite terrain (CGT = Eastern Ghats). (*b*) Profile from Superior craton (Archean, amphibolite facies) into Grenville belt (granulite facies). Diagram (*b*) also shows crustal structures modeled from gravity profiles. Both diagrams show gravity profiles at various positions along the fronts.

members of the gneissic or charnockitic suites. Some gneisses are younger intrusions, or at least were remobilized in a later event, but the age of this event relative to general deformation and/or collision is unclear. It is tempting to place the highest-grade metamorphic mobilization of the gneisses, intrusion of mafic and anorthositic bodies, and deformation and suturing all in one collisional event. Evidence for this concurrence, however, is absent. At least two, and possibly more, periods of deformation have been demonstrated by intersecting structures (e.g., Natarajan and Nanda, 1981). Polyphase metamorphism is also likely.

The time of suturing along the western margin of the belt can be somewhat constrained by the fact that the Godavari rift transects the Eastern Ghats. Middle Proterozoic sediments in the rift are metamorphosed at their eastern margin (Ch.

10), and thus some deformation along the Eastern Ghats front must have occurred after this time.

ABSOLUTE AGES

Only four rock suites have been accurately dated in the Eastern Ghats. They are:

The Chilka Lake anorthosite body at 1,400 to 1,300 m.y. (Sarkar et al., 1981).

The Kunavaram alkaline body at approximately 1,300 m.y. (Clark and Subbarao, 1971).

Granites on the eastern side of the Nellore schist belt, which yield ages ranging from 1,615 to 995 m.y. (Gupta et al., 1984). The ages of Gupta et al. include one whole-rock isochron with the 1,615 m.y. age and an initial $^{87}Sr/^{86}Sr$ ratio of 0.709 ± 0.005 and one combined whole-rock/ biotite age of 995 m.y. with an initial Sr isotopic

ratio of 0.735, clearly indicative of postemplacement reequilibration.
Sapphirine-bearing rocks at approximately 1,000
m.y. (Grew and Manton, 1986).

The timing of the granulite metamorphism is unclear, but the 1,400-m.y. age of the Chilka Lake
complex is consistent with a mid-Proterozoic deformation and metamorphism. Fission-track
studies on pegmatites in mica pegmatites east of
the Nellore schist belt are consistent with an event
at 1,600 m.y. (Parshad et al., 1979).

Various rock suites have been correlated with
Archean suites elsewhere in India based on lithologic evidence (e.g., Fermor, 1936). For example,
the Nellore schist belt is similar to some of the
coherent schist belts of the Western Dharwar craton (Ch. 2). No definite Archean dates are available on any rocks in the Eastern Ghats, however,
except for uncertain model ages of Perraju et al.
(1979).

GEOLOGY OF THE WESTERN BOUNDARY

The only detailed geologic study along the western margin of the Eastern Ghats belt was in the
Jeypore area (Fig. 5.3; Crookshank, 1938). The
contact between granulite-facies rocks in the Eastern Ghats and the amphibolite facies of Bastar, to
the west, is highly irregular. In particular, "apo-
physes" of charnockites extend into the Bastar
area (Fig. 5.3). Shearing and crushing are common both east and west of the contact zone, but
the charnockites just east of the contact appear to
be relatively unaffected. The major shear zone
shown in Figure 5.3 is linear and east of the western edge of the granulite terrain. Bodies of "granites" in the amphibolite terrain were attributed to
mobilization of components in the charnockite
terrain. In addition to magmatic activity, Crookshank proposed that metasomatic fluids injected
sodium into the border area and rocks to the west,
possibly as emanations from the charnockites.

CONFORMABLE MAFIC AND
FERRUGINOUS SUITES

Mafic rocks of granulite facies are common as
small, conformable layers and lenses throughout
the Eastern Ghats region. Large, mappable units
seem to occur mostly south of the Godavari River
and toward the western edge of the belt, primarily
in the Nellore and Khammam areas (Fig. 5.1).
Some rocks in these larger bodies are of lower (including greenschist) facies. It is important to dis-

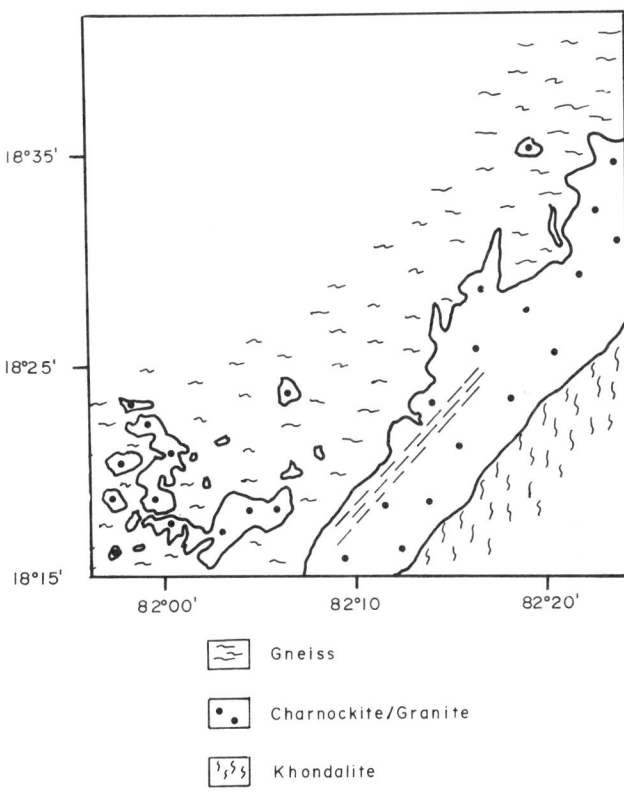

Fig. 5.3. Map of western boundary of Eastern Ghats in Bastar area (Crookshank,
1938). Eastern Ghats province is east of irregular line dividing gneiss (amphibolite facies) and charnockite. Shear zone is mapped
east of contact. Outliers of charnockite/
granite are mapped together because Crookshank regarded them as cogenetic.

tinguish these conformable mafic suites from superficially similar rock types associated with younger intrusive suites. Rocks of generally gabbroic composition in both suites can be metamorphosed to pyroxene granulites, commonly consisting of ortho- and clinopyroxenes, hornblende, and plagioclase. The distinguishing characteristics of the younger intrusive suite are cross-cutting relationships with khondalitic, charnockitic, or other wall rocks, although later deformation has caused conformability of many contacts in these younger suites; association with anorthositic bodies; and presence of chromite and other ultramafic assemblages. These younger suites are discussed later.

Nellore schist belt

The Nellore schist belt (Fig. 5.1) is a series of supracrustal rocks with many similarities to the older mafic schist belts of the Western Dharwar craton (Ch. 2), but its age has not been determined. A twofold subdivision based on metamorphic facies (Vasudevan and Rao, 1975) separated the younger greenschist facies of the western part of the belt from older upper amphibolite to granulite facies in the eastern part. Vasudevan and Rao proposed a fault separating the two suites, with an "oligomictic conglomerate" within the fault zone, presumably related to an unconformity at the base of the younger suite. Development of pegmatites and activity of feldspathizing solutions is also predominantly in the eastern part of the belt and in the gneissic terrain to the east (the Nellore mica belt; Babu, 1969, 1970). A stratigraphy indicated by Narayana Rao (1983) for the Nellore schist belt is shown in Table 5.1, including a lower sequence that is predominantly metasedimentary, and an upper part consisting mainly of metabasalts (Fermor, 1936).

The greenschist-facies rocks of the Nellore belt are primarily chloritic schists and phyllites and quartz-chlorite-sericite-biotite schists. Higher-grade assemblages include garnet-hornblende-biotite-muscovite schists ($\pm$ kyanite, staurolite, and sillimanite); mafic granulites containing amphibole, ortho- and clino-pyroxene, and plagioclase

($\pm$ garnet); and banded ferruginous quartzites and calc-silicate rocks. Parent rocks obviously include pelitic, cherty, and calcareous sediments; mafic volcanic rocks are also likely precursors for some of the amphibole-pyroxene rocks, but proof of this parentage is lacking. Babu (1970) estimated metamorphic conditions for the high-grade rocks of about 650° and 7.5 kb based on standard equilibrium relationships for minerals such as staurolite, kyanite, hornblende, and garnet.

Relative age relationships between rocks of the Nellore belt and the surrounding gneissic terrain are unclear, partly because of poor exposures. There is general conformity of structural trends in the gneisses and schists, but some sialic materials may be younger than the schists (Rama Rao and Narasingha Rao, 1976; Narayana Rao, 1982). Kanungo and Chetty (1978) showed that an anorthosite intrusive into the mafic rocks of the belt has been brought into substantial structural conformity with the mafic rocks.

Structural studies of Nellore rocks have been limited. Sastry and Vaidyanadhan (1968) indicated tight NE-SW folding early in the history of the area and open folding later, possibly along similar trends. Chetty (1983) showed three periods of deformation in the south-central part of the belt based on folding of quartzite layers (Fig. 5.4). The sequence of events proposed by Chetty is folding, detachment of fold hinges, and shearing and twisting of fold hinges. This sequence was regarded as a result of heterogeneous progressive simple shear leading to noncoaxial deformation.

Khammam Area

The Khammam area (Fig. 5.1) is south of the Godavari rift valley and is overlapped on the west by a series of mid-Proterozoic sediments (Pakhal sequence). The Chimalpahad anorthositic complex (Appavadhanulu et al., 1975) intrudes the mafic rocks.

Rocks in the Khammam belt are metasedimentary and metavolcanic types similar to those in the Nellore belt, including garnet-biotite schists, quartz-biotite schists and gneisses, garnet-kyanite-muscovite schists and gneisses (Kumar and

TABLE 5.1 *Stratigraphy of Nellore schist belt (Narayana Rao, 1983)*

Nellore Group	Kandra Formation Anthophyllite amphibolites, Hornblende amphibolites Kandukur Formation Metapelites, quartzites, banded iron ore quartzites, calc-silicate rocks, quartz-amphibole schists etc.

Prepared with permission from the Quarterly Journal of the Geological, Mining and Metallurgical Society of India, v. 55.

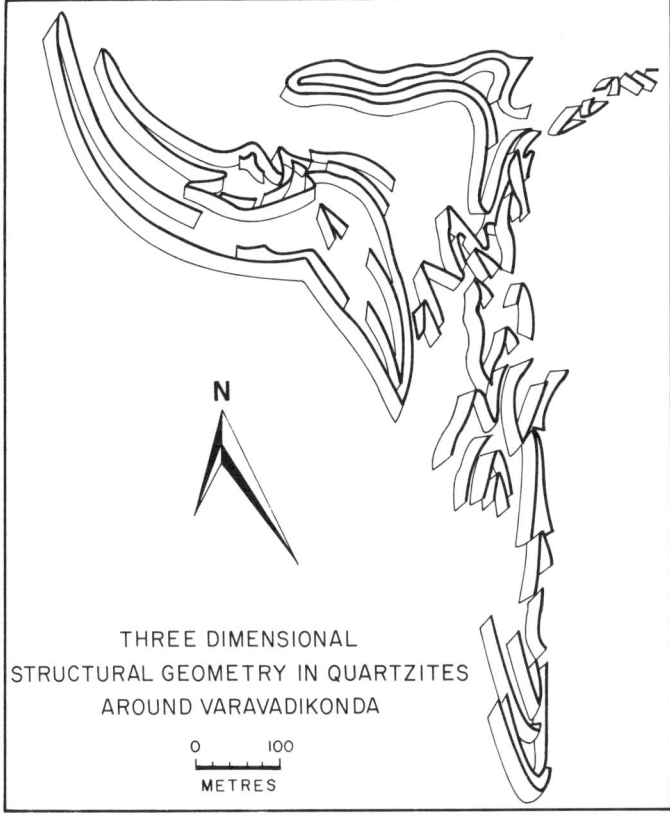

N

THREE DIMENSIONAL
STRUCTURAL GEOMETRY IN QUARTZITES
AROUND VARAVADIKONDA

0   100

METRES

Fig. 5.4. Deformation pattern near Varavadikonda (Chetty, 1983).

Somayajulu, 1981), calc-silicate rocks, banded iron formations, quartzites, and hornblende schists and gneisses. Amphibolites are abundant (Reddy, 1972). Talc-tremolite-actinolite schists, presumably metamorphosed ultramafic intrusions, occur as scattered conformable lenses. Peridotites have been reported by Ramamohana Rao et al. (1983). Late intrusive rocks include metamorphosed mafic dikes (now amphibolites), some granitic bodies, and pegmatites. Granulite facies does not appear to have been attained by the mafic rocks. The southeastern margin of the mafic belt is conformable with khondalites of the general Eastern Ghats terrain.

A summary of structural studies in the area by Hopgood (1968), Janardhan Rao and Satyanarayana (1972, 1975) and Subbaraju (1975) is shown as follows. No data are available to attach absolute ages to any of the events.

Oldest event:
1. Deposition of sedimentary/volcanic sequence.
2. NE-SW isoclinal folding.
3. Intrusion of mafic dikes.
4. Metamorphism, migmatization, and folding.

5. NE-SW folding and syntectonic intrusion of Chimalpahad anorthosite; also NE-SW and NW-SE shearing.
6. Intrusion of mafic and ultramafic rocks.
7. Low-grade metamorphism and additional NE-SW folding.
8. Open folding on roughly EW axes.

Hopgood (1968) placed these events within the framework of development of a gneiss dome in which metamorphic grade decreases outward to schists on the margins.

Other mafic schists and quartzites

In addition to the Nellore and Khammam areas, mafic schists calc-silicate rocks, and banded iron formations occur at a variety of locations. Somayajulu and Chandrasekhar (1978) described an area generally northeast of the Nellore belt that contains a suite of mafic granulites (hornblende, ortho- and clinopyroxene, plagioclase), khondalites, quartz-magnetite rocks, and garnetiferous quartzites conformably folded with charnockites and gneisses. The mafic granulites were regarded as metatholeiites in the sedimentary sequence.

Hypersthene and grunerite also occur in the magnetite quartzites.

KHONDALITES AND ASSOCIATED ROCKS

Khondalites form a coherent suite of rocks extending through much of the Eastern Ghats belt. They are intimately and conformably cofolded with charnockites, calc-silicate rocks, and mafic granulites. One of the principal areas in which khondalites have been studied is Visakhapatnam (including the Srikakulam and Garbham areas). Here, Rao (1969) proposed a stratigraphy for the khondalite section in which quartzitic rocks at the base grade upward into metapelites. Both suites contain intercalated calc-silicates, and manganese occurs in several horizons of spessartite-braunerite rocks.

The khondalite suite is nearly uniform throughout the Eastern Ghats. Mineralogically, the khondalites contain quartz, plagioclase, K-feldspar, garnet, and sillimanite. Biotite and cordierite are also common. Hypersthene occurs locally. Scattered graphite flakes occur in many rocks (Krishna Rao and Malleswara Rao, 1965). Sapphirine is present only locally (Nanda and Natarajan, 1977). Cordierite is present in areas of somewhat lower metamorphic grade than the common sillimanite-garnet ($\pm$ hypersthene) rocks without cordierite (Banerji, 1982). An inverse correlation between the abundances of orthopyroxene and garnet implies a reaction relationship. The garnets in khondalite are generally high in grossularite and low in pyrope components, in contrast with the almandine-pyrope garnets of associated charnockites (Subramaniam, 1967). Garnets tend to be paragenetically young, showing inclusions of all other minerals (Sahu, 1973).

Manganese ores are abundant in many areas (Krishna Rao, 1960, 1963; Roy, 1981) and occur as conformable horizons rich in such minerals as braunite, hausmannite, and possibly silicates such as rhodonite and spessartite. Secondary weathering has produced low-temperature enriched ores containing pyrolusite, etc. Relic oolitic textures of Mn minerals have been found in some places. Manganese is much more abundant in the khondalites of the Eastern Ghats than in equivalent rock types of the Granulite terrain to the south (Ch. 4).

The average composition of 12 khondalites (Divakara Rao, 1983) is shown in Table 5.2. This composition is notable for the excess Al_2O_3 characteristic of pelitic sediments. Total Fe is also high. The high local uranium contents reported by Satyanarayana and Krishna Rao (1974) are interesting in connection with the generally proposed high heat flow of the Indian shield (Ch. 1).

Zircons in the khondalites are invariably small, rounded, well sorted, and do not show overgrowths. Long axes are generally not parallel to the c crystal axis. They are presumably a sedimentary component that was not affected by metamorphism (Murthy and Siddiquie, 1964; Siddiquie and Viswanathan, 1969).

The khondalites are intimately interbedded with calc-silicate rocks (described shortly) and clearly represent a series of argillaceous, probably silty, sediments formed in an environment that also permitted the deposition of carbonate rocks and possibly some volcanism (now seen as enclaves of mafic granulite).

Calc-silicate rocks

Calc-silicate rocks are interbedded with khondalites at numerous places (C. Bhattacharyya, 1971c; Mukherjee and Rege, 1972b). The mineralogy is largely a result of prograde metamorphism, but some retrogressive effects have been found; for example, Choudhuri and Banerji (1974) described an assemblage of grossularite and quartz formed by retrogressive reaction of anorthite and wollastonite.

Nayudu and Jagadiswara Rao (1977) summarized the characteristics of 121 different localities of calc-silicates. The rocks occur commonly as lenticular bodies 50 to 800 m long and 5 to 200 m wide. They are more abundant in the northern part of the Eastern Ghats, associated with Mn-bearing rocks, than in the south. The principal mineral assemblage is salitic clinopyroxene-plagioclase-scapolite-calcite-wollastonite-microcline-quartz-sphene, an eight-phase assemblage presumably at equilibrium under granulite-facies conditions.

Sivaprakash (1981) studied calc-silicates in scattered outcrops near Visakhapatnam and subdivided them into three suites, all of which contain K-feldspar, wollastonite, calcite, and quartz. The other characteristic minerals and properties of the suites are (1) scapolite, diopside, andradite and sphene in a rock that has a high Fe^{+3}/Fe^{+2} ratio; (2) scapolite, diopside, grossularite, sphene and zoisite in a rock with high Ca content and high Mg/Fe ratio; and (3) scapolite, diopside, sphene, hornblende and magnetite in a rock with low Fe^{+3}/Fe^{+2} and Mg/Fe ratios. Most of the minerals were formed during prograde metamorphism, but zoisite developed during retrogression caused by influx of water.

Mafic granulites

Some mafic rocks in the khondalite terrain are clearly related to late- or syntectonic anorthositic suites or to diabasic intrusions, both of which were later metamorphosed to granulite facies. Other mafic rocks occurring as conformable lenses, however, may represent basaltic volcanism during sedimentation of the khondalite host

TABLE 5.2. *Compositions of rocks in Eastern Ghats*

	1	2	3	4	5	6	7	8
SiO_2	68.9	56.4	60.8	55.1	57.6	71.0	68.8	79.0
TiO_2	0.61	0.85	1.9	1.5	1.2	0.32	0.21	0.14
Al_2O_3	13.6	16.3	15.7	17.3	17.6	14.4	16.1	9.6
Fe_2O_3	2.0	1.9	3.0	1.5	2.1	2.3	0.78	0.34
FeO	3.4	10.4	6.0	7.5	8.0	6.3	3.7	2.1
MnO	0.06	0.22	0.20	0.50	0.30	0.42	0.24	0.21
MgO	1.6	3.5	2.3	5.1	5.0	0.26	1.3	0.81
CaO	3.4	5.6	3.4	7.2	2.2	2.5	1.5	0.32
Na_2O	1.3	3.1	2.4	2.6	2.1	0.33	2.2	2.8
K_2O	5.0	1.2	3.5	1.4	2.9	2.0	4.7	4.3
Rb		30	150					
Sr		1000	400					
Y		10	40					
Zr		200	300					
V		225	150					
Cr		80	30					
Ni		40	83					
Th								
U								

All abundances of major elements are in weight percent of oxides. Where no entry is made in the FeO position, the value at Fe_2O_3 is total iron calculated as ferric oxide. All abundances of trace elements are in parts per million by weight. Values are taken directly from quoted sources and not recalculated to 100%.

Table 5.2. Column References
1. Srikakulam. Charnockite (C. Bhattacharyya, 1972).
2. Visakhapatnam. Tonalitic charnockite (Sriramadas and Rao, 1979).
3. Visakhapatnam. Granodioritic charnockite (Sriramadas and Rao, 1979).
4. Chipurupalle. Basic charnockite (mafic granulite) (Perumal, 1975).
5. Average khondalite (Divakara Rao, 1983).
6. Chipurupalle. Khondalite (Perumal, 1975).
7. Chipurupalle. Leptynite (Perumal, 1975).
8. Chipurupalle. Leptynite (Perumal, 1975).

rocks. C. Bhattacharyya (1972) referred to these rocks in the Srikakulam area (Fig. 5.1) as "mafic charnockites" with a mineralogy of orthopyroxene, clinopyroxene (salite), plagioclase (with some reverse zoning) and garnet plus augen of biotite and K-feldspar. C. Bhattacharyya considered these rocks to be part of a magmatic suite with all compositional gradations to charnockite. Metamorphism of the rocks to granulite facies was shown by Blattner (1980) based on high F and Cl contents of coexisting amphibole and biotite. Halden et al. (1982), in their study of the Angul area (Figs. 5.1 and 5.5), regarded the precursor of the mafic granulite as the oldest material in the area, confirming volcanism during khondalite sedimentation.

Leptynites

As used in this book, the term "leptynite" refers to foliated garnet-quartz-feldspar rocks that lack sillimanite but may contain biotite and minor hypersthene (Subramaniam, 1962; Mukherjee and Rege, 1972a; C. Bhattacharyya, 1973). Generally, leptynites are adjacent to khondalites, commonly forming small border phases between khondalites and either charnockites or zones of migmatization. Massive bodies of "primary" (metasedimentary or metaigneous) leptynite are apparently absent. Formation of leptynites by further metamorphism/metasomatism of khondalite is consistent with other properties of the leptynites, including highly variable abundances of quartz, garnet, and feldspar; abundance of rounded zircons or overgrowths on zircons, presumably by inheritance of rounded zircons from the khondalites and later metamorphic overgrowth (Viswanathan, 1968); and similarity of garnet compositions (principally grossularite and almandine) in both khondalites and leptynites (Subramaniam, 1967).

The process that converts khondalites to leptynites is not clear. It cannot be simple isochemical metamorphism because such a process could

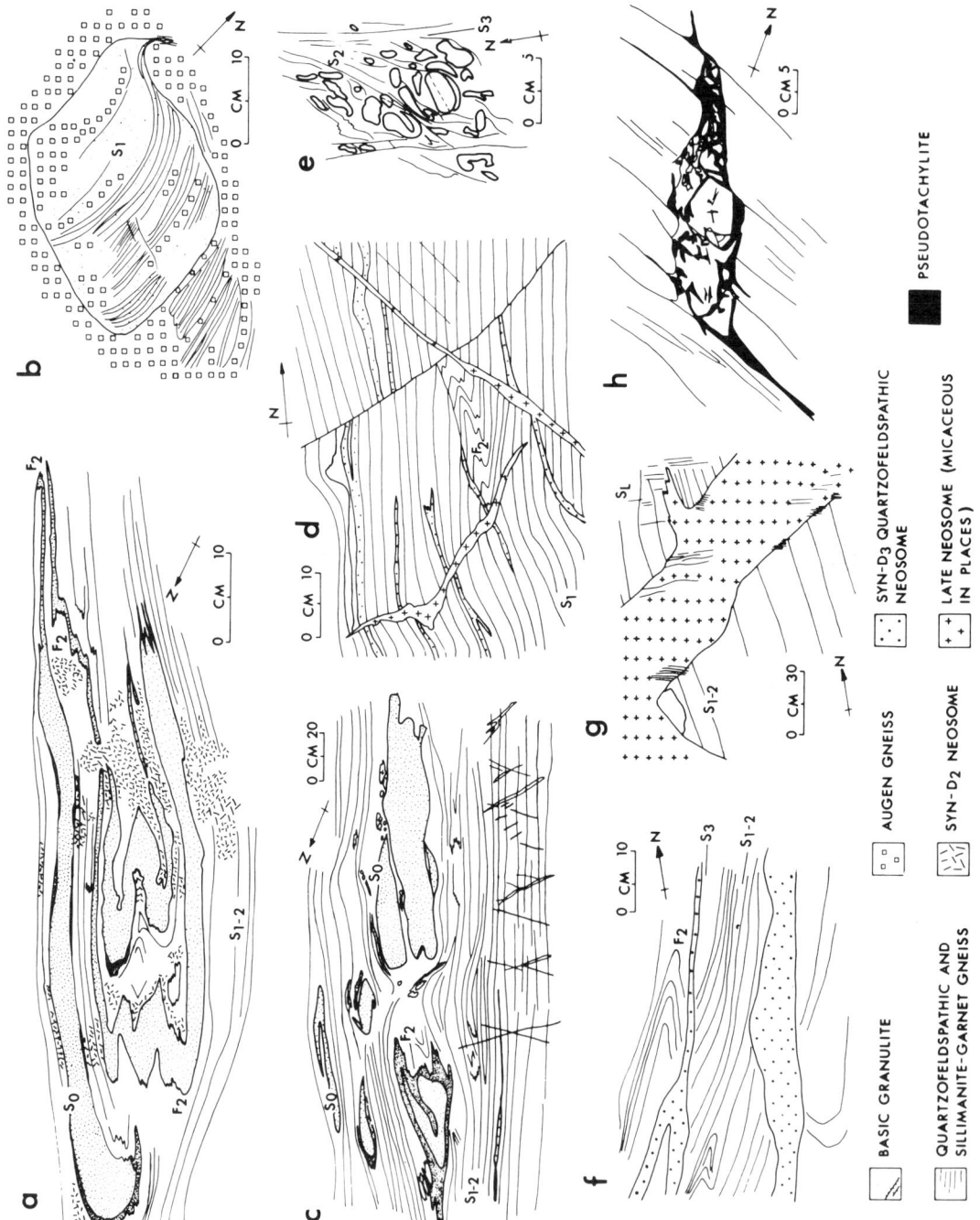

not eliminate the sillimanite in the original khondalite. Metasomatism could involve the addition of alkalies and possibly other ions to form additional feldspar and/or biotite at the expense of sillimanite. There is no indication that the temperatures and pressures of equilibration are different for the khondalites and leptynites.

CHARNOCKITES

Most of the studies conducted in the Eastern Ghats belt have not applied the concepts of metasomatic, *in situ* development of charnockites that have been demonstrated by a variety of studies in the Granulite terrain of southern India (Ch. 4). The only study of this type in the Eastern Ghats is by Halden et al. (1982) in the Angul area of Orissa (Fig. 5.5). Many of the studies in the Eastern Ghats have assumed that at least some of the "acid charnockites" represent magmatic intrusions or that there is a continuous "charnockite series" from acidic to basic compositions, possibly formed by magmatic differentiation.

Three principal regions of charnockite development have been most extensively studied (Fig. 5.1). The Visakhapatnam (including Srikakulam, Garbham, and Chipurupalle) area has been discussed earlier as a complex intermingling of gneisses and charnockites (Sriramadas et al., 1969; C. Bhattacharyya, 1972). The Srikakulam part of the Visakhapatman region is also the site of extensive intrusion by ultramafic rocks possibly related to an anorthosite suite. Another area of extensive study of charnockites is around Kondapalli, including Ganginemi, where charnockites, gneisses, and khondalites are intricately associated with small anorthosite masses (Srirama Rao, 1946, 1947). Similar complex interlayering also occurs south of the Kondapalli area, around Amaravathi, Kondavidu, and the northern part of the Guntur District (Sadashivaiah and Subbarayudu, 1970; Ramaswamy and Murty, 1973).

The charnockites of the Eastern Ghats have several consistent features in all areas. They are conformable with mafic granulites and members of the khondalite suite. Some cross-cutting charnockites have been reported, and early papers tended to attribute these bodies to injection of primary charnockite magmas (Srirama Rao, 1946, 1947). Other possibilities include metasomatic veining or possibly local anatexis. The charnockites may be associated with migmatization of gneisses (Dash and Paul, 1976). Paracharnockites, formed by metamorphism of metasedimentary parents, have been described locally (Rau and Adinarayana, 1968).

Contacts of the charnockites with other rocks are locally abrupt but more commonly gradational over short distances (centimeters to meters). Contacts with khondalites may be the sites of cordierite or sapphirine development (e.g., Nanda and Natarajan, 1977), although cordierite is somewhat more widely distributed than the contact zones.

The "acid charnockites" form a coherent suite with a consistent mineralogy, although the relative proportions of the minerals are quite variable. Mafic and felsic granulites appear to be separate suites, bimodal in abundances of SiO_2 and other components. Where intermediate varieties have been investigated, they appear to represent metasomatism (or ingestion?) of mafic granulite by felsic material (e.g., Subramaniam, 1967; Mall, 1973). Average compositions are reported in Table 5.2.

Common minerals in charnockites are quartz, K-feldspar, hypersthene, and plagioclase; minor components include clinopyroxene, garnet, and phlogopite. Garnets are much higher in almandine content and lower in grossularite content than those in adjoining khondalites and leptynites (Subramaniam, 1962, 1967). Garnet abundances tend to increase as the abundance of orthopyroxene decreases. The typical pyroxene is hypersthene (Leelanandam, 1967). Plagioclase tends to be more equidimensional than plagioclase grains in undoubted igneous rocks (C. Bhattacharyya, 1971b). Some reverse zoning has also been reported in plagioclase (Leelanandam, 1968). Myrmekite is developed locally (C. Bhattacharyya, 1971a). Zircons tend to be rounded, but the long axes of the grains are parallel to the c crystallographic axis (Viswanathan, 1969). High Cl and F contents have been reported for amphiboles and

---

Fig. 5.5. Relationships between structural elements and neosomes in Angul area (Halden et al., 1982). *(a)* Attenuated basic granulite layer(s) in sillimanite-garnet gneiss, folded and boudinaged in D2 with quartz-feldspar-biotite neosome in boudin necks. *(b)* Xenoliths of folded basic granulite in augen gneiss. *(c)* F2 folds in granulite-gneiss assemblage; the agmatitic appearance results from continued extension during D2. *(d)* Mica-bearing pegmatite following directions of the late conjugate fracture set that cuts late neosomes and D2 features in gneisses. *(e)* Augen in augen gneiss showing effects of overgrowth and rotation. S2 is poorly developed and isolated by stronger development of S3. *(f)* D3 neosome controlled by S3 and near parallel composite of S1 and S2. *(g)* late pegmatite affected by late WNW-ESE fracture cleavage. *(h)* Pseudotachylite veins cutting sillimanite-garnet gneiss near margin of dolerite dike.

biotites associated with charnockite development (Kamineni et al., 1982), indicating prograde metamorphism.

History of Angul area

Studies in the Angul area of Orissa (Fig. 5.1) showed a series of conformable khondalites, mafic granulites, and gneisses (including leptynites) in which charnockites have been developed (B. Bhattacharyya and De, 1964; Halden et al., 1982). Bhattacharyya and De regarded the charnockite and associated anorthositic rocks as syntectonic intrusions. Halden et al. described the charnockite development in terms of metasomatism and metamorphism, consistent with more recent work in the Granulite terrain (Ch. 4). The rock types recognized by Halden et al. (1982) include khondalite, mafic granulite, garnetiferous quartzofeldspathic gneiss, and augen gneiss with essentially the same mineralogy (except garnet) of the quartzofeldspathic gneiss but showing nonconformable (cross-cutting) relationships.

In addition to the preceding suites, quartzofeldspathic neosomes pervade and obliterate parts of all other rock types at Angul. The neosomes include pegmatite and aplite dikes and patches and bands of quartzofeldspathic material. Leptynite was regarded by Halden et al. (1982) as one end product of neosome development. Charnockite in conformable bands with gradational borders was regarded as another result. Garnet and orthopyroxene are apparently antipathetic. Relationships of particular importance recognized by Halden et al. are shown in Figure 5.5. The history of the Angul area outlined by Halden et. al. is as follows:

(Oldest event.) Development of (presumably igneous) precursors to the mafic granulites and sedimentation of the khondalite suite.

Injection of an igneous precursor to the augen gneisses during the first deformational phase. Some quartzofeldspathic metasomatism. This phase of deformation formed isoclinal folds with mineralogical banding caused by metamorphic differentiation. Metamorphism of all existing rock types to granulite facies.

Major compression to form mostly N-S folds with greatly attenuated limbs.

Major development of quartzofeldspathic neosomes. Charnockitization occurred during a late deformational event that formed open folds and widely spaced shears. Much of the charnockite is in the cores of anticlines formed at this time.

(Youngest event.) Open folds and kink folds with E-W axes.

Halden et al. (1982) proposed granulite metamorphism at two, and perhaps three, times. The only dated event is a Rb/Sr age of 854 m.y. on muscovite from a late pegmatite.

CONDITIONS OF METAMORPHISM AND DEFORMATION

Among the aspects of the history of the Eastern Ghats that are unknown are the number, duration, and age of the metamorphic and deformational events. Furthermore, there is no necessity that the different events be synchronous throughout the length of the belt. Nevertheless, some generalizations can be made.

At some time prior to the Middle Proterozoic, granulite-facies metamorphism affected at least part of the Eastern Ghats belt. The high-grade charnockites characteristic of upland massifs in the southern Indian Granulite terrain (Ch. 4) are absent from the Eastern Ghats, and thus metamorphic conditions of greater than about 8 kb and about 700°C were not attained at the level currently exposed in the region. This maximum of pressure and temperature is consistent with the presence of cordierite in some rocks. Estimates of pressure and temperature based on equilibration of the assemblage quartz-K-feldspar-Na feldspar-sillimanite-garnet are in the range of 6 to 7 kb and 600° to 700°C (Dhana Raju, 1977a; 1977b).

The conformability of charnockites and mafic granulites with preexisting khondalites in the major NE-SW structural trend of the belt strongly suggests that NW-SE compression occurred during the granulite formation. This event was also accompanied by metasomatism, development of quartzofeldspathic neosomes, and probably "drying" of the rocks by a CO_2 flux (Ch. 4). The close intermingling of charnockites, khondalites, leptynites, and other rock types shows that metamorphism of these various suites occurred at roughly the same temperature and total pressure. The mineralogical differences between the suites, therefore, must be the result of compositional differences of parent materials, caused not only by original compositions but also by variable effects of metasomatism and by local variations in fluid composition (particularly CO_2/H_2O) ratio.

Later events modified the granulite suite. Development of biotite and some hornblende in charnockites and mafic granulites may represent retrogressive metamorphism. It is also interesting to note that E-W fold axes postdate the N-S trends in most well-studied areas. A NW-SE trend is also the major structural orientation along the Mahanadi lineament, at the northern end of the belt (Ch. 10). It is not known whether there is any relationship between metamorphic retrogression and the E-W structures.

As discussed below, anorthositic/mafic intrusive rocks either accompanied, or formed shortly after, the granulite-facies metamorphism. The fact that many of these bodies are deformed shows that they were emplaced while tectonic activity was still in progress. Alkaline intrusive suites, discussed below, formed posttectonically.

ANORTHOSITES AND RELATED ROCKS

Suites of anorthosites and gabbros, partly accompanied by chromite-bearing ultramafic rocks, occur broadly throughout the Eastern Ghats (De, 1968; M. K. Bose, 1979). Some bodies occur along the thrust zone at the western margin of the belt. The anorthosite suites are generally deformed but do not show granulite-facies metamorphism. They were presumably emplaced after the major metamorphic event but before the cessation of compressive tectonism. Sarkar et al. (1981) stated that the Chilka Lake complex (Figs. 5.1 and 5.6) was emplaced during folding on E-W axes.

M. K. Bose (1979) listed major anorthosite occurrences in the Eastern Ghats and southern India. Two of his suites (Rampur; Banpur and Balugaon) and areas studied by Perraju (1973) were incorporated by Sarkar et al. (1981) in the Chilka Lake complex. This complex and the remaining suites of Bose are shown in Figure 5.1. Considerable effort has been made by workers in many areas to distinguish layered complexes from massif anorthosites (e.g. Ray and Bose, 1975), but this discrimination has not been wholly successful for many of the disconnected bodies. Small anorthosite bodies not shown in Figure 5.1 include one reported by Kanungo and Chetty (1978) in the Nellore area.

The anorthositic and related rocks at Kondapalli occur as veins, sills, and conformable sheets in a layered series of charnockites, khondalites, and mafic granulites (Sinha and Mall, 1974; Nanda and Natarajan, 1980; and S. K. Bose and Bose, 1982). The suite contains anorthosites, gabbros, norites, websterites, and chromite-bearing orthopyroxenites and serpentinized dunites (Table 5.3). Layering is found in some bodies and is conformable to the regional trend. S. K. Bose and Bose described a stratigraphy for one large sheet in which ultramafic rocks at the base grade upward through gabbros and norites into more silicic rocks.

Sinha and Mall (1974) proposed that the anorthositic suite at Kondapalli was affected by charnockite development, as shown by veining and by the fact that Mg-Fe distribution coefficients for orthopyroxene-clinopyroxene pairs are the same in both mafic granulites and the ultramafic members of the anorthosite suite. Nanda and Natarajan (1980) stated that anorthositic and related rocks occur only in anticlinal charnockite axes and are not intrusive into khondalites.

The Chilka Lake suite (Sarkar et al., 1981) covers an area of more than 1,200 km^2 (Fig. 5.6). Major rock types are anorthosite, leuconorite, norite, jotunite, and quartz mangerite. The rocks were apparently emplaced during cross-folding along E-W axes and postdate the intense deformation and granulite-facies metamorphism of the area. The massif contains an outer cover of quartz

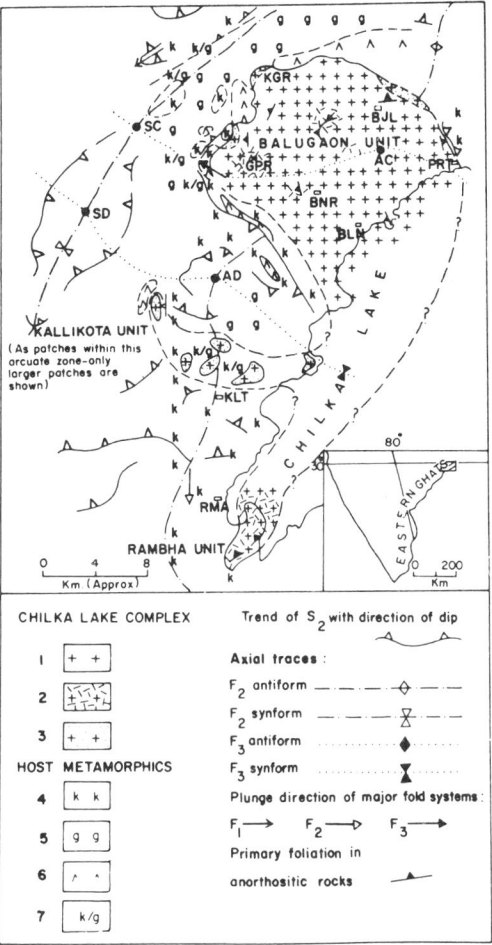

Fig. 5.6. Chilka Lake anorthositic complex (Sarkar et al., 1981). Lithologies include 1, largely anorthosite; 2, largely leuconorite/norite; 3, largely acidic rocks; 4, khondalite; 5, garnetiferous granite gneiss; 6, pyroxene gneiss and granulite (intermediate and acid charnockites and mafic granulites); 7, interaction zone of khondalite and garnetiferous granite gneiss. AC, culmination of antiformal axis (dome); SC, culmination of synformal axis; AD, depression of anticlinal axis. Locations include KGR, Keagora; BJL, Binjhala; GPR, Gohia Pahar; PRT, Pathakarta; BNR, Banpur; BLN, Belugaon; KLT, Kallikota; RMA, Rambha. Reproduced with permission from Lithos, v. 14, p. 94.

mangerite that is syntectonic with the regional E-W deformation, is intruded by the anorthosite core, and does not lie on differentiation trends of the anorthosite suite. The interior of the massif shows compositional layering and a continuous chemical gradation from anorthosites to norites to jotunites. This assemblage contains both xenoliths and autoliths, shows flow layering, and is presumably a differentiated magmatic suite. A five-point Rb/Sr isochron (four anorthosite and one granite) yielded an age of 1,312 $\pm$ 21 m.y.

TABLE 5.3. *Compositions of Kondapalli-Ganginemi suite (S. K. Bose and Bose, 1982)*

	1	2	3	4
SiO_2	49.1	50.6	50.3	65.1
TiO_2	0.75	0.04	1.0	0.09
Al_2O_3	6.1	29.6	16.7	22.2
Fe_2O_3	2.3	0.65	1.7	0.69
FeO	10.8	0.97	9.3	0.62
MnO	0.20	0.03	0.15	0.01
MgO	19.6	0.94	7.2	0.56
CaO	10.4	13.1	10.3	5.6
Na_2O	0.58	3.0	2.1	3.9
K_2O	0.12	0.65	0.36	0.69

All abundances of major elements are in weight percent of oxides. Where no entry is made in the FeO position, the value at Fe_2O_3 is total iron calculated as ferric oxide. All abundances of trace elements are in parts per million by weight. Values are taken directly from quoted sources and not recalculated to 100%.

Table 5.3. Column References
1. Pyroxenite.
2. Anorthosite.
3. Gabbro-norite.
4. Felsic rocks.

with an initial $^{87}Sr/^{86}Sr$ ratio of 0.70674. The four anorthosites yielded an age of 1,404 ± 89 m.y. with an initial ratio of 0.70661. This age presumably is younger than the major deformation and metamorphism in the area.

ALKALINE COMPLEXES

Posttectonic alkaline suites have been described at a number of locations throughout the Eastern Ghats (Fig. 5.1). All suites are diapiric and posttectonic, with the possible exception of foliated gneissic rocks in the earliest stages of the Elchuru body. The major bodies are at Koraput, Kunavaram, and Elchuru (M. K. Bose et al., 1982; Nag, 1983a). A number of small intrusions of carbonatite, syenite, and pyroxenite form sheetlike bodies in khondalites in the Visakhapatnam area (Ramam and Viswanathan, 1977). Choudhuri and Banerji (1976) attributed the development of apatite-bearing pegmatite veins in the same area to carbonatite intrusions not exposed at the surface. Numerous other small bodies occur in the Nellore and Khammam areas (Jayaram et al., 1972; Leelanandam and Krishna Reddy, 1981; S. Bhattacharyya et al., 1982; and Leelanandam and Ratnakar, 1983).

The Kunavaram complex is the largest alkaline suite in the Eastern Ghats (M. K. Bose et al., 1971; Subbarao, 1971). Syenites and related rocks occur throughout a belt elongated north-south with dimensions of 27 km by 5 km. The Kunavaram suite is the only one to be dated adequately, yielding a Rb/Sr whole-rock age of 1,265

± 85 m.y. with an initial $^{87}Sr/^{86}Sr$ of 0.7051 ± 0.0013 (Clark and Subbarao, 1971). A muscovite-biotite pair has a Rb/Sr age of 620 m.y., indicating some type of thermal event, presumably not simple uplift, at that time. The suite shows differentiation of alkalic diorites and monzodiorites to various types of syenites. The final product is nepheline syenite and minor amounts of carbonatite. Mafic and ultramafic schlieren are present locally. The Koraput suite follows a similar differentiation trend to that at Kunavaram, but mafic members of the assemblage are more abundant and less alkalic (M. K. Bose, 1970).

The Elchuru suite has several different outcrop areas, with the major one apparently being a ring dike. The differentiation trend is more complicated than those for Koraput and Kunavaram, with mafic rocks appearing at two times (M. K. Bose et al., 1982). Nag (1983b) summarized the differentiation path as follows: alkaline gneiss, apparently syntectonic; mafic alkalic rocks; hypersolvus nepheline syenite; subsolvus nepheline syenite; lamprophyre dikes; and carbonatite. Madhavan and Leelanandam (1979) showed a complete gradation between hypersolvus and subsolvus syenites.

The northernmost alkaline suite in the Eastern Ghats province is directly south of the Sukinda thrust. The Baradangua body (Sahu, 1976) consists primarily of nepheline syenite containing calcite, sphene, apatite, and ilmenomagnetite. Sahu estimated a crystallization temperature of about 700° based on nepheline composition.

All of the nepheline syenites of the Eastern Ghats belt have similar properties. On a quartz-nepheline-kalsilite diagram (Fig. 5.7), most of the whole-rock samples plot in the alkali feldspar field, with a few rocks on the cotectic between feldspar and nepheline. The nephelines all have similar compositions, but there is a broad range of compositions for the alkali feldspars. This relationship is consistent with the concept that the alkali suites formed by fractional crystallization toward an end product consisting of nepheline syenite magma.

EVOLUTION OF THE EASTERN GHATS

The general history of the Eastern Ghats belt appears to contain the following significant events. The oldest activity was the deposition of silty and shaly sediments (now khondalites), carbonates (now calc-silicates), and intercalated basalts (now mafic granulites). At various places along what is now the western edge of the belt, there was deposition of mafic volcanosedimentary sequences that now occur as schist belts (Nellore, Khammam). The age relationships between these mafic suites and the more silicic khondalite suite are unknown, as are the absolute ages of either one.

Major E-W compression caused the prevailing

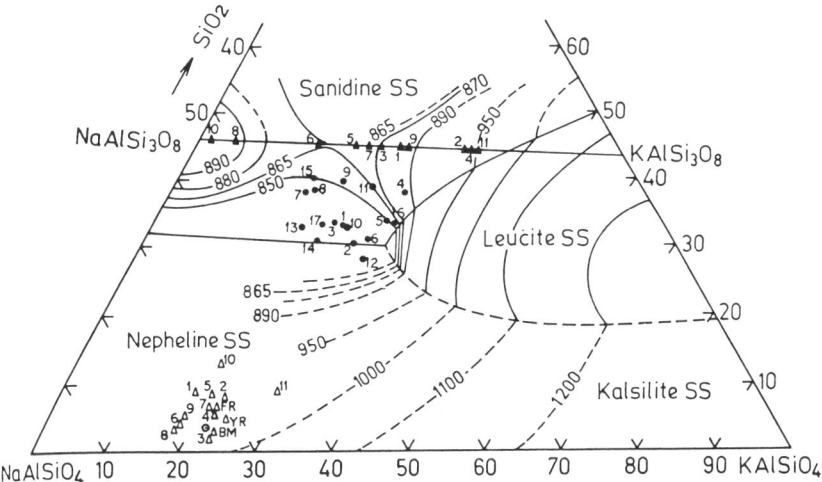

Fig. 5.7. Quartz-nepheline-kalsilite diagram (omitting quartz apex) showing relationships among compositions of nepheline (open triangles), whole rocks (solid circles), and alkali feldspars (solid triangles) in alkaline rocks of Eastern Ghats (Nag, 1983a). Numbers 1 through 9 label coexisting phases for rocks in the Eastern Ghats, and higher numbers refer to nepheline syenites elsewhere. Letter symbols (Em, YR, and FR) refer to analyzed nephelines from other areas. Diagram shows contours on liquidus surface at 1,000 kg/cm². Reproduced with permission from Neues Jahrbuch fur Mineralogie, Abhandlungen, v. 148, p. 107.

NE-SW strikes in the belt. Several stages of compression probably occurred. Granulite-facies metamorphism presumably accompanied one or more of these deformational events, associated with metasomatism, CO_2 fluxing, etc. The ages of deformation and metamorphism are unknown, but they may have coincided with emplacement of granites in the Middle Proterozoic.

Following compression, open cross-folding on E-W axes occurred at about the same time as emplacement of anorthositic and related magmatic suites. The time relations of these events are unclear, although the Chilka Lake massif has been proposed to be synchronous with the cross-folding. It is also not clear whether the anorthosites were emplaced during or after granulite-facies metamorphism, or whether different bodies have different relationships to the granulites. The Chilka Lake body crystallized 1,300 to 1,400 m.y. age.

Intrusion by alkali magmas, and possibly dolerites, was mostly posttectonic. The Kunavaram alkali suite is 1,285 m.y. old.

Formation of the western margin of the Eastern Ghats belt was probably caused by some type of continent-continent collision. The age of this event relative to other deformational events is unknown, although presumably it did not precede granulite metamorphism.

The similarities between the Eastern Ghats and the southern Indian Granulite terrain are clear. Both areas show granulite-facies metamorphism, intrusion of anorthositic suites, and late formation of alkalic suites. The proportion of mafic schists is smaller in both areas than in the Dhar-war craton. The Eastern Ghats, however, are characterized by the following features that distinguish them from the southern Granulite terrain: close intermingling of granulites and the khondalite suite, and a greater abundance of khondalites; intense deformation of the terrain, apparently both before, during, and after metamorphism; a steep gravity gradient where the margin of the belt is in contact with amphibolite-facies rocks; and abundance of manganese formations. The implication of these differences is that the Eastern Ghats granulite suites were developed during an intense deformational (perhaps collisional) event, for which no evidence has yet been found in the southern Granulite terrain.

REFERENCES

Appavadhanulu, K., Setti, D. N., Badrinarayanan, S., and Subba Raju, M. (1975). The Chimalpahad meta-anorthosite complex, Khammam District, Andhra Pradesh. Geol. Surv. India Misc. Publ. 23, Part 2, 267–278.

Babu, V. R. R. M. (1969). Temperatures of formation of pegmatites of Nellore mica-belt, Andhra Pradesh, India. Econ. Geol., v. 64, 66–71.

———— (1970). Petrology of metamorphic rocks of almandine-amphibolite facies in Saidapuram-Podalakuru area, Nellore District, Andhra Pradesh, India. Tschermak's Mineral. Petrogr. Mitt., v. 14, 171–194.

Banerji, P. K. (1982). The khondalites of Orissa, India—A case history of confusing terminology. Geol. Soc. India J., v. 23, 155–159.

Bhattacharyya, B., and De, A. (1964). The sequence of deformation, metamorphism, and igneous intrusion in the area around Angul (Orissa). Geol. Soc. India J., v. 5, 159–171.

Bhattacharyya, C. (1971a). Myrmekite from the charnockitic rocks of the Eastern Ghats, India. Geol. Mag., v. 108, 433–438.

—— (1971b). Size and shape analysis of plagioclase in the charnockite series of the Eastern Ghats, Srikakulam District, India. Neues Jahrb. Mineral. Monatsh., no. 2, 58–65.

—— (1971c). Chemistry and metamorphic facies of the calc-granulites from the area around Garbham, Srikakulam District, Andhra Pradesh. Geol. Min. Metall. Soc. India Quart. J., v. 43, 109–112.

—— (1972). Granitization in relation to evolution of the charnockite series from the Eastern Ghat hills, Srikakulam District, Andhra Pradesh, India. Neues Jahrb. Mineral. Monatsh., no. 5, 220–240.

—— (1973). Petrology of the leptynites and garnetiferous granite-gneisses around Garbham, Srikakulam District, Andhra Pradesh. Geol. Soc. India J., v. 14, 113–123.

Bhattacharyya, S., Bose, S. K., and Dutta, H. S. (1982). Petrology of the alkalic suite of Errakonda Hill near Chimakurti, Prakasam District, Andhra Pradesh. Ind. J. Earth Sci., v. 9, 84–89.

Blattner, P. (1980). Chlorine and fluorine in apatite, biotite and hornblende of basic charnockites from Kondapalli, India. Neues Jahrb. Mineral. Monatsh., no. 6, 283–288.

Bose, M. K. (1970). Petrology of the intrusive alkalic suite of Koraput, Orissa. Geol. Soc. India J., v. 11, 99–126.

—— (1979). On the Eastern Ghats Precambrian granulite belt and associated anorthosites. Ind. J. Earth Sci., v. 6, 200–219.

——, Chakravarti, S., and Sarkar, A. (1971). Petrological observations on alkali syenites of Kunavaram, A. P., India. Geol. Mag., v. 108, 273–280.

——, Ghosh Roy, A. K., and Czygan, W. (1982). K-Rb relations in the alkaline suites of the Eastern Ghats Precambrian belt, India. Lithos, v. 15, 77–84.

Bose, S. K., and Bose, M. K. (1982). Petrology and geochemistry of the Kondapalle-Ganginemi igneous complex, Krishna District, Andhra Pradesh. Ind. J. Earth Sci., v. 9, 150–166.

Chetty, T. R. K. (1983). Structural geology of schistose rocks around Kalichedu-Degapudi area of Nellore schist belt, Andhra Pradesh, South India. Unpubl. Ph.D. Thesis, Sri Venkateswara Univ., Tirupati, 280 p.

Choudhuri, R., and Banerji, K. C. (1974). On some calc-silicate rocks around Sitarampuram, Visakhapatnam District, Andhra Pradesh. Geol. Soc. India J., v. 15, 48–57.

——, and —— (1976). On the occurrence, em-

placement and origin of the apatite deposits of Kasipatnam in Visakhapatnam District, Andhra Pradesh. Proc. Ind. Natl.. Sci. Acad., Part A., v. 42, 387–406.

Clark, G. S., and Subbarao, K. V. (1971). Rb-Sr isotopic age of the Kunavaram series—A group of alkaline rocks from India. Can. J. Earth Sci., v. 8, 1597–1602.

Crookshank, H. (1938). The western margin of the Eastern Ghats in southern Jeypore. Geol. Surv. India Records, v. 73, 398–434.

Dash, B., and Paul, A. K. (1976). Migmatites associated with the Precambrian rocks of of a part of Eastern Ghats. Geol. Surv. India Misc. Publ. 23, Part 2, 335–345.

De, A. (1968). Anorthosites of the Eastern Ghats, India. In Origin of Anorthosite and Related Rocks (ed. Y. W. Isachsen), N.Y. State Mus. Sci. Serv., Mem. 18, 425–434.

Dhana Raju, R. (1977a). P-T conditions of granulite-upper amphibolite facies metamorphism in the Precambrian granitic rocks of the Chipurupalle-Razam area of the Eastern Ghats. Geol. Soc. India J., v. 18, 281–287.

—— (1977b). Metamorphism of the granitic rocks of Chipurupalle-Razam area, Srikakulam District, Andhra Pradesh. Geol., Min. Metall. Soc. India Bull. 49, 35–54.

Divakara Rao, V. (1983). Khondalites from the Eastern Ghats granulite belt. Geophys. Res. Bull. (Hyderabad), v. 21, 233–242.

Fermor, L. L. (1936). An attempt at the correlation of the ancient schistose formations of India. Geol. Surv. India Mem. 70, Part 2, 1–217.

Grew, E. S., and Manton, W. I. (1986). A new correlation of sapphirine granulites in the Indo-Antarctic metamorphic terrain: Late Proterozoic dates from the Eastern Ghats. Precamb. Res., v. 33, 123–137.

Gupta, J. N., Pandey, B. K., Chabria, T., Banerjee, D. C., and Jayaram, J. M. V. (1984). Rb-Sr geochronological studies on the granites of Vinukonda and Kanagiri, Prakasam District, Andhra Pradesh, India. Precamb. Res., v. 26, 105–109.

Halden, N. M., Bowes, D. R., and Dash, B. (1982). Structural evolution of migmatites in granulite-facies terrane: Precambrian crystalline complex of Angul, Orissa, India. Roy. Soc. Edinburgh Trans., Earth Sci., v. 73, 109–118.

Hopgood, A. M. (1968). The structural geology and tectonic history of Pre-Cambrian rocks exposed in the Kinarsani River, eastern Andhra Pradesh. Ind. Geosci. Assoc. J., v. 8, 13–34.

Janardan Rao, Y., and Satyanarayana, K. (1972). Structural studies in the Precambrian rocks of Yellambailu-Mailaram area, Khammam District (A. P.). Ind. Acad. Geosci. J., v. 14, 9–23.

——, and —— (1975). Petrology and geochemistry of Yellambailu-Palvoncha area,

Khammam District, A. P. Chayanica Geologica, v. 1, 212–229.

Jayaram, M. S., Devendranath, T., and Sarma, S. R. (1972). The nepheline syenite belt of Khammam District, Andhra Pradesh. Geol. Soc. India J., v. 13, 418–420.

Kaila, K. L., and Bhatia, S. C. (1981). Gravity study along Kavali-Udipi deep seismic sounding profile in the Indian peninsular shield: Some inferences about origin of anorthosites and Eastern Ghats orogeny. Tectonophysics, v. 79, 129–143.

Kamineni, D. C., Bonardi, M., and Rao, A. T. (1982). Halogen-bearing minerals from Airport Hill, Visakhapatnam, India. Amer. Mineral., v. 67, 1001–1004.

Kanungo, D. N., and Chetty, T. R. K. (1978). Anorthosite body in the Nellore mica-pegmatite belt of eastern India. Geol. Soc. India J., v. 19, 87–90.

Krishna Rao, J. S. R. (1960). Structure and stratigraphy of the manganese deposits of Visakhapatnam and Srikakulam Districts, India. Econ. Geol., v. 55, 827–834.

—— (1963). Microscopic examination of manganese ores of Srikakulam and Visakhapatnam (Vizagapatnam) Districts, Andhra Pradesh, India. Econ. Geol., v. 58, 434–440.

——, and Malleswara Rao, V. (1965). Occurrence and origin of graphite in parts of Eastern Ghats, South India. Econ. Geol. v. 60, 1046–1051.

Kumar, D. R., and Somayajulu, P. V. (1981). Stratigraphy and petrology of the garnet-mica-kyanite schist belt of Kothagudem, Khammam District, Andhra Pradesh. Geoviews, v. 9, 111–128.

Leelanandam, C. (1967). Chemical study of pyroxenes from the charnockitic rocks of Kondapalli (Andhra Pradesh), India, with emphasis on the distribution of elements in coexisting pyroxenes. Mineral. Mag., v. 36, 153–179.

—— (1968). Zoned plagioclase from the charnockites of Kondapalli, Krishna District, Andhra Pradesh, India. Mineral. Mag., v. 36, 805–815.

——, and Krishna Reddy, K. (1981). The Uppalapadu alkaline pluton, Prakasam District, Andhra Pradesh. Geol. Soc. India J., v. 22, 39–45.

——, and Ratnakar, J. (1983). The Purimetla alkaline pluton, Prakasam District, Andhra Pradesh. Geol., Min. Metall. Soc. India Quart. J., v. 55, 14–30.

Madhavan, V. D., and Leelanandam, C. (1979). Hypersolvus and subsolvus rocks of the Elchuru alkaline pluton, Prakasam District, Andhra Pradesh, India. Neues Jahrb. Mineral. Abh., v. 136, 276–286.

Mall, A. P. (1973). Distribution of elements in coexisting ferromagnesian minerals from ultra-basics of Kondapalle and Ganginemi, Andhra Pradesh, India. Neues Jahrb. Mineral. Monatsh., nos. 7/8, 323–336.

Mukherjee, A., and Rege, S. M. (1972a). Facies transition and growth of hypersthene in some high grade metamorphic rocks from the Eastern Ghats, India. Neues Jahrb. Mineral. Monatsh., no. 3, 116–132.

——, and —— (1972b). Stability of wollastonite in the granulite facies; Some evidences from the Eastern Ghats, India. Neues Jahrb. Mineral. Abh., v. 118, 23–42.

Murthy, M. V. N., and Siddiquie, H. N. (1964). Studies on zircons from some garnetiferous sillimanite gneisses (khondalites) from Orissa and Andhra Pradesh, India. J. Geol., v. 72, 123–127.

——, Viswanathan, T. V., and Roy Choudhury, S. (1971). The Eastern Ghats Group. Geol. Surv. India Records, v. 101, Part 2, 15–42.

Nag, S. (1983a). The alkaline rocks of Eastern Ghats orogenic belt, India. Neues Jahrb. Mineral. Abh., v. 148, 97–112.

—— (1983b). Geochemical study on Proterozoic mafic alkaline rocks of Elchuru, Prakasam District, Andhra Pradesh, India. Neues Jahrb. Mineral. Abh., v. 147, 217–227.

Nanda, J. K., and Natarajan, V. (1977). A note on the sapphirine bearing rocks of Anakapalle area, Visakhapatnam District, Andhra Pradesh. Ind. Minerals, v. 31, no. 3, 27–29.

——, and —— (1980). Anorthosites and related rocks of the Kondapalli Hills, Andhra Pradesh. Geol. Surv. India Records, v. 113, Part 5, 57–67.

Narayana Rao, M. (1982). Petrology and chemistry of some quartzo-feldspathic gneisses from the Chundi-Malekonda area, Prakasam District, Andhra Pradesh. Geol., Min. Metall. Soc. India Quart. J., v. 54, 39–47.

—— (1983). Lithostratigraphy of the Precambrian rocks of the Nellore schist belt. Geol., Min. Metall. Soc. India Quart. J., v. 55, 83–89.

Narayanaswamy, S. (1975). Proposal for charnockite-khondalite system in the Archaean shield of peninsular India. Geol. Surv. India Misc. Publ. 23, Part 1, 1–16.

Natarajan, V., and Nanda, J. K. (1981). Large scale basin and dome structures in the high grade metamorphics near Visakhapatnam, South India. Geol. Soc. India J., v. 22, 584–592.

Nayudu, R. T., and Jagadiswara Rao, R. (1977). Calc-silicate rocks of Andhra Pradesh. Panjab Univ. (Chandigarh) Centre Adv. Studies Geol., Publ. 11, 11A–16A.

Parshad, R., Lal, N., and Nagpaul, K. K. (1979). Tectonic uplift of the Nellore mica belt, India, as revealed by fisson track dating technique. Geol. Soc. India J. v. 20, 31–36.

Perraju, P. (1973). Anorthosites of Puri District,

Orissa. Geol. Surv. India Records, v. 105, Part 2, 101–116.

———, Kovach, A., and Svingor, E. (1979). Rubidium-strontium ages of some rocks from parts of the Eastern Ghats in Orissa and Andhra Pradesh, India. Geol. Soc. India J., v. 20, 290–296.

Perumal, N. V. A. A. (1975). A petrological study of charnockites and associated rocks east of Chipurupalle, Srikakulam District, Andhra Pradesh. Geol. Soc. India J., v. 16, 29–36.

Radhakrishna, B. P., and Naqvi, S. M. (1986). Precambrian continental crust of India and its evolution. J. Geol, v. 94, 145–166.

Rama Rao, P., and Narasingha Rao, A. (1976). Geology of Kanagiri-Gogulapalle area, Prakasam District, Andhra Pradesh. Ind. Acad. Geosci. J., v. 19, 15–25.

Ramam, P. K., and Viswanathan, T. V. (1977). Carbonatite complex near Borra, Visakhapatnam District, Andhra Pradesh. Geol. Soc. India J., v. 18, 605–610.

Ramamohana Rao, T., Prasada Rao, V. S. N., Sairampersmal, D., and Ranga Rao (1983). Peridotite from the Archaean belt of Khammam, Andhra Pradesh. Curr. Sci., v. 52, 1139–1141.

Ramaswamy, A., and Murty, M. S. (1973). The charnockite series of Amaravathi, Guntur District, Andhra Pradesh, South India, Geol. Mag., v. 110, 171–184.

Rao, G. V. (1969). The geology and manganese ore deposits of parts of Visakhapatnam (Vizagapatnam) manganese belt, Srikakulam District, Andhra Pradesh. Geol. Surv. India Bull., ser. A, no. 35, 129 p.

Rau, R. V. R., and Adinarayana, D. (1968). Paracharnockites from Anantagiri, Eastern Ghats, Andhra Pradesh. Curr. Sci., v. 37, 226–228.

Ray, S., and Bose, M. K. (1975). Tectonic and petrological evolution of the Eastern Ghats Precambrian belt. Chayanica Geologica, v. 1, 1–13.

Reddy, B. R. G. (1972). Trace element studies of amphibolites from the Precambrian Tanikalla metamorphic complex, east of Khammam, Andhra Pradesh. Ind. Acad. Sci. Proc., Sect. B, v. 76, 153–164.

Roy, S. (1981). Manganese Deposits. Academic Press, London, 488 p.

Sadashivaiah, M. S., and Subbarayudu, G. V. (1970). Pelitic hornfels xenoliths in the garnetiferous granulites and granitic rocks of the Kondavidu area, Guntur District, Andhra Pradesh. Geol. Soc. India J., v. 11, 1–16.

Sahu, K. N. (1973). Garnets from khondalites of Tapang, Orissa. Geol., Min. Metall. Soc. India Quart. J., v. 45, 211–214.

——— (1976). Petrological observations on nepheline syenites of Baradangua, Dhenkanal, Orissa. Geol. Soc. India J., v. 17, 484–489.

Sarkar, A. N., Bhanumathi, L., and Balasubrahmanyan, M. N. (1981). Petrology geochemistry and geochronology of the Chilka Lake igneous complex, Orissa State, India. Lithos, v. 14, 93–111.

Sastry, A. V. R., and Vaidyanadhan, R. (1968). Structure and petrography of the quartz-magnetite and associated rocks of Vemparala area, Nellore District, Andhra Pradesh. Geol. Soc. India J., v. 9, 49–57.

Satyanarayana, B., and Krishna Rao, J. S. R. (1974). Occurrence of high uranium-bearing granitic gneisses near Narsipatnam, Vishakhapatnam District, A. P. Geol. Soc. India J., v. 15, 97–98.

Siddiquie, H. H., and Viswanathan, T. V. (1969). Zircons of khondalites. Geol. Surv. India Misc. Publ. 9, 202–213.

Sinha, R. C., and Mall, A. P. (1974). The chromite bearing ultrabasics of Kondapalle and Ganginemi (A. P.) and the petrogenetic relations between associated "charnockite series" of rocks. Ind. Natl. Sci. Acad. Proc., Part A, v. 40, 46–56.

Sivaprakash, C. (1981). Petrology of calc-silicate rocks from Koduriu, Andhra Pradesh, India. Contrib. Mineral. Petrol., v. 77, 121–128.

Somayajulu, P. C., and Chandrasekhar, K. (1978). Geology and status of the iron formation and petrology of the associated mafic granulites, Addanki, Prakasham District, Andhra Pradesh. Ind. Acad. Geosci. J., v. 21, 49–65.

Srirama Rao, M. (1946). Geology and petrography of the Bezwada and Kondapalli hill ranges; Part I. Bezwada gneiss and associated rocks. Ind. Acad. Sci. Proc., Sect. A, v. 24, 199–215.

——— (1947). Geology and petrography of the Bezwada and Kondapalli hill ranges; Part II. Charnockites and associated rocks and chromite. Ind. Acad. Sci. Proc., Sect. A, v. 26, 133–166.

Sriramadas, A., and Rao, A. T. (1979). Charnockites of Visakhapatnam, Andhra Pradesh. Geol. Soc. India J., v. 20, 512–516.

———, Rao, K. S. R., and Rao, A. T. (1969). Aluminous augite in the pyroxenite of the charnockite rocks of Vishakapatnam District, Andhra Pradesh. Ind. Natl. Sci. Acad. Proc., Sect. B, v. 70, 15–27.

Subbaraju, K. (1975). Some aspects of the schistose rocks of Khammam District, Andhra Pradesh. Ind. Mineralogist, v. 16, 35–42.

Subbarao, K. V. (1971). The Kunavaram series—A group of alkaline rocks, Khammam District, Andhra Pradesh, India. J. Petrol., v. 12, 621–641.

Subrahmanyam, C. (1983). An overview of gravity anomalies, Precambrian metamorphic terrains and their boundary relationships in the southern Indian shield. In Precambrian of South India (eds. S. M. Naqvi and J. J. W. Rogers), Geol. Soc. India Mem. 4, 553–570.

Subramaniam, A. P. (1962). Pyroxenes and garnets from charnockites and associated granulites. In Petrologic Studies; A volume in Honor of A. F. Buddington (ed. A. E. J. Engel, H. L. James, and B. F. Leonard), Geol. Soc. Amer., New York, 21–36.

—— (1967). Charnockites and granulites of southern India; A review. Dansk Geol. Foren. Meddel., v. 17, 479–493.

Vasudevan, D., and Rao, T. M. (1975). The high grade schistose rocks of the Nellore schist belt, Andhra Pradesh and their geologic evolution. Ind. Mineralogist, v. 16, 43–47.

Viswanathan, T. V. (1968). Zircon studies of charnockites and associated country rocks from Phulbani District, Orissa. Geol. Surv. India Misc. Publ. 9, 185–201.

—— (1969). The granulitic rocks of the Indian Precambrian shield. Geol. Surv. India Mem. 100, 37–66.

6. BHANDARA CRATON

The rectangular portion of the Indian shield bounded by the Eastern Ghats, the Mahanadi and Godavari rifts, and the Narmada-Son lineament is designated as the Bhandara craton (Fig. 6.1). It contains a vast tract of granites and gneisses with engulfed and overlying supracrustal rocks of the Dongargarh, Sakoli, Sausar (including Chilpighat and Sonawani), Bengpal, Sukma, and Bailadila (Iron Ore) suites. These older rocks are overlain by eight unmetamorphosed, Late Proterozoic basins, including the major ones of Chattisgarh and Bastar (Fig. 6.1; Table 6.1).

Early work in India, summarized by Fermor (1936), correlated supracrustal rocks in the Bhandara area with those in many other regions, including the Dharwar belts (Ch. 2). Later work, however, has shown that most of the supracrustal suites of the Bhandara craton apparently are much younger than the Dharwar rocks. The major schistose supracrustal rocks of the Bhandara craton occur (Fig. 6.1) near Nagpur and west of the Chattisgarh basin; along the Narmada-Son lineament near Jabalpur, separated by Deccan basalts from the first area; and in the area west of the Eastern Ghats and south and east of the Bastar basin, near Bijapur and Bastar. There seems to be little similarity between the rocks of these diverse areas.

The Bhandara craton has three significant features:

It contains the Satpura orogenic belt, which is mostly east-west and intersects the dominantly north-south structural trend of the shield at nearly a right angle.

It contains an area known as the "Bhandara triangle," where three sets of planar, penetrative features intersect each other and which contains supracrustal rocks in three separate basins.

It contains abundant manganese ores in the supracrustal rocks of the Sausar Group. These ores are sedimentary, nonvolcanogenic deposits.

Early workers considered that all crystalline rocks of the Bhandara craton are Archean, but limited recent geochronological work has failed to detect any Archean ages. Lithologic features of the Bengpal, Sukma and Bailadila (Iron Ore) suites

(near Bijapur and Bastar), however, are similar to those of Archean belts elsewhere.

GNEISS AND GRANITE

A vast area of the Bhandara craton consists of granites and granitic gneisses of uncertain age and genesis. All of the rocks investigated thus far are intrusive into at least one set of supracrustal rocks, and the sialic "basements" that have been described in the Dharwar and other cratons have not been found in the Bhandara craton. This absence is particularly intriguing in view of the fact that the sediments are mostly platformal and have been derived largely from sialic terrains, indicating that some continental basement must have existed in the area at the time of their formation.

Gneisses similar to the Peninsular Gneiss of the Dharwar craton have been studied in several areas (Crookshank, 1963; Shukla and Anandalwar, 1973; Rajarajan, 1976). These gneisses are predominantly biotite bearing and were classified by Rajarajan into three groups: fine-grained biotite gneiss, porphyritic and streaky biotite gneiss, and fine-grained gneissose granites. In some places these gneisses contain garnet and sillimanite. Enclaves of hornblende- and pyroxene-bearing gneisses are also present. Some gneissic suites appear to be older than, and form the basement for, associated supracrustal suites. Other gneisses appear to intrude the supracrustal rocks. Detailed studies of the relationships are generally not available.

The Bhandara Triangle contains the Tirodi gneissic suite, a name given by Narayanaswamy et al. (1963) to hybrid gneisses and associated rocks exposed between the Sausar and Sakoli outcrops and as enclaves within the Sausar belt itself (Fig. 6.1). The gneiss has a metamorphic age of $1,525 \pm 70$ m.y. with an initial $^{87}Sr/^{86}Sr$ of 0.7148 $\pm$ 0.0033 (Sarkar et al., 1986). The main constituents of the Tirodi suite are biotite-muscovite and biotite gneisses, hornblende-biotite schists and gneisses, epidote and biotite quartzites, calc-silicate gneisses, quartz-biotite granulites, and feldspathic biotite-muscovite schists. Garnet and sillimanite occur in areas metamorphosed to higher grades. Small manganese ore deposits

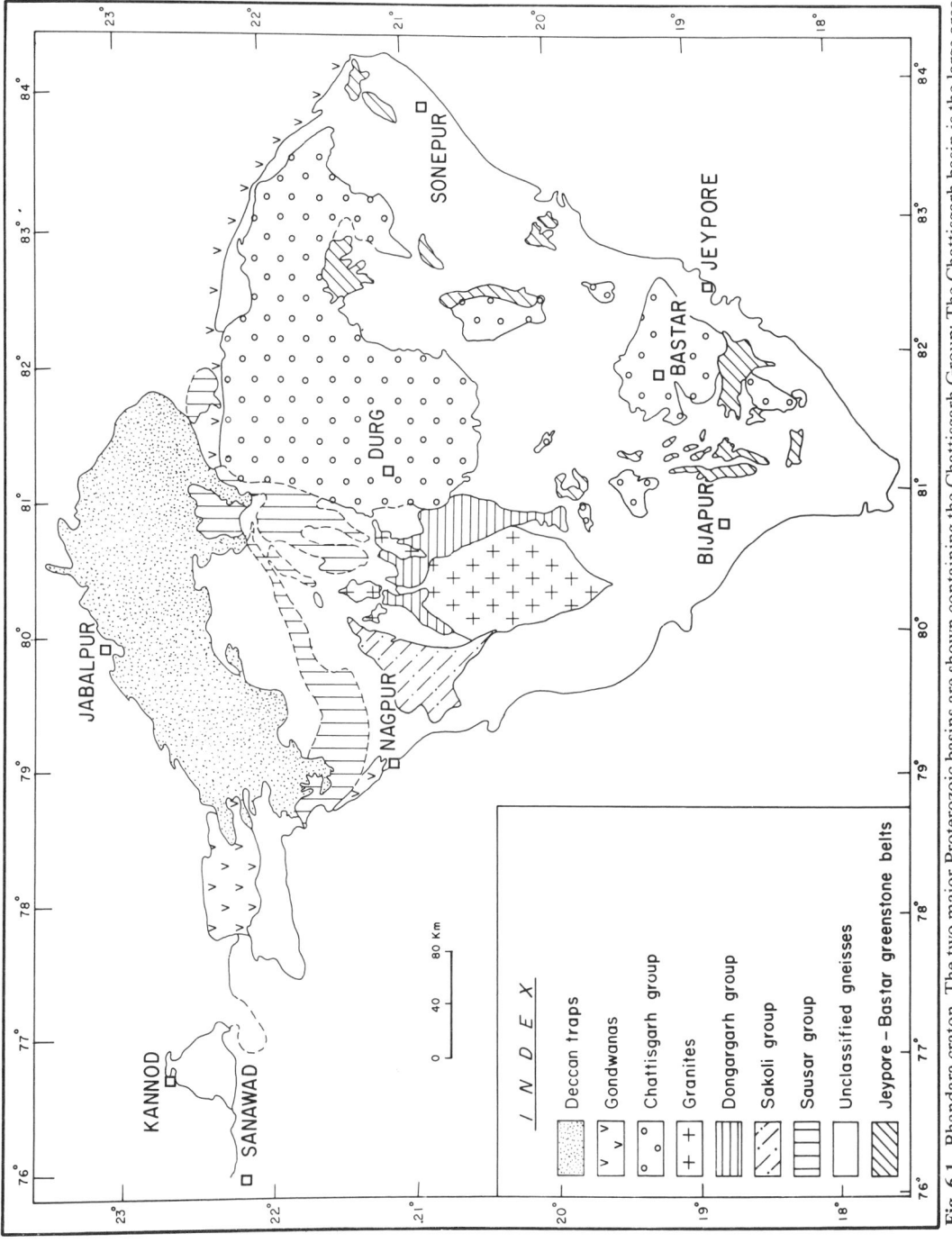

Fig. 6.1. Bhandara craton. The two major Proterozoic basins are shown containing the Chattisgarh Group: The Chattisgarh basin is the large area around, and northeast of, Durg; the Bastar basin is around the city of Bastar. The Dongargarh Granite is the large granite mass intrusive into the Dongargarh Group.

121

TABLE 6.1. *General stratigraphy of Bhandara craton*

Nagpur, Bhandara, Chindwara (west of Chattisgarh basin)	Jeypore, Bastar (east, southeast, and south of Chattisgarh basin)
Chattisgarh and Bastar basins	
-- Unconformity --	
Sausar Group (?? 950 m.y. ??)	Bailadila Group (age ?)
Dongargarh Supergroup (2,200 m.y.?)	Bengpal Group (age ?)
Sakoli Group (2,200 m.y.?)	Sukma Group (age ?)
Basement gneisses and granites ?	

occur in the Nagpur area. The gneisses are coarse to medium grained, banded, commonly porphyroblastic, and were regarded as paragneisses by Narayanaswamy and Venkatesh (1973).

A large granitic body in the southern part of the Bhandara triangle is referred to as the Dongargarh Granite. It is intrusive into some of the supracrustal suites and appears to represent a relatively young body similar to the Closepet Granite of the Eastern Dharwar craton (Ch. 2). Very little information is available on the granite.

SUPRACRUSTAL ROCKS BETWEEN NAGPUR AND WEST SIDE OF CHATTISGARH BASIN

The three major supracrustal suites of the Bhandara triangle are Sakoli, Dongargarh and Sausar (Fig. 6.1). The rocks in the Dongargarh belt are significantly different from those of the other suites and have been designated as the "Dongargarh Supergroup" by Sarkar (1957, 1983). These various suites are described in this section.

Dongargarh Supergroup

The Dongargarh Supergroup was regarded as part of the Sakoli Group in older classifications but is now considered to be a separate suite (Sarkar, 1957, 1983). Sarkar et al. (1981) subdivided the Dongargarh Supergroup into the Amgaon, Nandgaon, and Khairagarh Groups (Table 6.2). The belt of Dongargarh rocks is about 90 km wide and 150 km long, extending northeasterly between the Sakoli synclinorium on the west and the Chattisgarh basin on the east (Fig. 6.1). The Tirodi gneisses (discussed previously) separate the Dongargarh belt from the Sakoli and Sausar Groups.

The Amgaon Group is composed mainly of quartz-feldspar-biotite gneisses and metamorphosed sedimentary rocks interbedded with metabasalts and minor other volcanic rocks. The supracrustal rocks include quartz-sericite schists and quartzites $\pm$ garnet, $\pm$ epidote, $\pm$ biotite. All rocks have been metamorphosed to amphibolite facies. Sarkar et al. (1981) proposed an "Amgaon Orogeny" that occurred about 2,300 m.y. ago, at which time granites and gneisses in the Amgaon

TABLE 6.2. *Stratigraphy of Dongargarh Supergroup (Sarkar et al., 1981)*

Chattisgarh Supergroup	Raipur Group
	Chandarpur Sandstone
--Unconformity --	
	Khairagarh orogenic phase (c.900 Ma ?)
Khairagarh Group	Mangikhuta volcanics
	Karutola Formation
	Sitagota volcanics (1367 Ma)
	(Intertrappean shale) (1686 Ma)
	Bortalao Formation
	Basal shale (1534 Ma)
--Unconformity --	
	Dongargarh Granite (< c.2200 Ma)
Nandgaon Group	Pitepani volcanics
	Bijli rhyolites (c.2200 Ma)
	Amgaon orogeny, metamorphism, and granitization (> c.2300 Ma)?
Amgaon Group	Quartz-sericite schist, feldspathic quartzite, garnet-epidote quartzite, hornblende biotite quartzite, quartz-feldspar biotite gneiss, hornblende schist, and amphibolite

Prepared with permission from the Indian Journal of Earth Sciences, v. 8, p. 134.

suite were developed by syntectonic granitization. The gneisses were regarded as granitized metasediments because of their high contents of Al_2O_3, Fe_2O_3, CaO, Na_2O, and K_2O and low contents of SiO_2, MgO and FeO.

The Nandgaon Group consists of the Bijli Rhyolite and the Pitapani Formation. The Bijli Rhyolite has a total thickness of about 4,500 m and contains rhyolites plus rhyolitic conglomerates, sandstones, shales, and tuffs. The suite contains inclusions of Amgaon amphibolites and quartzites. Sarkar et al. (1981) suggested that the Bijli Rhyolite is post-Sakoli in age and presented an eight-point Rb/Sr isochron with an age of 2,180 ± 25 m.y. and an initial $^{87}Sr/^{86}Sr$ ratio of 0.7057 ± 0.0015.

The Bijli Rhyolite (Jafri, 1981) is mostly porphyritic. Phenocrysta are quartz, with embayed margins, subhedral and sericitized K-feldspar, and albite. The matrix is a microcrystalline mass of quartz, feldspar, biotite, sericite, and opaque minerals and shows flow layering around the phenocrysts. The Pitapani volcanic suite overlies the Bijli Rhyolite and also contains inclusions of the rhyolites. Rock types vary from quartz-normative to olivine-normative tholeiites (Jafri, 1981). Phenocrysts are plagioclase (andesine to labradorite) and pyroxene, and the matrix consists of plagioclase, chlorite, clinopyroxene and opaque minerals ± interstitial glass.

The Nandgaon Group is intruded by the Dongargarh Granite, which has a seven-point Rb/Sr isochron age of 2,270 ± 90 m.y. with an initial $^{87}Sr/^{86}Sr$ ratio of 0.7092 ± 0.0054 (Sarkar et al., 1981). The error on the initial ratio is extremely large, but if the ratio is valid, then the granite is presumably derived from crustal material. The Dongargarh Granite consists of granophyres, porphyritic microgranites, and porphyritic to coarse-grained granites. Although field evidence clearly shows that the Dongargarh Granite is intrusive into, and thus younger than, the Bijli Rhyolite, radiometric age determinations show a reverse relationship (2,270 m.y. for the granite and 2,180 m.y. for the rhyolite). Sarkar et al. explained the isotopic data as resulting from contamination of the granite with older rock.

The Nandgaon Group and Dongargarh Granite are overlain by the Khairagarh Group, which consists of shales, sandstones, and volcanic rocks (Sarkar et al., 1981). The basal unit is the Bortalao shale and sandstone, with a basal conglomerate containing pebbles of quartz and granite overlain by quartzite. The Bortalao Formation is overlain by the Sitagota volcanic suite, consisting of quartz- to olivine-normative tholeiites. The Sitagota suite is overlain by the Karutola Sandstone, which is a fine- to coarse-grained ferruginous orthoquartzite. The Mangikhuta quartz-normative tholeiites are the youngest rocks of the Dongargarh Supergroup.

The Dongargarh Supergroup has been affected by at least three phases of complex and tight folding whose ages are mostly unknown. The first phase was originally designated as the Sakoli orogeny but has been named the Amgaon orogeny by Sarkar et al. (1981). It produced isoclinal folds in the Amgaon Group. The second phase affected the Bijli Rhyolite and older rocks and has been designated as the Nandgaon orogeny. The third phase affected the Bortalao Formation and older rocks and is referred to as the Khairagarh orogeny.

Sakoli Group

The Sakoli Group occurs in a large synclinorium (Fig. 6.1). Major rock types are largely metapelitic and include muscovite-quartz-garnet-biotite schists ± staurolite ± chlorite with interlayered amphibolites and banded hematite quartzites. The metamorphism of the Sakoli Group appears to have been at a somewhat lower grade than that of the Sausar Group (discussed below), with the highest-grade rocks being kyanite bearing (Sengupta, 1965). The Sakoli suite contains very little mafic/ultramafic volcanic material.

The Sakoli rocks have undergone two phases of deformation (Sengupta, 1965). The first phase mainly produced isoclinal folds of the bedding planes (S1) and an axial plane schistosity (S2) defined by preferred orientation of micas and oriented quartz grains. The second phase of deformation folded both S1 and S2 and shows steep dips. A metamorphic high extending through the center of the belt is represented by a strip of kyanite schist, and a Barrovian zonation of metamorphic grade decreases toward either side of this strip. First-generation folding was accompanied by development of schistosity and crystallization of mica and quartz. Kyanite formed shortly afterward, along with minor garnet and staurolite outside of the kyanite zone. Major garnet crystallization occurred during the second phase of folding, followed shortly by staurolite. Crystallization of plagioclase occurred during all metamorphic and deformational phases. After progressive metamorphism, the Sakoli rocks were subjected to retrogression.

Sausar Group

Metamorphosed supracrustal rocks designated as the Sausar Group form an arcuate belt about 32 km wide and 210 km long (Fig. 6.1; Fermor, 1909, 1936; Narayanaswamy et al., 1963). The Sausar Group is composed primarily of metamorphosed sandy, shaly, and calcareous sediments and manganese ores and is the major Mn-producing suite in India. Volcanic rocks are virtually absent. Calcareous formations (Lohangi and Bichua) are better developed in the north and west, and argilla-

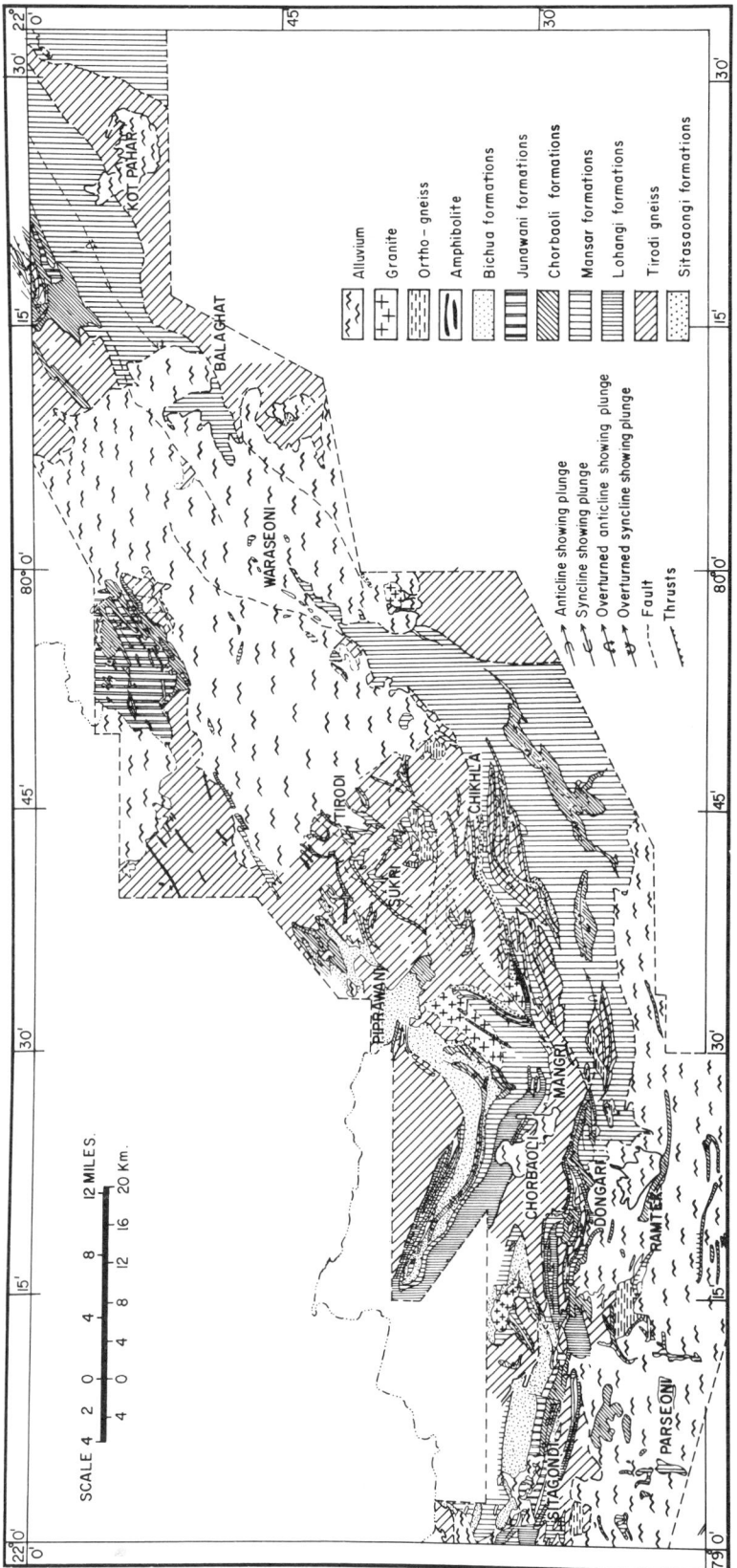

Fig. 6.2. Sausar Group (Narayanaswamy et al., 1963).

Alluvium
Granite
Ortho – gneiss
Amphibolite
Bichua formations
Junawani formations
Chorbaoli formations
Mansar formations
Lohangi formations
Tirodi gneiss
Sitasaongi formations

Anticline showing plunge
Syncline showing plunge
Overturned anticline showing plunge
Overturned syncline showing plunge
Fault
Thrusts

SCALE 4 2 0
4 0 4 8 12 16 20 Km.
4 0 4 8 12 MILES.

124

ceous formations (Mansar and Chorbaoli) in the south and east of the belt (Fig. 6.2; Table 6.3). Because of lateral facies changes, the sequence and names of formations differ slightly from place to place (Roy, 1973). Relationships among high-grade metamorphic equivalents of the Sausar Group in the northwestern part of the outcrop area (Satpura Range) are discussed in the next section.

The stratigraphic terminology of the Sausar Group is controversial. Two old stratigraphic terms (Chilpighat and Sonawani) were once used in the Sausar area. Narayanaswamy et al. (1963) proposed that the phyllites and sericitic schists of the Chilpighat series can be correlated with the muscovite schists of the Mansar and Sitasaongi Formations of the Sausar Group. Similarly, marbles, calc-granulites, and quartz-muscovite schists of the Sonawani suite have been correlated with calcareous rocks of the Sausar Group. The term "Chilpighat," however, is still used as equivalent to all or part of the Sausar Group (Rao, 1979).

As noted in the introduction, the Sausar rocks were once thought to be the oldest formations of central India, with all gneissic rocks intrusive into them. The only actual ages that can be used to date the Sausar rocks, however, are limited radiometric ages for the Satpura orogeny, which deformed and metamorphosed them. This orogeny has been dated by Rb-Sr methods on the Tirodi Gneiss as 1,525 m.y. (Sarkar et al., 1986). In addition, U-Th-Pb methods on pegmatites (Holmes, 1955) and K-Ar methods on micas (Sarkar et al., 1981) yield closing ages in the range of 1,000 to 850 m.y.

The grade of metamorphism in the Satpura belt gradually increases westward and northward. Near Balaghat (Fig. 6.2) rocks are primarily in greenschist facies. Farther west are amphibolite-facies suites marked by the presence of garnet, staurolite, and kyanite. Farthest to the north-northwest is a granulite-facies assemblage characterized by sillimanite, clinopyroxene and iron-bearing garnet.

Structural studies of the Sausar Group have been conducted by numerous workers (Straczek et al., 1956; Narayanaswamy et al., 1963; Basu and Sarkar, 1966; and Narayanaswamy and Venkatesh, 1973). Narayanaswamy et al. described the general structure, including the Satpura Range. The southern part consists of isoclinal folds that are overturned or recumbent toward the north. Some of the folds were described as small nappes showing northward transport directions and southerly dips. This area is bordered on the north by gneisses of the Tirodi suite that apparently constitute the core of the orogen ("central crystalline axis"). The gneisses contain intricately infolded fragments of schist belts, manganese ores, and other supracrustal suites. Structures in this core are extremely complex and generally show northeast-southwest strikes and variable dips. North of the gneissic core of the orogen, in the Satpura Range, the belt shows extensive development of southward-directed nappes. The folding and thrusting are so complex that the area consists of detached sheets of schists, gneisses, and manganese ores. One nappe (Deolapar) was proposed to have moved 4 to 5 km southward.

The entire Sausar area has been refolded (along axes subparallel to the original fold axes) or cross-

TABLE 6.3. *Stratigraphy of Sausar Group (modified from Narayanaswamy et al., 1963)*

Bichua Formation	Dolomitic, serpentine-bearing marble; calc-silicate granulites; occurs in all areas except south and east
Junewani Formation	Muscovite-biotite quartz schists, granulites, and gneisses; index minerals are garnet, staurolite, sillimanite, and kyanite; widespread, lenticular, locally interbedded with Bichua Formation
Chorbaoli Formation	Quartzites, micaceous and feldspathic quartz schists, and local autoclastic quartz conglomerates; sporadic garnet, sillimanite, and kyanite; widespread except in center of belt
Mansar Formation	Micaceous schists and phyllites, commonly with garnet; most widespread formation in the Sausar Group; contains manganese ores, including gondites
Lohangi Formation	Three interdigitating members; pinches out to south and east into Sitasaongi Formation; contains manganese ores
Lohangi member	Calcitic and dolomitic marbles
Utekata member	Calc-silicate granulites and gneisses
Kadhikhera member	Quartz-biotite granulite and gneiss
Sitasaongi Formation	Quartz-muscovite-feldspar schists and intercalated quartzites; local kyanite and garnet
Tirodi Gneiss	Biotite gneiss with minor amphibolite, calcareous gneisses and schists; garnet common; locally porphyroblastic; mainly in center of belt

---Unconformity --

Older gneisses, amphibolites, etc.

folded (along axes at high angles to the original folds). The most intensely refolded area is in the crystalline core. Late, potassic granites were emplaced along structures formed during the refolding. The granites in the area have not been well studied, but possibly correlative plutons in the Jabalpur area (Ghose and Gupta, 1980) are diapiric, clearly posttectonic, and have very high total alkali contents and K_2O contents (up to 6.7%) and virtually no MgO. These properties are consistent with those of postorogenic alkali granites of other areas.

Metamorphism of the Sausar Group was apparently roughly synchronous with the various stages of deformation (Narayanaswamy et al., 1963). Nappes and other faults offset some isograds, but other isograds transect folds and faults, showing an interconnected series of movement and metamorphism.

Detailed structural studies have been made locally. Sarkar et al. (1977) proposed that Sausar rocks in the Chikla area (Fig. 6.2) have undergone three phases of deformation and four phases of metamorphism. The first deformational phase produced isoclinal folds, axial plane schistosity, and mineral and micropucker lineations. Metamorphism formed phyllites and caused the growth of garnet-I, garnet-II, and staurolite-I. The second phase of deformation did not produce major structures but generated superposed pucker folds and crenulation cleavage accompanied by the growth of garnet-III. The third deformational phase formed open folds with steeply dipping axial planes. Metamorphic growth of staurolite-II and -III and kyanite-I occurred during, and slightly before, the last folding. Agarwal (1974, 1975) also mapped three different phases of folding in the area north of Nagpur (Fig. 6.1). These rocks were metamorphosed to almandine amphibolite facies and migmatized. The first deformation produced major recumbent folds, including the Deolapar nappe.

High-grade metamorphic rocks were described in the Ramtek area (Fig. 6.2; Basu and Sarkar, 1966). The lower unit consists of coarse-grained quartzites, conglomerates, and manganiferous rocks. The manganese rocks consist either of an assemblage of pyroxmangite, garnet, rhodochrosite, and quartz ± apatite, or an assemblage of bustamite, rhodochrosite, and quartz. Above this unit, much of the section consists of calcareous rocks. Calc-gneisses contain calcite and/or dolomite and have a variable mineralogy of epidote, diopside, hornblende, oligoclase, microcline, garnet, biotite, actinolite, and quartz. Diopside is commonly hedenbergitic, the garnet is grossularite, and plagioclase is andesine/oligoclase (Bhaskara Rao and Ananthapadmanabhan, 1970).

The Mansar Formation is the uppermost unit in the Ramtek area and consists of gondites, pelitic schists, and manganese ores. In its present usage, the term "gondite" refers to rocks composed largely of garnets (rich in spessartite or almandine) and quartz with variable amounts of rhodonite, blandfordite (a variety of aegerine-acmite), brown manganiferous pyroxene (a variety of aegerine/augite), and minor manganiferous minerals. Manganese-free accessory minerals include apatite, microcline, sericite, and calcite. Among the minor manganese minerals, tirodite is a manganiferous amphibole consisting of a mixture of cummingtonite, soda tremolite, and manganotremolite; and juddite is a combination of glaucophane and magnesio-riebeckite. The gondites are regarded as the products of granulite-facies metamorphism of manganiferous sediments.

SUPRACRUSTAL ROCKS OF THE JABALPUR AREA

A small belt of supracrustal rocks is exposed west of Jabalpur, near the Narmada-Son valley. This exposure is separated from the main Sausar suite in the Nagpur-Bhandara-Balaghat area by an area of Deccan basalts (Fig. 6.1). The rocks of the Jabalpur area, which are highly deformed and metamorphosed, were originally correlated with the Bijawar suite of the Aravalli craton (Ch. 9) or the Gangpur suite of the Singhbum craton (Ch. 7). Crookshank (1936) correlated the Jabalpur rocks with the Sausar Group on lithological similarities, and Agarwal (1976, 1977) strengthened the correlation by comparing metamorphic grade, structure, and the emplacement pattern of mafic dikes. All correlations are tentative, however, because no radioactive age data are available for the Jabalpur region.

The Jabalpur rocks consist of conglomerates, phyllites, mica schists, calcitic and dolomitic marbles, and banded ferruginous rocks associated with manganese ores. Mafic rocks are at least partly intrusive (Khanna, 1979). The foliation strikes parallel to the general ENE-WSW trend of the Satpura belt, and three distinct phases of deformation have been recorded (Agarwal, 1976, 1977). The marbles, in particular, show an early phase of tight folding overprinted by a later phase of gentle folds with NNW-trending axes, perpendicular to the Satpura trend. All rocks have been metamorphosed in a Barrovian sequence to garnet-, staurolite-, and sillimanite-bearing grades.

SUPRACRUSTAL ROCKS EAST AND SOUTHEAST OF THE BASTAR BASIN

Seven scattered and isolated patches of supracrustal rocks occur east and southeast of the Bastar basin in the general region of Bijapur and Bastar (Figs. 6.1 and 6.3; Crookshank, 1963). No general name has been applied to this suite because of uncertainty of correlation between individual out-

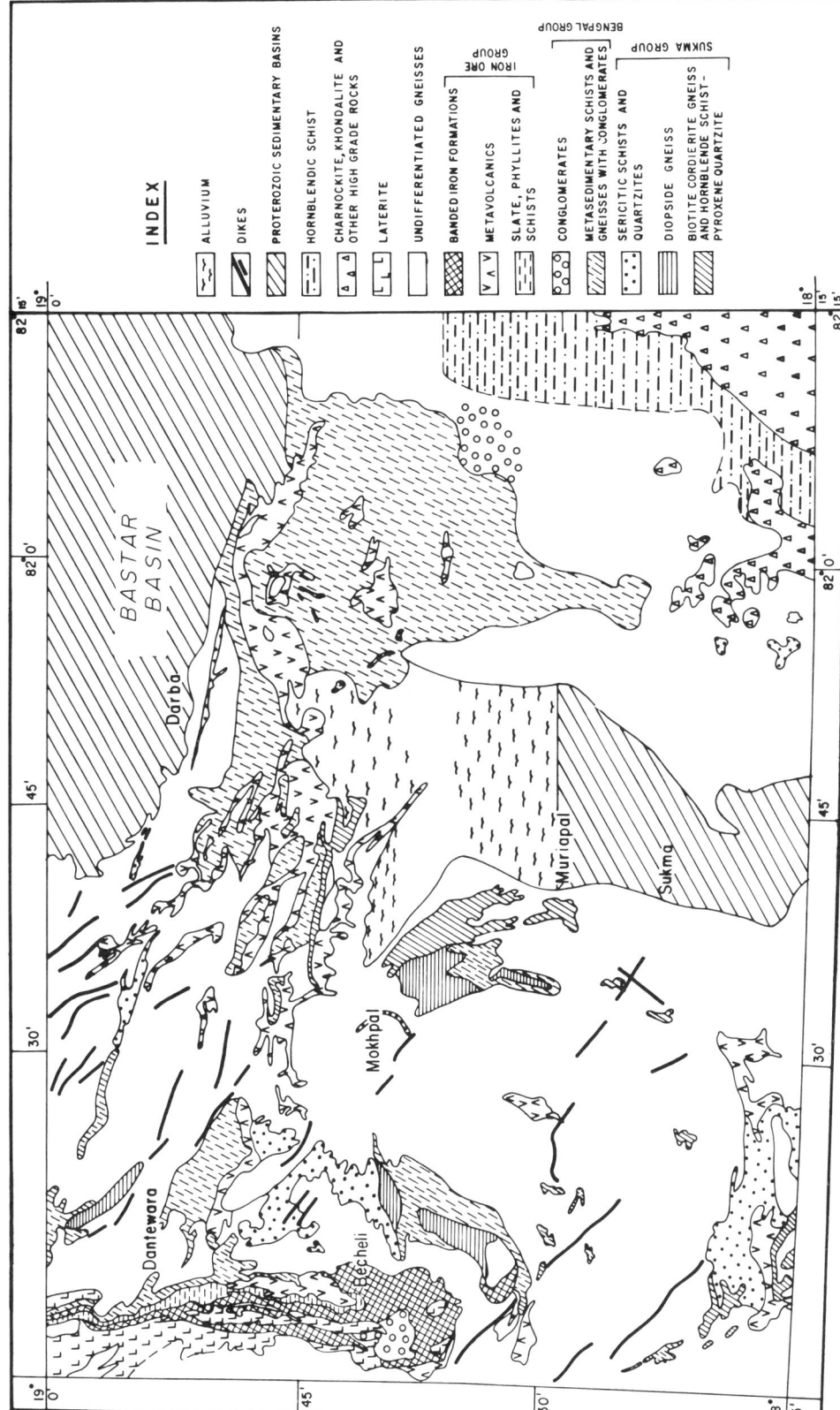

INDEX

ALLUVIUM
DIKES
PROTEROZOIC SEDIMENTARY BASINS
HORNBLENDIC SCHIST
CHARNOCKITE, KHONDALITE AND OTHER HIGH GRADE ROCKS
LATERITE
UNDIFFERENTIATED GNEISSES
BANDED IRON FORMATIONS   IRON ORE GROUP
METAVOLCANICS
SLATE, PHYLLITES AND SCHISTS
CONGLOMERATES   BENGAL GROUP
METASEDIMENTARY SCHISTS AND GNEISSES WITH CONGLOMERATES
SERICITIC SCHISTS AND QUARTZITES   SUKMA GROUP
DIOPSIDE GNEISS
BIOTITE CORDIERITE GNEISS AND HORNBLENDE SCHIST-PYROXENE QUARTZITE

BASTAR BASIN

Darba

Muriapal

Sukma

Mokhpal

Dantewara

Recheli

Fig. 6.3. Jeypore-Bastar area (Crookshank, 1963). The Iron Ore Group is also referred to as the Bailadila Group.

TABLE 6.4. *Stratigraphy of Jeypore area (Crookshank, 1963)*

Purana	Upper: Limestones, purple shales, and slates
	Lower: Pale sandstones and shales, purple shales, quartzites, grits, conglomerates
--Unconformity --	
Igneous rocks	Dolerite dykes
	Granite and pegmatite
	Charnockites
	Greenstones and granite-gneiss
--Unconformity --	
Khondalites	(Position uncertain)
--Unconformity --	
Bailadila (Iron Ore) Series	Banded hematite-quartzites, grunerite-quartzites and white quartzites
--Unconformity --	
Bengpal Series	Ferruginous schists, schistose conglomerates, biotite-hornblende-quartzites, shales, slates
	Slates, schists, phyllites, grunerite-garnet-schists, magnetite-quartzites, garnet-biotite-gneiss, with basaltic flows and tuffs
	Sericite-quartzites, andalustite-gneiss, banded magnetite-quartzites, grunerite-schists, and quartzites with intercalated basalt flows
--Line of division uncertain --	
Sukma Series	Sillimanite-quartzites, grunerite-schists magnetite and diopside-quartzites, hornblende-schists, biotite-cordierite gneiss, etc.

crops. The rocks in these suites are mostly sandy, clayey and calcareous metasediments and mafic volcanic rocks and iron formations (Mukharya, 1975; Dutta et al., 1979). The succession of rocks proposed by Crookshank is shown in Table 6.4 with minor modifications to conform to modern stratigraphic terminology. No age data are available, but Ghosh et al. (1977) suggested that mafic lavas are the oldest rocks in the area and are succeeded upward by sedimentary rocks of the Bengpal and Sukma Groups. Crookshank correlated the supracrustal rocks with the Dharwar schist belts (Ch. 2), and Table 6.4 shows considerable lithological resemblance of the supracrustal rocks with some parts of the Dharwar belts and also with the Iron Ore Series of the Singhbhum craton (Ch. 7).

Division between the Sukma and Bengpal Groups is essentially based on the difference in metamorphic grade (Table 6.4; Fig. 6.3). The Sukma Group is sillimanite bearing and the Bengpal Group contains andalusite. Crookshank (1963) regarded diopsidic quartzites and pyroxene gneisses as diagnostic of the Sukma Group.

According to Crookshank (1963), the Sukma Group has five lithologic assemblages: biotite-cordierite gneiss, diopside-hornblende gneiss/pyroxene gneiss/diopside quartzite, hornblende schist, magnetite quartzite and grunerite schist, and quartzite. No stratigraphic sequence can be established because of isolation of the outcrops. The biotite-cordierite gneiss is medium grained and

consists largely of quartz, irregular masses of cordierite (partly or completely pinitized), biotite, and muscovite. Low-silica varieties of the rock consist of coarse intergrowths of cordierite with orthoclase, chlorite, anthophyllite, muscovite, and biotite; inclusions of picotite occur in the cordierite. The diopside-hornblende gneiss/pyroxene gneiss/diopside quartzite is composed primarily of diopside and quartz. The diopside is locally altered to hornblende or actinolite, and some is serpentinized. Major minerals in the hornblende schist are green hornblende and quartz, with minor garnet, biotite, and labradorite, probably formed by metamorphism of calcareous/magnesian sediments. Quartzites contain magnetite and minor sillimanite and fuchsite.

The principal constituents of the Bengpal Group are andalusite-bearing gneisses and schists with biotite and muscovite. Sillimanite and garnet occur at high metamorphic grades. Basaltic and tuffaceous rocks are abundant in the Bengpal sequence, in contrast to the virtual absence of metavolcanic material in the Sakoli and Sausar Groups, west of the Chattisgarh basin. The mafic rocks consist primarily of fine-grained hornblende and plagioclase and show amygdaloidal flow bases.

The Sukma and Bengpal Groups are severely deformed. South of Bastar, Chatterjee (1970) found that horizontal shearing had produced mostly steep, NNE-plunging overturned folds with axial planes dipping steeply ENE. To the west, the fold geometry changes to N-S-trending, nonplunging folds. Based on preferred orientation of micas and quartz, Chatterjee inferred that the major phase of flexural slip folding was associated with rotational movements with unrestricted transport, whereas nonrotational movements with restricted transport occurred later in the deformational process.

Relationships between the Sukma and Bengpal Groups and the Bailadila (Iron Ore) Group (Fig. 6.1; Table 6.4) are controversial. Although the Bailadila Group appears, at some places, to overlie the other suites unconformably, it has also been proposed that quartzites at the base of the Bailadila Group are continuous with the Bengpal Group (P. K. Ghosh, as cited by Crookshank, 1963). The most abundant rock of the Bailadila sequence is banded hematite quartzite, which consists of roughly equal amounts of iron ores and quartz plus minor amphiboles of the magnesioriebeckite/riebeckite series (Chatterjee, 1969).

According to Chatterjee (1964), the Bailadila Group has been folded twice, with deformation less severe than that of the Sukma and Bengpal Groups. The major folds are asymmetric and open, with subhorizontal axes and axial planes dipping steeply eastward. This simple structure has been complicated by cross-folding, which formed gentle flexures with steep axial planes.

SATPURA OROGENY

The Satpura Range extends ENE-WSW across a large part of central India (Figs. 1.1 and 6.1). Much of it is covered by a thick section of Deccan basalts, but one portion in the western part of the Bhandara craton contains the Sausar Group, which has been extensively deformed and metamorphosed. The age of the Satpura orogeny is essentially unknown. As discussed earlier, the Tirodi Gneiss has a metamorphic age of 1,525 m.y. Also, the Satpura belt may have formed a southern margin for Vindhyan sedimentation about 1,400 m.y. ago (Ch. 9). The Satpura orogeny is most likely to have occurred at some time in the Middle Proterozoic.

Two lines of evidence indicate that the Satpura belt is the result of a typical compressive orogeny. One is the presence of recumbent folds and overthrusts along both the northern and southern margins of the belt. The nappe structure is particularly well developed in the north, where southward movement has occurred. The symmetrical nature of the Sausar orogen is remarkable, with thrust and fold vergence from both sides toward a crystalline core zone. A second line of evidence is the increase in metamorphic grade toward the northwest, with granulite-facies rocks near the northern margin.

The preceding features are characteristic of approximately north-south compression during the Satpura orogeny. Stable blocks north and south of the orogen (in its present orientation) were apparently compressed against a mobile belt. The pattern in the Sausar suite, however, is different from that in most mobile belts in showing fold and thust vergence toward the center of the orogen instead of away from it. Possibly the mobile belt contained fragments or extensive areas of stable crust that were overridden during compression.

On a broad scale, the stable masses on each side of the orogenic belt were presumably either the Bhandara craton and the Bundelkhand massif of the Aravalli craton (Ch. 9) or, according to Radhakrishna and Naqvi (1986), the Dharwar-Singhbhum protocontinent and the Aravalli-Bundelkhand protocontinent. One problem with either interpretation is the presence of the Narmada rift (Ch. 10) between the Satpura belt and the Bundelkhand massif. There are two possible relationships between the Satpura belt and the Narmada rift:

The rift is older than the Satpura belt. The Satpura orogeny occurred during the closure of the rift basin, which would be associated with compression perpendicular to the rift margin. In this case, the present topographic expression of the rift is the result of reactivation of a former rift area.

The rift is younger than the Satpura orogeny. In this case, the Satpura event could be an intracratonic deformation, which would be consistent with the dominantly siliceous and platformal nature of the deformed sedimentary suites. The rift could have formed preferentially along the orogenic belt because of the tendency of new continental rifts to develop in areas that have recently been tectonically thickened (Ch. 10).

The mild deformation of Vindhyan sediments on the southern side of the Vindhyan basin in the Aravalli craton (Ch. 9) is consistent with either interpretation of the Satpura orogeny and may have resulted from the same forces that caused the Satpura compression. Deformation of the Vindhyan sequences is very limited, and most of the basin is completely unaffected by compression. This lack of deformation leads to some doubt about the time relationship between the Satpura orogeny and Vindhyan sedimentation, which is commonly placed in the age range of 1,400 to 900 m.y. ago (Ch. 9). The Satpura orogeny might be older than Vindhyan time, thus explaining the lack of deformation of Vindhyan rocks.

If the Satpura orogeny is largely intracratonic, then it may be possible to correlate supracrustal rocks of the Bhandara craton with rocks farther north in India. A simple map interpretation indicates that the Satpura trend may join the Chotanagpur-Singhbhum-Dhalbhum trend in Bihar and Orissa and the Aravalli-Delhi and Bijawar trends in western Madhya Pradesh and Gujarat (Fig. 1.5; Chs. 7 and 9; Radhakrishna and Naqvi, 1986). Thus, it is possible that the Sausar, Delhi, and Bijawar suites formed a semicircular geosyncline around the massif consisting of the Banded Gneissic Complex and Bundelkhand area of the Aravalli craton (Ch. 9). A time span of this magnitude (more than 1,000 m.y.) is partly supported by the possibility that the Aravalli and Delhi suites represent a continuous sedimentary interval (Ch. 9). All the rock suites involved in this basin are platformal sedimentary sequences with only minor volcanic component, which strengthens the correlations and also further indicates the intracratonic nature of the basin. The entire Aravalli-Bijawar-Sausar-Delhi assemblage might then have been deformed in a Satpura-Delhi orogeny.

LATE PROTEROZOIC BASINS

The Bhandara craton contains two major and six minor, detached, Proterozoic basins containing unmetamorphosed supracrustal rocks (Fig. 6.1). The major basins are the Chattisgarh, containing the Chattisgarh Supergroup, and the Bastar basin, containing the Indravati Group.

The Chattisgarh basin has a saucer shape, covers about 36,000 sq. km, and contains conglomerates, orthoquartzites, and other sandstones, shales, limestones, cherts, and dolomites (King, 1885; Dutt, 1964; Schnitzer, 1971, 1977). The aggregate thickness of the sedimentary sequence is more than 1,500 m. Early studies of the basin divided the sedimentary sequence into a lower, Chandarpur series and an upper, Raipur series. A stratigraphy proposed by Dutt (1964) is shown in Table 6.5, with the entire sequence being regarded as the Raipur series of five formations. Schnitzer (1971) later readopted the older terminology, naming the entire sequence the Chattisgarh Supergroup, containing a lower Chandarpur Group and an upper Raipur Group of five different cy-

cles. Kreuzer et al. (1977) found K/Ar ages of 700 to 750 m.y. for glauconites of the Chandarpur suite.

The Chattisgarh basin contains typical platform sediments. Dolomitization was common (Adyalkar and Dube, 1978). Authigenesis occurred under shallow-water conditions (Srivastava and Schnitzer, 1976). Phosphates occur in both the Chandarpur Group and Charmuria Formation of the Raipur Group (Adyalkar et al., 1975). Stromatolites occur in the limestones of the Raipur Group.

The supracrustal rocks of the Bastar basin (Table 6.6) were designated the Indravati Group by Dutt (1964) and are lithologically similar to those of the Chattisgarh basin. The Bastar suite has been tentatively correlated with the Kurnool Group of the Cuddapah basin. Many geologists have presumed that the isolated Late Proterozoic basins of the Bhandara craton were formerly connected as one basin that has since been partly removed by erosion (Ahmad, 1958), but the validity of this conclusion is unknown.

MINERAL DEPOSITS

The principal mineral deposits of the Bhandara craton include manganese in the Sausar Group, iron in the Bailadila Group, and copper in the Malanjkhand granite near Balaghat (Muktinath et al., 1979; Sretharam, 1979). Other economic materials include tin and kyanite. Base metals have been described by Subrahmanyam and De (1979).

Manganese

The manganese ore deposits in the Sausar Group have been studied extensively (Narayanaswamy et al., 1963, and references cited therein; a series of papers by Supriya Roy, including Roy, 1973, plus other references in Roy, 1981; and papers by Fuchs, 1970, 1971). As discussed previously, the

TABLE 6.5. *Stratigraphy of Chattisgarh basin (Dutt, 1964)*

Stage	Description
Raipur (450m)	Greenish gray, shaly limestone, fine grained in the lower part and purple in the upper (seen in the area around Raipur and at shallow depth in wells at Bhilai)
Khairagarh (Variable)	Current bedded subarkose with 10–15% feldspar. Outcrop is arcuate
Gunderdehi (180m)	Splintery calcareous shale, with thin sandstone laminae near the top
Charmuria (300m)	Gray, fine-grained, thin-bedded limestones; becomes shaly towards the top. The Mahanadi follows the junction of this with the lower sandstone between Dhamtari and Mohdi
Chandarpur (300m)	Medium feldspathic sandstone or subarkose with conglomerate at the base. Shale layers in the upper part

TABLE 6.6. *Stratigraphy of Indravati Group in Bastar basin (Dutt, 1964)*

Jagdalpur Formation (200 to 250 m)	Upper purple shale with quartzite intercalations
	Purple shale with interlaminations of purple limestone
	Greenish gray and purple cryptocrystalline limestone
	Banded purple shale with intercalations of quartzites and limestones
Kanger Formation (0 to 140 m)	Gray laminated limestone
Tirathgarh Formation	Purple shale with platy quartzite intercalations
	Subarkose and oothoquartzite
	Basal conglomerate

--Unconformity--

Sausar Group is a miogeosynclinal orthoquartz-ite-shale-carbonate sequence. Igneous rocks are virtually absent except for minor occurrences of late- and posttectonic granites. The manganese deposits are stratigraphically confined to pelitic rocks in the Mansar Formation and, on a more minor scale, in the Lohangi Formation and the Tirodi gneisses that adjoin the Sausar belt. The pelitic rocks of the Mansar Formation have been metamorphosed from greenschist to almandine amphibolite facies in different parts of the belt.

The manganese formations are individual beds of manganese oxides and silicates; the silicates are commonly referred to as gondites. The gondites and oxide ores are intimately interbedded among themselves and with the enclosing pelitic meta-sediments. The oxide ores show sedimentary bedding, are concordant with surrounding sediments (Roy, 1973), and have been co-folded at various scales with the adjacent formations (Agarwal, 1974). These manganese-rich rocks were apparently deposited as manganiferous sediments in a platformal, nonvolcanogenic environment.

Roy (1981) discussed the stability ranges and parageneses of manganese minerals at different grades of metamorphism. Generally, low-temperature manganiferous gels produced pyrolusite, cryptomelane, manganite, and coronadite. Braunite was the first mineral to form at greenschist facies and is stable at all metamorphic grades. Bixbyite commonly followed or accompanied braunite and occurs in the garnet zone and higher facies. Jacobsite and vredenburgite first appeared in the staurolite zone and continued into the sillimanite zone. Hausmanite is characteristic of sillimanite-grade metamorphism. Hollandite and pyrolusite formed during waning stages of the metamorphic sequence.

Iron

Large iron deposits occur in the Bailadila Iron Ore Group of the Bailadila Range (Fig. 6.3; Chatterjee, 1964, 1969). The major deposits occur near the top of the range, extending in a north-south direction for about 45 km. They are directly un-derlain by ferruginous slates and, farther down in the section, by sericitic-feldspathic quartzites and conglomerates. Metabasalts overlap the quartzites in some locations.

Two types of iron formations are present: coarse and wavy-banded hematite-chert rocks, and finely and evenly laminated martite-magnetite-chert rocks. The deposits range from thin lenses to large bodies extending up to 2 to 5 km long and a few hundred meters wide. Their long dimensions are generally parallel to bedding planes. Oolites are present in laminated deposits; the individual ooids are composed of concentric rings of hematite with small cores of quartz. Other features include colloform banding and automorphic crystallization of specular hematite (Chatterjee, 1964). The deposits obviously are bedded sediments.

REFERENCES

Adyalkar, P. G., and Dube, V. N. (1978). Dolomitization in the northern part of the Chhattisgarh basin, Bilaspur District, Madhya Pradesh. Geol. Soc. India J., v. 19, 69–73.
———, Phadtre, P. N., and Ramanna, K. (1975). On the occurrence of phosphatic limestone in Chattisgarh basin of eastern Madhya Pradesh. Geol. Soc. India J., v. 16, 494–495.
Agarwal, V. N. (1974). Tectonics of the manganese ore bodies associated with Sausar Group, Nagpur District, Maharashtra, India. Ind. Natl. Sci. Acad. Proc., Part A, v. 40, 101–111.
——— (1975). Fold interference patterns in the Sausar Group, northern Nagpur District, Maharashtra, India. Geol. Soc. India J., v. 16, 176–187.
——— (1976). Tectonics of the marble rocks of Jabalpur, Madhya Pradesh. Geol. Soc. India J., v. 17, 194–200.
——— (1977). Metamorphic history of the Precambrian rocks of the Narmada section, southwest of Jabalpur, M.P. Geol. Soc. India J., v. 18, 493–499.
Ahmad, F. (1958). Paleogeography of central

India. Geol. Surv. India Records, v. 87, 530–548.

Basu, K. K., and Sarkar, S.N. (1966). Stratigraphy and structure of the Sausar series in the Mahuli-Ramtek-Junawani area, Nagpur District, Maharashtra. Geol., Min. Metall. Soc. India Quart. J., v. 38, 77–105.

Bhaskara Rao, B., and Ananthapadmanabhan, P. (1970). Petrology of the calc-gneisses and marbles of the Sausar Group in Khumari-Mogra area, Nagpur District, Maharashtra. Geol., Min. Metall. Soc. India Quart. J., v. 42, 117–126.

Chatterjee, A. (1964). Geology, mineralogy and genesis of iron ores of some deposits of the Bailadila Range, Bastar District, M. P. Geol., Min. Metall. Soc. India Quart. J., v. 36, no. 2, 57–72.

—— (1969). Mineralogy and stability relations of magnesio-riebeckite amphiboles from metamorphosed iron formation of Bailadila, Bastar District, Madhya Pradesh. Geol., Min. Metall. Soc. India Quart. J., v. 41, no. 1, 25–36.

—— (1970). Structure, tectonics and metamorphism in a part of South Bastar (M. P.). Geol., Min. Metall. Soc. India Quart. J., v. 42, no. 2, 75–96.

Crookshank, H. (1936). Geology of the northern slopes of the Satpuras between the Morand and Sher rivers. Geol. Surv. India Mem. 71, Part 2, 173–381.

—— (1963). Geology of southern Bastar and Jeypore from the Bailadila Range to the Eastern Ghats. Geol. Surv. India Mem. 87, 150 p.

Dutt, N. V. B. S. (1964). Suggested succession of the Purana formations of Chattisgarh. Geol. Surv. India Records, v. 93, Part 2, 143–148.

Dutta, S. M., Mishra, V. P., Dutta, N. K., and Pandhare, S. A. (1979). Precambrian geology of a part of Narainpur and Kondagaon Tahsils, Bastar District, with special reference to Rowghat iron ore deposits. Geol. Surv. India Spec. Publ. 3, 55–67.

Fermor, L. L. (1909). The manganese ore deposits of India. Geol. Surv. India Mem. 37, 235–364.

—— (1936). An attempt at the correlation of the ancient schistose formations of peninsular India. Geol. Surv. India Mem. 70, Part 1, 1–217.

Fuchs, H. D. (1970). Stratigraphische und petrographische Untersuchungen im Kristallinen des zentralindischen Manganerz-Gurtels; das Gebiet des Bawanthari Flusses. Neues Jahrb. Geol. Paleontol. Abh., v. 136, 262–302.

—— (1971). Tektonische Untersuchungen in dem zentralindischen Manganerz-Gurtel des zudlichen Bawanthari Fluss Gebietes. Geol. Rundschau, v. 60, 569–588.

Ghose, N. C., and Gupta, S. D. (1980). Chemistry of the Precambrian Madan Mahal granites, Jabalpur, India. Recent. Res. Geol., v. 5, 276–287.

Ghosh, P. K., Prasad, U., and Banerjee, A. K. (1977). Geology and mineral occurrences of part of the Abuj Mar area, Bastar District, Madhya Pradesh. Geol. Surv. India Records, v. 108, Part 2, 182–188.

Holmes, A. (1955). Dating the Precambrians of peninsular India and Ceylon. Geol. Assoc. Canada Proc., v. 7, 81–106.

Jafri, S. H. (1981). Geochemistry of volcanic rocks from a part of Dongargarh Supergroup, east of Bhandara triangle, central India. Unpublished Ph. D. Thesis, Aligarh Muslim Univ., Aligarh.

Khanna, V. K. (1979). Geochemistry of basic rocks in the Ghamapur-Lalmati area of Jabalpur, Madhya Pradesh. Geol. Surv. India Spec. Publ. 3, 155–174.

King, W. (1885). Sketch of the geological work in the Chattisgarh division of the Central Provinces. Geol. Surv. India Records, v. 18, 169–200.

Kreuzer, H., Harre, W., and Kuersten, M. (1977). K/Ar dates of two glauconites from the Chandarpur series (Chhattisgarh/India); On the stratigraphic status of the late Precambrian basins in central India. Geologisches Jahrb., Ser. B, no. 28, 23–36.

Mukharya, I. L. (1975). Metamorphism and petrogenesis of the Dalli-Rajhara and Ari Dongri iron formations, Madhya Pradesh. Geol. Soc. India J., v. 16, 441–449.

Muktinath, Mathur, S. M., Sharma, R. S., Neelakantam, S., Narang, J. L., and Sonakia, A. (1979). Geology and mineralization in Malanjkhand area, Balaghat District, M. P. Geol. Surv. India Spec. Publ. 3, 203–207.

Narayanaswamy, S., and Venkatesh, V. (1973). The geology and manganese ore deposits of the manganese belt in Madhya Pradesh and adjoining parts of Maharashtra; Part IV. The geology and manganese deposits of northern Bhandara District, Maharashtra. Geol. Surv. India Bull., Ser. A, no. 22, Part 4, 183 p.

——, Chakravarty, S. C., Vemban, N. A., Shukla, K. D., Subramanyam, M. R., Venkatesh, V., Rao, G. V., Anandalwar, M. A., and Nagarajaiah, R. A. (1963). The geology and manganese ore deposits of the manganese belt in Madhya Pradesh and adjoining parts of Maharashtra; Part I. General introduction. Geol. Surv. India Bull., Ser. A, no. 22, Part I, 69 p.

Radhakrishna, B. P., and Naqvi, S. M. (1986). Precambrian continental crust of India and its evolution. J. Geol., v. 94, 145–166.

Rajarajan, K. (1976). Epizonal granitic complex in Bhandara and Chanda Districts, Maharashtra. Geol. Surv. India Misc. Publ., 23, Part 2, 347–363.

Rao, G. V. (1979). The correlation of the Dongargarh, Chilpi and Sausar Groups. Geol. Surv. India Spec. Publ. 3, 9–15.

Roy, S. (1973). Genetic studies on the Precambrian manganese formations of India with particular reference to the effects of metamorphism. In Genesis of Precambrian Iron and Manganese Deposits, UNESCO Earth Sci. Series, v. 9, 229–242.

——— (1981). Manganese Deposits. Academic Press, London, 458 p.

Sarkar, S. N. (1957). Stratigraphy and tectonics of the Dongargarh system, a new system in the Precambrian of Bhandara-Drug-Balaghat area, Bombay and Madhya Pradesh. Ind. Inst. Tech. J. Sci. Engineering Res., Kharagpur, v. 1, 237–268.

——— (1983). Present status of Precambrian stratigraphy and geochronology of peninsular India—A synopsis. Ind. J. Earth Sci., v. 10, 104–106.

———, Gautam, K. V. V. S., and Roy, S. (1977). Structural analysis of a part of the Sausar Group rocks in Chikla, Sitekere area, Bhandara District, Maharashtra. Geol. Soc. India J., v. 18, 627–643.

———, Gopalan, K., and Trivedi, J. R. (1981). New data on the geochronology of the Precambrians of Bhandara-Drug, central India. Ind. J. Earth Sci., v. 8, 131–151.

———, Trivedi, J. R., and Gopalan, K. (1986). Rb-Sr whole-rock and mineral isochron age of the Tirodi Gneiss, Sausar Group, Bhandara District, Maharashtra. Geol. Soc. India J., v. 27, 30–37.

Schnitzer, W. A. (1971). Die jungprakambrium Indiens ("Purana System"); Neugliederung, Stromatolithenfuhrung und lithofazielle Vergleich. Erlanger Geol. Abh., No. 85, 44 p.

——— (1977). Distribution of stromatolites and stromatolitic reefs in the Precambrian of India. In Fossil Algae; Recent Results and Developments (ed. E. Fluegel). Springer Verlag, Berlin, 101–106.

Sengupta, A. (1965). Some aspects of metamorphism of Sakoli series around Gangajhiri, Bhandara District, Maharashtra. Geol. Soc. India J., v. 6, 1–17.

Shukla, K. D., and Anandalwar, M. A. (1973). The geology and manganese ore deposits of the manganese belt in Madhya Pradesh and adjoining parts of Maharashtra; Part VII, The geology and manganese ore deposits of the Balaghat-Ukwa area, Balaghat District, Madhya Pradesh. Geol. Surv. India Bull., Ser. A, no. 22, Part 7, 65 p.

Sretharam, R. (1979). Mineragraphic studies of the copper ore from Malanjkhand deposit, Balaghat District, M. P. Geol. Surv. India Spec. Publ. 3, 141–151.

Srivastava, N. K., and Schnitzer, W. A. (1976). Authigenic minerals in the Precambrian sedimentary rocks of the Chattisgarh System (central India). Neues Jahrb. Mineral. Abh., v. 126, 221–230.

Straczek, J. A., Narayanaswami, S., Subramanyam, M. P., Shukla, K. D., Vemban, N. A., Chakravarty, S. C., and Venkatesh, V. (1956). Manganese ore deposits of Madhya Pradesh, India. 20th Internat. Geol. Cong. Rept., Mexico, Symposium on Manganese, Part IV, 63–96.

Subrahmanyam, B., and De, S. (1979). On some results of geophysical investigations for base metals over the Archaeans of central India. Geol. Surv. India Spec. Publ. 3, 273–277.

7. SINGHBHUM CRATON

The Singhbhum craton is bounded by the Mahanadi graben (and Sukinda thrust), Narmada-Son lineament, the Indo-Gangetic plain, and the east coast of India (Fig. 7.1). To the northeast, the Precambrian rocks of the craton are covered by the Garo-Rajmahal plateau basalts but may continue under the alluvium in Bangladesh and connect with Precambrian suites in the Assam area (Megahalaya and Shillong plateau; Figs. 1.1 and 8.1). A. N. Sarkar (1982a, 1982b) recognized three geologic provinces in the craton: (1) a Singhbhum-Iron Ore-Orissa nucleus south of the Singhbum thrust zone, which we designate the Singhbhum nucleus; (2) the Singhbhum-Dhalbhum mobile belt, north of the Singhbum thrust and partly bordered on the north by the Dalma thrust; and (3) the Chotanagpur-Satpura belt of gneisses and granites north and northwest of the mobile belt, which we refer to simply as the Chotanagpur terrain (Fig. 7.1).

The Singhbhum craton is the most intensely studied area of the Indian Precambrian, and almost all aspects of the geology are heatedly debated. The nature of the Singhbhum thrust is a typical example. S. N. Sarkar and Saha (1962) proposed that the Iron Ore Series south of the thrust is older than similar rocks north of the thrust and referred to these northern rocks as the Singhbhum Group. Other workers, however, found no clear structural hiatus between the Iron Ore and Singhbhum Groups (Iyengar and Alwar, 1965; D. Mukhopadhyay et al., 1975, 1980; Bose and Chakraborty, 1981; Iyengar and Murthy, 1982). Another example is the question of whether the Singhbhum Granite is completely intrusive into the Iron Ore Group (S. N. Sarkar and Saha, 1962, 1983) or whether parts of the Singhbhum Granite should be considered as a basement for a younger Iron Ore Supergroup (Iyengar and Murthy, 1982). Thus, various published maps and stratigraphic sections are greatly different from each other in almost all parts of the Singhbhum craton. A very generalized classification of lithologic suites is shown in Table 7.1.

The lithology of the Iron Ore Group is identical to that of the Bababudan Group of the Western Dharwar craton (Ch. 2). Both suites appear to have developed on tonalitic basement about 3,000 m.y. ago. Lithologies similar to those of the Iron Ore Group also occur in the Jeypore-Bastar area of the Bhandara craton (Ch. 6).

Radiometrically determined isochron ages are sparse for rocks of the Singhbhum craton. The principal ones are listed as follows, together with comments on conflicting data.

The Older Metamorphic Group (OMG) (Fig. 7.2) has been dated by both Rb/Sr and Sm/Nd methods. S. N. Sarkar et al. (1979) obtained a Rb/Sr age of 3,200 ± 85 m.y. with an initial $^{87}Sr/^{86}Sr$ ratio of 0.7018 ± 0.0003. Possibly correlative rocks yielded a Rb/Sr age of 3,180 ± 85 m.y. with an initial $^{87}Sr/^{86}Sr$ ratio of 0.703 ± 0.0003. A. R. Basu et al. (1981) obtained a Sm/Nd age of about 3,775 ± 89 m.y.; the isochron, however, is based on samples from different outcrop areas, which makes it essentially a two-point graph. Baksi et al. (1984) used $^{39}Ar/^{40}Ar$ techniques to demonstrate that rocks of the OMG were last heated about 3,400 m.y. ago.

The Singhbhum Granite has been dated at two places by Rb/Sr methods (S. N. Sarkar et al., 1979). A four-point isochron from a border with the OMG gneisses yielded an age of 2,910 ± 250 m.y. with an initial $^{87}Sr/^{86}Sr$ ratio of 0.7023 ± 0.0025. A five-point isochron for another set of samples yielded an age of 2,950 ± 200 m.y. with an initial ratio of 0.711 ± 0.009. The errors on these ages are large.

The Simlipal complex has been dated by Rb/Sr methods as 2,084 ± 70 m.y. old with an initial $^{87}Sr/^{86}Sr$ ratio of 0.745 (Iyengar et al., 1981a).

Uranium mineralization in the Singhbhum thrust zone has been dated as about 1,600 m.y. old by Pb isotopic methods (Krishna Rao et al., 1979).

The Newer Dolerites have been dated as ranging from about 1,500 to 1,000 m.y. old by K/Ar methods, involving total, rather than incremental, Ar emission (S. N. Sarkar and Saha, 1983).

Other ages of events and rock suites have been obtained by a variety of methods, and we can summarize age relationships in the Singhbhum craton as follows. The Singhbhum nucleus in the southern part of the craton is apparently the oldest area. Activity probably extended from early Archean until about 2,700 m.y. ago. The oldest event in the Singhbhum-Dhalbhum mobile belt is approximated as about 2,200 m.y. ago. The age of

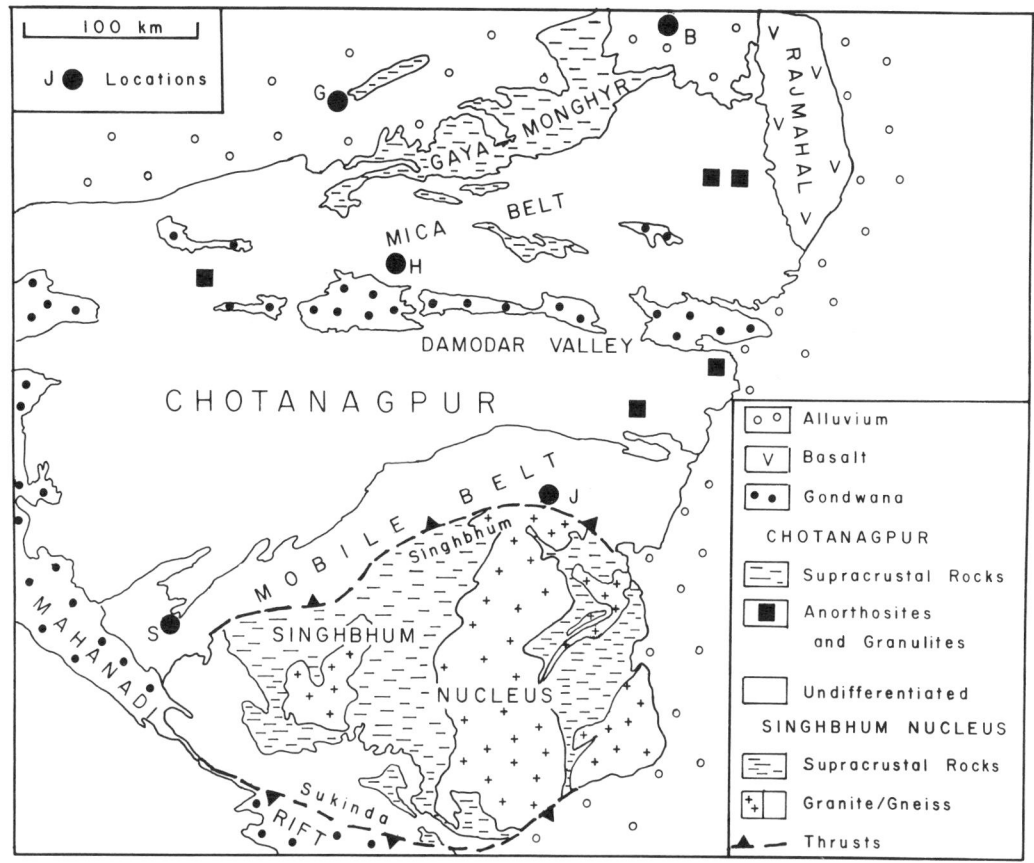

Fig. 7.1. Singhbhum craton. G, Gaya; B, Bhagalpur; H, Hazaribagh; J, Jamshedpur; S, Sundargarh.

TABLE 7.1. *General stratigraphy of Singhbhum craton*

Age(in b.y.)	Singhbhum Nucleus	Singhbhum-Dhalbhum Mobile Belt	Chotanagpur Belt
0.9–1.6	Newer dolerites		Syn- to late- and posttectonic granites/gneisses
1.5	Kolhan Group		
1.6	Mayurbhanj Granite; gabbro/ anorthosite	Chakradharpur granite; gabbro/anorthosite	Gabbro/anorthosite
	Ultramafic intrusions		
	------------ Unconformity --------------		
	Dhanjori Group	Dalma lavas	
	------------ Unconformity --------------		
	Singhbhum Group (Chaibasa formation)	Singhbhum Group	
	------------ Unconformity --------------		
2.9	Singhbhum Granite; Iron Ore Group		Orthogneisses
	------------ Unconformity --------------		
3.8	Older Metamorphic Group gneisses		
	Older Metamorphic Group supracrustal rocks		
	Basement?	Basement?	Basement?

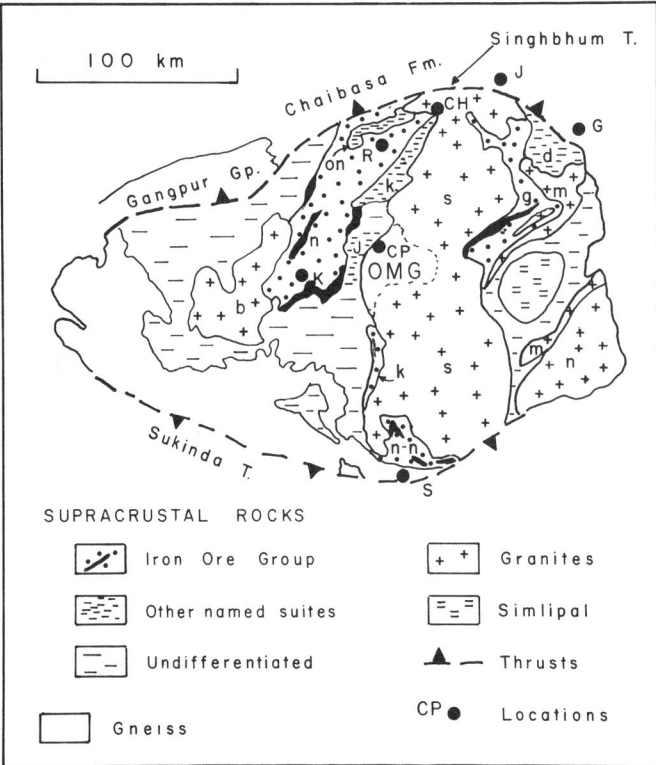

Fig. 7.2. Singhbhum nucleus. Locations include J, Jamshedpur (and Tatanagar); G, Ghatsila; CH, Chaibasa; R, Roro igneous suite; CP, Champua; K, Koira; S, Sukinda ultramafic suite. OMG is the major area of enclaves of Older Metamorphic Group in the Singhbhum Granite. Singhbhum and Sukinda thrusts are located at the northern and southern margins of the nucleus. The Gangpur Group and Chaibasa Formation of the mobile belt are shown north of the Singhbhum thrust.

Basins of the Iron Ore Group include n, Naumandi (Noamundi); g, Gorumahisani; k, Keonjhar; n-n, Nausahi-Nilgiri. Major bands of iron formation are shown in black.

Designated areas of other supracrustal rocks include on, Ongarbira; k, Kolhan; j, Jagannathpura.

Areas of granite include s, Singhbhum; m, Mayurbhanj; b, Bonai; n, Nilgiri.

This map is generalized from the very different interpretations of A. K. Banerji (1977), Mazumdar (1978), Iyengar and Murthy (1982) and S. N. Sarkar and Saha (1983).

the Chotanagpur suite is not well constrained but may include events ranging from late Archean until the Phanerozoic (A. N. Sarkar, 1982a). The age of the orogenic activity northwest of the Singhbhum-Dhalbhum mobile zone is also unclear; it is commonly regarded as Satpura age, about 950 m.y. ago (Holmes, 1955), although the Satpura orogeny actually may be significantly older (Ch. 6).

MAJOR STRUCTURES

The Singhbhum craton contains three major thrust belts: the Dalma thrust in the north of the Singhbhum-Dhalbhum mobile belt, the Singhbhum thrust between the mobile belt and the Singhbhum nucleus, and the Sukinda thrust, on the southern margin of the craton (Fig. 7.1). In ad-

dition to these thrust zones, a major rift/graben system occurs along the Damodar valley in the Chotanagpur gneissic terrain. The eastern end of the Damodar graben is buried by alluvium, and the western end intersects the western extension of the Mahanadi rift system.

The most spectacular feature of the Singhbhum craton is a 200-km-long arcuate thrust zone that is referred to as the Singhbhum thrust, Singhbhum shear zone, or Copper Belt thrust. In the west, the zone is more than 25 km wide and contains at least three major thrust slices. These slices are compressed into a narrow zone of 1-km width in the center of the thrust belt, but the zone widens to about 5 km at its eastern end (A. N. Sarkar, 1982a). Verma et al. (1984) regarded the Singhbhum thrust as a deep feature penetrating into the mantle. The thrust zone contains mafic and ultramafic rocks. The presence of glaucophane schists has been reported (S. N. Sarkar and Saha, 1977, 1983) but not confirmed.

S. N. Sarkar and Saha (1977, 1983) regarded the Singhbhum thrust zone as separating a southern terrain stabilized about 2,900 m.y. ago from a northern province containing younger rocks. They explained the Singhbhum thrust as the expression of an intraplate thrust zone that was intermittently active from about 2,300 m.y. ago until the Satpura orogeny. Instead of an intraplate model, A. N. Sarkar (1982a, 1982b) proposed that the Singhbhum thrust was formed by the collision of old terrains of the Singhbhum and Chotanagpur microplates with subduction of oceanic lithosphere.

Both the Singhbhum and Dalma thrusts have supracrustal rocks on both sides of their exposed traces (Fig. 7.1). For the major Singhbhum thrust, there are three possible interpretations: rocks south of the thrust are part of the Singhbhum nucleus, either older or younger than the Singhbhum Granite, and are wholly separate from rocks north of the thrust; rocks north and south of the thrust are all part of the same prethrust assemblage, with only limited movement on the thrust; and rocks south of the thrust are foreland molasse shed from the highlands in the mobile belt, and the thrust advanced over its own molasse.

The southward-dipping Sukinda thrust, on the southern edge of the Singhbhum craton, forms a major join between granulites of the Eastern Ghats belt and lower-grade rocks of the Iron Ore Group. The Sukinda thrust is north of the Mahanadi rift zone in the east and forms the northern border of the rift toward the west. The thrust dips steeply to the south.

The Singhbhum craton shows broad isostatic compensation. Local Bouguer gravity highs occur over relatively mafic supracrustal rocks and lows over granite/gneiss terrains (Verma et al., 1984). High heat flow values occur in the Singhbhum thrust zone (Rao and Rao, 1974; M. L. Gupta, 1982). These values are consistent with reported high concentrations of radioactive elements (Rao and Rao, 1974; Rao et al., 1976).

OLDER METAMORPHIC GROUP (OMG)

A major part of the Singhbhum nucleus is occupied by the Singhbhum Granite complex, which covers an area of about 10,000 sq. km (Fig. 7.2). Relics of the OMG occur within the batholithic complex. The type area of the OMG is west of Champua (Fig. 7.2), where the suite consists of medium-grained mica schists, quartzites, calc-silicate rocks, and para- and orthoamphibolites (S. N. Sarkar and Saha, 1977, 1983). These supracrustal rocks also contain gneisses, either as part of the OMG or as early phases of the Singhbhum

Granite suite. The gneisses of the OMG enclaves have been dated as 3,800 m.y. old by Sm/Nd techniques (A. R. Basu et al., 1981) and 3,200 m.y. old by Rb/Sr techniques (S. N. Sarkar et al., 1979), but both dates are questionable.

Relationships of the OMG with other rock suites are unclear. S. N. Sarkar and Saha (1983) regarded the OMG as unconformably overlain by the Iron Ore Group (IOG). These suites are not in contact, however, except along one roof pendant within the Singhbhum Granite, where contact relationships are not clear. The wide distribution of OMG supracrustal and gneissic enclaves in the Singhbhum Granite complex indicates that the OMG must have occupied a large part of the area now invaded by granite. This widespread unit may have been a basal part of the supracrustal suite now recognized as the IOG. Iyengar and Murthy (1982) grouped the Gorumahisahni Group of the IOG (east of the Singhbhum Granite) with the OMG as the Badampahar Group, underlying their Dhanjori Group (Fig. 7.2). They regarded gneisses in the OMG as older phases of the Singhbhum Granite.

Structural relationships among OMG supracrustal rocks, apparently older gneissic enclaves, and the Singhbhum Granite have been investigated in several areas around Champua (Fig. 7.2). S. N. Sarkar and Saha (1977, 1983) described a large, near-vertical, appressed, megascopic fold of bedding surfaces in paraamphibolites, mica schists, and thin bands of fuchsitic and garnetiferous quartzites. Amphibolite-facies metamorphism occurred during the development of schistosity. A second generation of mesoscopic, moderately appressed folds occurs locally. Bedding foliation and first-generation mineral lineation in the OMG supracrustal rocks are continuous with foliation and lineation in adjacent biotitic gneisses.

Supracrustal, including amphibolitic, rocks of the OMG locally show migmatization and in situ granitization adjacent to gneisses. Complete gradation occurs along some contacts from amphibolites to tonalitic and granodioritic gneisses, suggesting that the gneisses may be partly metasomatic. S. N. Sarkar and Saha (1983) reported numerous examples of structural discordance between well-foliated tonalitic gneisses and younger, more granitic phases of the Singhbhum Granite suite. Thus, there may have been a major deformational event between original formation of the OMG supracrustal/gneiss suite and emplacement of the major part of the Singhbhum Granite.

The OMG tonalite/trondhjemite gneisses consist of plagioclase (An_{20-40}), quartz, biotite and minor hornblende (Table 7.2). The major-element composition could have been the result of partial melting of amphibolites (Saha and Ray,

TABLE 7.2. *Compositions of Older Metamorphic Group*

	1	2	3	4
SiO_2	67.2	68.1	52.5	53.6
TiO_2	0.08	0.10	1.3	0.16
Al_2O_3	15.7	15.7	14.0	14.6
Fe_2O_3	1.5	1.4	2.1	2.1
FeO	3.0	2.7	9.0	8.2
MnO	0.12	0.10	0.19	0.45
MgO	1.2	1.1	8.1	6.6
CaO	3.4	3.8	9.1	9.7
Na_2O	5.3	5.2	2.7	3.5
K_2O	1.5	1.6	0.92	0.85

All abundances of major elements are in weight percent of oxides. Where no entry is made in the FeO position, the value at Fe_2O_3 is total iron calculated as ferric oxide. All abundances of trace elements are in parts per million by weight. Values are taken directly from quoted sources and not recalculated to 100%.

Table 7.2. Column References
1. Biotite gneiss (Saha, 1979).
2. Tonalite gneiss (Saha and Ray, 1984).
3. Orthoamphibolite (Saha and Ray, 1984).
4. Paraamphibolite (Saha and Ray, 1984).

1984). Trace element characteristics include highly variable K/Rb ratios (100 to 500), high Sr contents, low Th and U contents, and Ba/Sr ratios near 1.0 (Ray et al., 1980). Saha and Ray (1984) found moderate light REE enrichment, minor heavy REE enrichment, $La_N/Yb_N = 26.7$, and a gently sloping REE pattern without Eu anomaly (Fig. 7.3).

The hornblendic schists and amphibolites of the OMG consist of hornblende and plagioclase

with minor iron ore minerals, epidote, sphene, and apatite (Table 7.2). No komatiites are present, and the low K_2O contents and other features suggest a generally tholeiitic composition. Their REE patterns (Fig. 7.3) indicate weak fractionation in light REE and minor negative Eu anomaly. Enrichment in lithophilic elements is suggested by high Rb/Sr and Ba/Sr ratios and a low K/Rb ratio (Table 7.2). Saha and Ray (1984) proposed an origin by 20% to 30% partial melting of low-K basalts or 5% to 15% partial melting of LIL-enriched peridotite.

SUMMARY OF COHERENT SUPRACRUSTAL SUITES SOUTH OF CHOTANAGPUR TERRAIN

Schistose rocks apparently formed from preexisting volcanic and sedimentary accumulations are widely distributed in basins around the Singhbhum Granite (Fig. 7.2). Some geologists have correlated presumed older members of these suites with OMG enclaves in the granite massif. No radiometric dates are available on any of the supracrustal rock suites, and this fact, plus the isolation of the various basins, renders stratigraphic correlation virtually impossible. The major basins are Gorumahisani, Nausahi (Nausahi-Nilgiri), Keonjhar, Naumandi (Noamundi), Dhanjori, Simlipal, Kolhan, and Gangpur. An area on the northern side of the Singhbhum Granite, between the granite and the Singhbhum thrust zone, contains supracrustal rocks of uncertain correlation.

According to the summary by S. N. Sarkar and Saha (1983), the Iron Ore Group (IOG) is regarded as the oldest coherent supracrustal suite in the area. It occurs in the Gorumahisani, Nausahi,

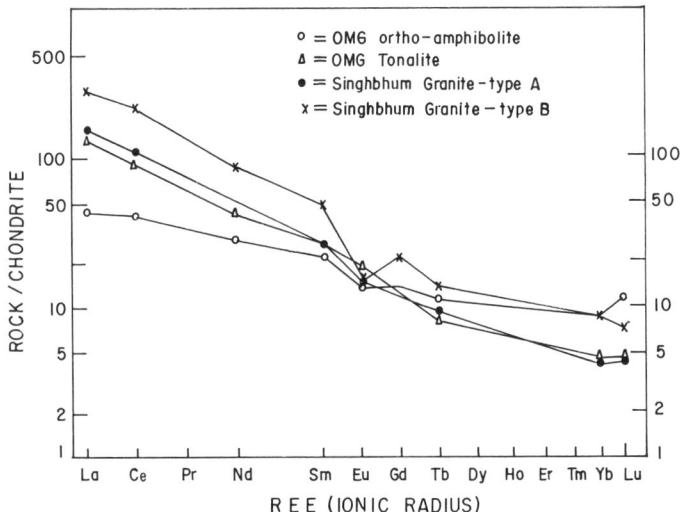

Fig. 7.3. REE patterns in Older Metamorphic Group and associated rocks (Saha and Ray, 1984).

and Naumundi basins. In addition, metasedimentary rocks on the northern side of the massif and south of the Singhbhum shear zone have been correlated with the IOG by some workers. The term "Chaibasa Formation" has been used to designate metasedimentary rocks presumed to be in the lower part of the Singhbhum-Dhalbhum mobile belt, particularly south of the shear zone, and is thus possibly correlative with the IOG. Metasedimentary rocks in isolated outcrops south of the granite massif, and extending into the Sukinda valley, have also been referred to as IOG.

Several complications arise with respect to the term "Iron Ore Group." Some workers (e.g., Iyengar and Alwar, 1965; Iyengar and Murthy, 1982) separated the lower sequence in the Gorumahisani basin as the Badampahar Group and regarded it as at least partly correlative with the Older Metamorphic Group enclaves in the Singhbhum Granite. In addition, they considered the entire overlying supracrustal sequence around the Singhbhum Granite and in the Singhbhum mobile belt to be one suite, to which the term "Dhanjori" has been applied. Many other stratigraphic terms are used for groups within the Iron Ore Supergroup.

The Dhanjori suite is a sequence of dominantly metavolcanic rocks in the Dhanjori basin (Fig. 7.2). Some writers (e.g., Iyengar and Alwar, 1965) have attempted to correlate the Dhanjori volcanic rocks with the igneous suite of the Simlipal complex. The Simlipal complex, however, shows many characteristics of a caldera collapse or cryptovolcanic structure, possibly even an astrobleme. Another possible correlative with the Dhanjori suite is a sequence of metavolcanic rocks on the western side of the Singhbhum Granite that has been referred to as the Jagannathpura lavas, which occurs in a broad area south and east of the Naumandi basin.

Four principal stratigraphic terms have been used for rock suites within the Singhbhum-Dhalbhum mobile belt. The major part of the sedimentary sequence has commonly been subdivided into two formations. The lower one is the Chaibasa Formation, consisting of mica schists and quartzites, and the upper one is the Dhalbhum Formation, consisting of sericitic schists and phyllites with some unusual "carbon phyllites." The Dalma volcanic suite occurs in the mobile belt. The fourth suite in the mobile belt is the Gangpur Formation, which occurs in an anticlinal or overturned synclinal dome in the western part of the belt. The Gangpur Formation is lithologically distinct from other supracrustal rocks in the area in its content of manganese silicate rocks (gondites).

The Kolhan supracrustal suite, on the northwestern side of the massif, is a sequence of compositionally mature quartzites that have undergone very little deformation. It appears to be younger than other supracrustal suites and has no direct correlative elsewhere.

IRON ORE GROUP (IOG)

S. N. Sarkar and Saha (1962, 1977) regarded the Naumandi (Noamundi) basin on the western side of the massif (Fig. 7.2; Table 7.3) as the type area of the IOG and correlated it with sequences in the Nausahi basin, on the south of the massif, and the Gorumahisani basin, on the eastern side. They regarded the Naumandi basin as a NNE-plunging synclinorium overturned toward the southeast. Murthy (1975a), however, considered the eastern limb to be upright, leading to a different stratigraphic interpretation. The IOG of the Naumandi basin is intruded by the Bonai Granite (Fig. 7.2). The age of the granite is unknown, although it is speculatively regarded as younger than the Singhbhum Granite (Prasada Rao et al., 1964; D. Mukhopadhyay, 1976; Bose and Chakraborty, 1981; Iyengar and Murthy, 1982).

Metasedimentary rocks along the northern border of the Singhbhum nucleus bear a controversial stratigraphic relationship to the type IOG. Bose and Chakraborty (1981) grouped rocks both north and south of the Singhbhum thrust zone as one suite. D. Mukhopadhyay et al. (1975) inferred that the sedimentary assemblage on the northern border of the Singhbhum Granite was deposited unconformably on the massif and correlated these suites with the IOG. S. N. Sarkar and Saha (1983) regarded metasedimentary assemblages in this northern zone as younger than the IOG and correlated them with the Chaibasa Formation within the mobile belt. Dunn and Dey (1942) grouped the IOG and OMG suites and correlated the IOG with the Chaibasa Formation of the Singhbhum-Dhalbhum mobile belt, although the Chaibasa Formation does not contain iron ores.

Detailed structural work has been done on the Malangatoli iron ore deposits of the Naumandi basin. The structure can be interpreted as the result of three periods of folding (A. K. Chatterjee and Mukherji, 1981). The F1 folds are generally inclined to reclined with axes plunging N to NNE at 20° and axial planes dipping steeply westward. These first-generation folds have been deformed by upright and asymmetric open folds (F2) along axes subparallel to the F1 folds and plunging NNE to NE. Third-generation (F3) folds trend east-west and cause a series of canoe folds by intersecting with the F2 structures.

The Ongarbira belt (Fig. 7.2; A. K. Gupta et al., 1981) has uncertain relationships to the IOG and/or Chaibasa Formation. It is exposed in a westward-plunging synclinorium overturned to the south and consists of basalt, quartzite, and related sedimentary rocks overlain by a suite of mafic and

TABLE 7.3. *Stratigraphy on west side of Singhbhum nucleus (S. N. Sarkar and Saha, 1977)*

Koira-Noamundi-Champua Area		Chaibasa-Saraikela-Chakradharpur Area	
Newer dolerites		Chakradharpur Granite Gneiss	
		--------Singhbhum Orogeny (c.1550–850 Ma) --------	
		Newer Dolerites (c.1600–950 Ma)	
		Jojohatu ultrabasic intrusives	
Kolhan	Shale	Kolhan	Shale
Group	Limestone	Group	Limestone
(c.1500–	Sandstone	(c.1500–	Sandstone-
1600 Ma)	conglomerate	1600 Ma)	conglomerate
--------------------------Unconformity--------------------------			
Jagannathpur lava (c.1600–1700 Ma)			
--------------------------Unconformity--------------------------		----------------------- Unconformity-----------------------	
Singhbhum Granite (c.2950 Ma)		Singhbhum Granite (c.2950 Ma)	
------------------------ Iron Ore Orogeny -----------------------		----------------------Iron Ore Orogeny ----------------------	
Iron	Upper shales with volcanics		Phyllites with volcanics
Ore	Banded haematite jasper with	Iron	Quartzite
Group	iron ore	Ore	Phyllite
	Lower shales	Group	Ongarbira lava flow
	Mafic lavas with tuffs		Several bands of orthoquartzites
	Sandstone and conglomerate		with minor arkose and
			conglomerate alternating with phyllite.
--------------------------Unconformity--------------------------		----------------------- Unconformity-----------------------	
Older metamorphic gneiss (c.3200 Ma)		Older metamorphic banded gneisses and	
-----------------Older Metamorphic Orogeny-----------------		amphibolites (related to O.M. orogeny) as relics	
O. M. Group: Banded calc-gneisses, mica schists,		within Singhbhum Granite	
quartz-schists, garnetiferous quartzites			

Prepared with permission from the Indian Journal of Earth Sciences, S. Ray Volume.

ultramafic volcanic rocks, some of which have pillow structures. Gupta et al. regarded the belt as a separate basin, but Iyengar and Murthy (1982) classified it with the Dhanjori Group, partly because of a lack of iron ores.

Correlation of the Naumandi IOG with suites in the Gorumahisani basin, on the east of the Singhbhum massif, is unclear. S. N. Sarkar and Saha (1977) regarded the suites in the Goruma-hisani basin as IOG (Table 7.4), but an older age was assigned by Prasada Rao et al. (1964), Iyengar and Alwar (1965), A. K. Banerji (1975), and Iyengar and Murthy (1982). The only basis for correlation is an approximate lithologic similarity of the Dhanjori volcanic suite, which overlies the Gorumahisani rocks, and the Ragunathpura volcanic suite, designated by Iyengar and Alwar as underlying the IOG of the Naumandi basin.

Rocks presumably correlative with the IOG have been divided into six sequences in the Nausahi basin (Prasada Rao et al., 1964). The base rests on an unconformity above a sequence of volcanic rocks. The first unit overlying the unconformity consists of mica schists, chlorite schists, quartzites, and interbedded hornblende schists and amphibolites that may represent metavolcanic rocks. Calcareous sedimentary rocks are ab-

sent, and the entire sequence resembles that of the IOG. The lower part of the second sequence contains ferruginous clastic sediments (principally shales), and the upper part is highly siliceous, consisting of banded black chert, jasper-bearing rocks, and quartzites. The third sequence is unconformable at its base and consists of conglomerates, sandstones, shales, and slates with mafic volcanic flow rocks and tuffs in the upper part. The fourth sequence consists of shales and banded hematite jasper rocks that lie unconformably over folded rocks of the third sequence and were regarded as typical IOG by Dunn (1940). Granites that have been designated as the Bonai Granite intrude the fourth sequence, but their correlation is uncertain. The fifth sequence consists of clastic rocks. The sixth sequence lies with moderate dips on steeply folded rocks of the fifth sequence and contains clastic rocks with some transition to calcareous schists and siliceous limestones toward the top.

Rocks correlated with the IOG occur at other locations. In the Sukinda valley, north of the Sukinda thrust, the suite consists of quartzites, intraformational conglomerates, banded iron formations, and tuffs. Folds trend roughly parallel to the Sukinda thrust and Mahanadi graben. Chromite-bearing ultramafic rocks (described later) intruded the IOG and are presumably related to the Sukinda thrust. Possible IOG correlatives metamorphosed to greenschist facies also occur in the Keonjhar District, where the stratigraphic sequence is basal metapelitic and volcanic rocks, banded iron formations, upper metapelites and volcanic rocks, and conglomeratic quartzite (Acharya, 1964).

Both iron and manganese are widespread in the IOG (A. K. Banerji, 1977). The iron formations contain abundant sedimentary structures, such as current markings, penecontemporaneous slumps, and banding (Rai et al., 1980; Pandey and Chatterjee, 1984). Initial precipitation of magnetite in a silica gel and later segregation into bands has also been proposed (Majumder and Chakraborty, 1977, 1979; Majumder et al., 1979). Stromatolites and other evidence of organic activity are abundant. Compositions of argillaceous members of the assemblage are consistent with a marine environment of deposition (A. Mukhopadhyay and Chanda, 1972; Murthy, 1975b).

All features of the metasedimentary rocks of the IOG are consistent with deposition in a platformal, cratonic area. It is not clear whether this environment means that the Singhbhum Granite complex existed, at least partly, before deposition of the IOG sediments and formed the basement for the suite. The basement may be either poorly known gneissic terrains south of the Nausahi

basin, with the Singhbhum Granite intrusive into both the older basement and the IOG, or the OMG, which may have been widely distributed before engulfment in the granites.

Four suites of volcanic rocks south of the Singhbhum thrust zone have been associated with various classifications of the IOG (Table 7.5). The Naumandi basin is a synclinorium, with the Bonai volcanic suite on the western limb and the Dongapasi volcanic suite along the eastern limb. The Ongarbira belt occurs in the questionably correlated suites north and northwest of the Singhbhum Granite. The Gorumahisani igneous rocks occur in the Gorumahisani basin.

The Bonai metavolcanic rocks consist of variable amounts of mafic lavas and tuffs with minor silicic volcaniclastic interbeds. Picritic rocks show pillow structures (Bose, 1982). P. K. Banerjee (1982) regarded the suite as island arc basalt. The Dongapasi volcanic rocks consist of olivine- and hypersthene-normative tholeiites (P. K. Banerjee, 1982) and some ultramafic flows (A. K. Gupta et al., 1981), and P. K. Banerjee has interpreted the suite as platformal, with compositions intermediate between those of modern continental plateau basalts and oceanic tholeiites. The Ongarbira suite consists of both mafic and ultramafic rocks. The Gorumahisani suite is partly volcanic but primarily intrusive (P. K. Banerjee, 1982); mafic and ultramafic rocks are both present, some with garnet peridotite inclusions, and P. K. Banerjee

TABLE 7.4. *Stratigraphy on east side of Singhbhum nucleus (S. N. Sarkar and Saha, 1977)*

Newer Dolerite Dykes (c.1600–c.950 Ma)	Amjhor differentiated sill
Mayurbhanj Granite (c.1200 Ma)	Romapahari Aplo-granite Course ferrohastingsite-biotite granite Biotite hornblende microgranite; locally granophyre
---Singhbhum Orogeny---	
Gabbro-norite-anorthosite with pockets of magnetite (c.1400–1500 Ma) and grading to pyroxenite, picrite and peridotite	
Dhanjori Group (Simlipal)	Three lava formations (c.1600–1700 Ma) alternating with two sandstone-quartzite horizons Basal quartzite-conglomerate with black phyllites
---Unconformity---	
Singhbhum Group	Mica schists and orthoquartzites with orthohornblende schists
---Unconformity---	
Singhbhum Granite	Mainly biotite granodiorite (c.2950 Ma), adamellite, also hornblende-epidote granodiorite and diorite
--Iron Ore Orogeny--	
Iron Ore Group	Phyllites and tuffs, chlorite-actinolite schist, epidiorite, banded chert, banded hematite quartzite with hematite ore (> c.3000 Ma)
----------------------------------Older Metamorphic Orogeny----------------------------------	
Older Metamorphic Group	Hornblende schist and amphibolite (c.3200 Ma) mostly granitized

Prepared with permission from the Indian Journal of Earth Sciences, S. Ray Volume.

TABLE 7.5. *Compositions of volcanic rocks in Singhbhum nucleus (P. K. Banerjee, 1982)*

	1	2	3	4
SiO_2	46.1	44.0	42.3	47.4
TiO_2	1.2	0.60	0.06	1.2
Al_2O_3	15.5	10.9	9.8	10.5
Fe_2O_3	5.8	4.4	4.9	21.0
FeO	9.2	8.1	6.0	
MnO	0.03	0.03	0.09	0.20
MgO	5.7	20.0	28.8	8.0
CaO	8.4	6.8	1.2	9.8
Na_2O	3.2	0.24	0.29	0.57
K_2O	0.32	0.07	0.06	1.6
Rb				
Sr	400	10	10	40
Y				
Zr	80	10	10	20
V	200	40	30	200
Cr	100	1200	1800	150
Ni	30	200	200	100
Th				
U				

All abundances of major elements are in weight percent of oxides. Where no entry is made in the FeO position, the value at Fe_2O_3 is total iron calculated as ferric oxide. All abundances of trace elements are in parts per million by weight. Values are taken directly from quoted sources and not recalculated to 100%.

Table 7.5. Column References
1. Near Bonai. Ultramafic rocks.
2. Gorumahisani basin. Schistose ultramafic rocks.
3. Gorumahisani basin. Massive ultramafic rocks.
4. Ongarbira belt. Metavolcanic rocks.

suggested that the suite represents a subvolcanic assemblage intrusive into, and flowing over, the IOG.

SINGHBHUM GRANITE COMPLEX

The Singhbhum Granite complex occupies about 10,000 sq km and consists of 12 separate, domal or sheetlike, arcuate, magmatic bodies (Saha, 1972). According to Saha, these 12 units were emplaced in three successive, but closely related, phases in an area that was mainly occupied by preexisting rocks of the OMG. Iyengar and Murthy (1982), however, proposed that the Singhbhum suite represents various magmatic events well separated in time. Sequential emplacement was also shown by B. K. Chattopadhyay (1983).

Chronologic relationships between the Singhbhum Granite and the surrounding rocks are unclear. S. N. Sarkar and Saha (1977, 1983) proposed that the granite is intrusive into the IOG, which went through an early phase of folding in

the Chaibasa region before intrusion. Isochron ages (Rb/Sr) for two sets of samples indicate that at least parts of the granite suite crystallized about 2,900 to 3,000 m.y. ago (S. N. Sarkar et al., 1979). The reported initial $^{87}Sr/^{86}Sr$ ratios are 0.711 and 0.7023, a spread so large that no conclusions can be drawn.

Rocks of the complex range from adamellites to granites (Saha and Rao, 1971; Saha, 1979; Table 7.6). The rocks within each of the three phases of intrusion proposed by Saha show a limited range of composition, and there is a general increase in abundance of K-feldspar and muscovite, decrease in ferromagnesian mineral content, and increase in K_2O and SiO_2 from the older (phase I) to the younger phases (II and III). Trace-element abundances are also consistent within phases (Roonwal, 1972; Saha et al., 1973b). Phases I and II are characterized by gently sloping REE patterns with slightly depleted heavy REE concentrations and very limited negative or absent Eu anomalies (Saha and Ray, 1984). In phase III, light REE are

TABLE 7.6. *Compositions of Singhbhum Granite (Saha, 1979)*

	1	2	3	4
SiO_2	71.5	70.1	73.2	74.8
TiO_2	0.15	0.15	0.1	0.2
Al_2O_3	15.0	15.4	14.3	15.5
Fe_2O_3	1.0	0.9	1.1	0.4
FeO	1.7	1.3	1.0	0.6
MnO				0.1
MgO	1.0	1.2	0.5	1.4
CaO	2.8	3.1	2.5	0.9
Na_2O	4.9	4.8	4.3	4.2
K_2O	1.7	2.8	3.0	3.8
Rb				
Sr	470	717	494	42
Y				
Zr				
V	24	37	22	5
Cr	7	14	8	
Ni	11	23	34	10
Th				
U				

All abundances of major elements are in weight percent of oxides. Where no entry is made in the FeO position, the value at Fe_2O_3 is total iron calculated as ferric oxide. All abundances of trace elements are in parts per million by weight. Values are taken directly from quoted sources and not recalculated to 100%.

Table 7.6. Column References
1. Magmatic phase I.
2. Magmatic phase II.
3. Magmatic phase III.
4. Granodiorite-adamellite (near Chaibasa).

enriched, and the heavy REE pattern is flat; Eu anomalies are moderate to strong.

Ray et al. (1980) estimated temperatures and oxygen fugacities of the granite by study of biotites. Conditions for phase II were $\log_{10} O_2 = -15$ and temperature of 590°. Conditions for phase III granites were $\log_{10} O_2 = -15.5$ to -16.7 and temperatures in the range of 530° to 730°.

SINGHBHUM GROUP

The Singhbhum Group refers to rocks exposed in the Singhbhum-Dhalbhum mobile belt and adjacent areas (Figs. 7.1, 7.4). In many terminologies, it is restricted to the area north of the Singhbhum thrust and south of the Dalma thrust or Chotanagpur gneissic terrain. The regional east-west strike of the belt is an approximate continuation of the trend of the Satpura belt in the Bhandara craton (Ch. 6). Early workers proposed that the Satpura and Singhbhum trends were a continuous orogenic belt (Krishnan, 1953; Holmes, 1955). In addition to rocks of the Singhbhum Group, the mobile belt also contains the Gangpur Group, regarded as younger than the Singhbhum rocks, and several intrusive bodies.

The Singhbhum Group consists mostly of mica schists, hornblende schists, quartzose schists and granulites, chloritic schists, and ortho- and paraamphibolites. Metavolcanic rocks are dominantly mafic to ultramafic. Stratigraphic relationships have led to a great deal of controversy and confusing terminology. The principal difficulty is caused by the Singhbhum thrust zone, particularly its occurrence within an area consisting wholly of supracrustal rocks. Thus, for example, it is not clear whether rocks south of the thrust but north of the Singhbhum Granite massif are correlatable with rocks north of the thrust, implying limited movement on the thrust; correlatable with other rocks in supracrustal basins around the Singhbhum Granite; or correlatable both with rocks in the mobile belt, north of the thrust, and in other areas around the granite massif, thus forming a large geosynclinal area surrounding the Singhbhum Granite. Another difficulty related to the thrust zone is the possibility that the main Singhbhum thrust is simply the sole thrust for a complex melange of disparate blocks now occurring to the north.

A stratigraphic section for the mobile belt is shown in Table 7.7. Different stratigraphic concepts include the following. S. N. Sarkar and Saha (1962, 1977, 1983) divided the Singhbhum Group into a lower Chaibasa Formation and upper Dhalbhum Formation, overlain by Dalma volcanic rocks. They grouped all rocks south of the thrust zone as an older Iron Ore series. Iyengar and Alwar (1965), Mazumdar (1978), and Iyengar and Murthy (1982) regarded the rocks of the mobile belt as equivalent to the Dhanjori Group (Fig.

7.2), implying correlation across a thrust zone of minor displacement. A. N. Sarkar (1982b) placed the Dalma volcanic suite as a concordant member of the Singhbhum Group, interbedded with the metapelitic rocks. A. K. Gupta et al. (1980), however, regarded the Dalma suite as discordant, and D. S. Bhattacharyya and Bhattacharyya (1970) used magnetic anomalies to demonstrate that the Dalma suite extends to a very great depth, consistent with a discordant relationship.

Structure and metamorphism

Dunn (1929), Dunn and Dey (1942), and S. N. Sarkar and Saha (1962) proposed that the Singhbhum Group north of the Singhbhum thrust consists of an anticlinorium of Singhbhum Group rocks south of a synclinorium of Dalma volcanic rocks, all overturned to the south. Second-generation cross-folding occurred in the western and southeastern parts of the belt and continued after thrusting had stopped. In this view, the Dalma thrust was younger than the folding. The two generations of folding plus the thrusting were considered to be three deformational episodes (S. N. Sarkar and Saha, 1977).

Four generations of folding have been proposed in the western part of the mobile belt (A. N. Sarkar and Bhattacharyya, 1978a, 1978b, 1978c; A. N. Sarkar, 1982a, 1982b) and in subsurface studies of the thrust zone (A. Basu et al., 1979). Not all of the deformational patterns are found throughout the belt. The first deformation developed subhorizontal folds (F1) on bedding planes (S1) with schistosity having a general east-west strike. These folds occur throughout the belt except in the south, where they have been destroyed by younger deformation. In the second phase of deformation, F1 folds and axial plane schistosity were refolded about steeply plunging fold axes (F2) at a high angle to F1. Regional structures shown, for example, by the Dalma volcanic band (Fig. 7.4) are F2 folds, and the F2 folding is most intense in the Dalma suite (A. N. Sarkar and Bhattacharyya, 1978a, 1978b, 1978c). The F3 folds were superimposed on F1 and F2. They are subhorizontal, with axial planes dipping slightly northward and striking ENE-WSW to E-W. They dispersed all earlier structures and are accompanied by northerly dipping cleavage that is less intense in the northern part of the belt. The fourth phase of deformation is represented by intense shearing in the southern part of the belt, associated with the development of the Singhbhum thrust.

Relationships between deformation and metamorphism are controversial. We describe studies in different parts of the belt.

The most extensive work has been done in the western part of the mobile belt. Ray (1976) stated that deformation was continuous throughout the

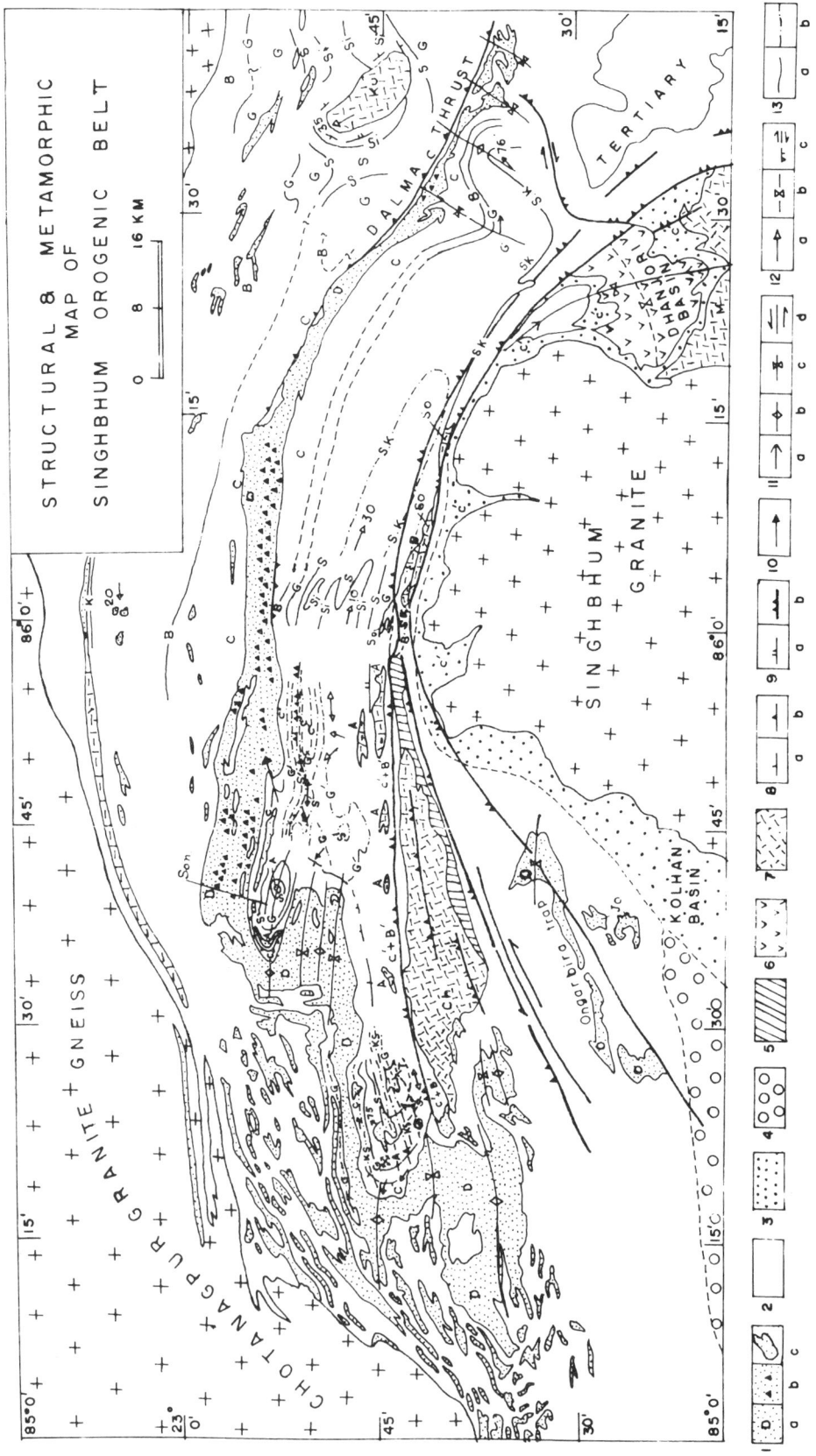

STRUCTURAL & METAMORPHIC MAP OF SINGHBHUM OROGENIC BELT

0   8   16 KM

144

TABLE 7.7. *Stratigraphy of mobile belt (A. N. Sarkar, 1982a)*

	Stratigraphic Units	Approximate Age in m.y.	Geologic Events
	Metadolerites (Newer Dolerites?), Arkasani Granite Gneiss, Soda granite, Chakradharpur Granite Gneiss (later and predominant phase), Chotanagpur Granite Gneiss (later phase)	c.850	F_3 cycle
	Amphibolites	c.1170	F_2 cycle
	Kuilapal Granite Gneiss, Tebo migmatites, Chotanagpur Granite Gneiss (earlier phase), Chakradharpur Granite Gneiss (earlier phase)	c.1550	F_1 cycle
Singhbhum Group	Mica schists, phyllites, and quartzites (mainly) with Dalma metavolcanics and subvolcanics, carbon-phyllite, purple phyllite, volcanic breccia, pyroclasts (cumulates?), chert, crystalline limestone, and calc-silicates	2000–1700	Sedimentation and volcanism

evolution of the belt but was not related to metamorphic episodes. In other studies, three cycles of metamorphism (M1, M2, and M3) have been correlated with the first three generations of deformation (D. S. Bhattacharyya et al., 1976; A. N. Sarkar, 1982a, 1982b). Minerals formed during and after M1 metamorphism include chlorite, muscovite, biotite, and the first generations of garnet, staurolite, and andalusite. The M2 event included rotation of garnet and staurolite porphyroblasts and bending and folding of mica without further prograde mineral growth; some retrogression occurred. The M3 cycle caused further development of chlorite, muscovite, biotite, garnet, kyanite, staurolite, and andalusite. During M1 metamorphism, conditions varied from low pressure and temperature at the western end of the belt to temperatures of about 660° and pressures of 3.3 kb farther east, where anatexis of pelitic

rocks occurred. Anatexis also occurred in other parts of the mobile belt.

In the eastern part of the mobile belt, around Ghatsila (Fig. 7.2), one prograde and one retrograde metamorphic episode have been proposed (Naha, 1965). They have been correlated with the M1 and M2 phases of the western part of the belt. Evidence for the single phase of metamorphism consists of a correspondence between the boundaries of Barrovian zones and the trends of fold axes, indicating metamorphism during one major deformational event. This interpretation was challenged by D. Mukhopadhyay et al. (1975), who described superposition of folding and synto postkinematic development of biotite, garnet, staurolite, and kyanite.

The northern part of the belt, north of the Dalma volcanic suite, contains up to three cycles of metamorphism and deformation (S. K. Ghosh,

1a	—Dalma Ophiolite	10	—F1 axis
1b	—Dalma Ophiolite with Ultramafic Fragments	11a	—F2 axis
1c	—Exotic Masses	11b	—Antiformal axial trace of F2
2	—Flysch	11c	—Synformal axial trace of F2
3	—Molasse	11d	—Sinistral F2 shear
4	—Platform Sediments	12a	—F3 axis
5	—Mafic and Ultramafic Rocks of Singhbhum Thrust Zone	12b	—Synformal axial trace of F3
6	—Dhanjori Lava	12c	—Composite dextral shear overthrust of F3
7	—Granites and Migmatites	13a	—M1 isograd
A	—Arkansani Granophyre	13b	—M3 isograd
Ch	—Chakradharpur Granite Gneiss	C	—Chlorite Zone
Ku	—Kuilapal Granite/Migmatite	B	—Biotite Zone
M	—Mayurbhanj Granite	G	—Garnet Zone
So	—Soda Granite	S	—Staurolite Zone
T	—Tebo Migmatite	K	—Kyanite Zone
8a	—Generalized S2	Si	—Sillimanite Zone
8b	—Early thrust	Jo	—Jojohatu
9a	—Generalized S3	O	—Ongarbira Basalts
9b	—Late thrust	Son	—Sonapet Valley
		T	—Tebo

(C', B', and G' represent corresponding minerals of M3)

Reproduced with permission from Tectonophysics, v. 86, pp. 367–368.

Fig. 7.4. Singhbhum mobile belt (A. N. Sarkar, 1982a).

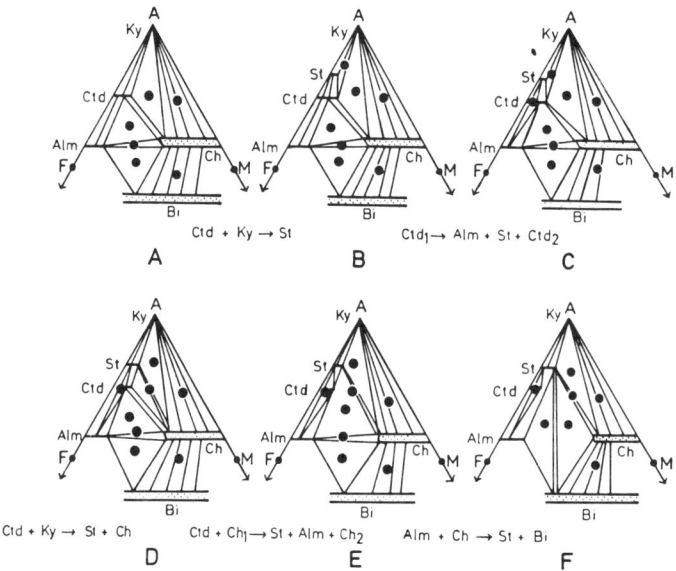

Fig. 7.5. AFM projection showing prograde changes in metamorphic rocks of Sini area (north of Chaibasa; Fig. 7.2) in Singhbhum mobile belt (Lal and Singh, 1978). Increasing degree of metamorphism yields mineral assemblages in diagrams A (lowest) to F (highest). Solid circles show observed assemblages. Principal reactions from one zone to the next are shown below the diagrams. Reproduced with permission from Neues Jahrbuch fur Mineralogie, Abhandlungen, v. 131, p. 324.

1963; A. N. Sarkar, 1977). Metamorphism culminated in the development of garnet, staurolite, and sillimanite and possibly the development of anatectic granites. This conclusion contrasts with that of K. L. Chakraborthy and Sen (1967), who found only one period of deformation and metamorphism in areas just south of the Dalma suite.

In the central part of the mobile belt (Fig. 7.4), Lal and Singh (1977, 1978) and Lal and Ackermand (1979) observed two phases of deformation and two phases of medium-pressure, Barrovian metamorphism (M1 and M2). A chemographic diagram of the various metamorphic zones is shown in Figure 7.5. The M1 phase was synchronous with the first deformation and continued after deformation ceased. Lal and Singh (1978) estimated the following temperatures and pressures at various isograds: 500° and 4.5 kb at the transition from the chlorite-phengite-biotite zone to the almandine-chlorite-muscovite zone, 520° and 4.75 kb at the appearance of the staurolite-biotite zone, and 670° and 6 kb at the appearance of the sillimanite-almandine-biotite zone. The various zones show a decrease in metamorphic grade outward from an elongated area north of the thrust belt. All rocks south of the thrust belt were regarded as being in chlorite grade.

Kyanite-grade assemblages immediately north of the Singhbhum thrust (A. K. Banerji, 1981b) occur predominantly in kyanite-quartz bodies. They are associated with muscovite-biotite-chlorite schist, garnet-staurolite-mica schist, and minor chert and quartzite. Metamorphic condi-

tions have been estimated as 450° to 500° and 5 to 6 kb. The presence of these rocks along the thrust zone contrasts with the more general observations of low-grade rocks on both sides of the thrust.

Variation in metamorphic grade across the Singhbhum thrust zone is obviously important in determining the history of the thrust. High-grade rocks (but below granulite facies) generally occur north of the thrust. Metamorphic gradients, however, decrease southward toward the thrust, and there is little difference in grade between rocks immediately north and immediately south of the thrust, with most of the assemblages in greenschist facies (A. N. Sarkar, 1982b).

Dalma volcanic suite

The Dalma volcanic suite and other intrusive and extrusive bodies form a major part of the Singhbhum mobile belt (Fig. 7.4). Present evidence indicates that at least some of the bodies represent a tectonically emplaced ophiolite assemblage. The volcanic suite is associated with a sequence of sedimentary rocks that is somewhat different from other pelitic schists in the mobile belt.

The Dalma igneous rocks are mafic to ultramafic. The lower part of the section consists of tholeiites, alkali basalts, andesites, and komatiites interbedded with tuffaceous sediments, tuffs, and a variety of sedimentary rocks. This horizon is overlain by high-Mg vitric tuffs with quench textures, and most of the komatiitic rocks in the belt appear to be pyroclastic. The upper unit is largely a high-Fe, low-K, pillowed tholeiite. Not all of the rocks show quench textures, and some cumulate texture has been recognized, indicating a mixture of intrusive and extrusive activity. The presence of pillow structures and other features indicates that most of the volcanism was subaqueous.

The metasedimentary rocks associated with the Dalma igneous suite include quartzites, siltstones, "carbon phyllites," "purple phyllites," and minor cherts and limestones. These rocks are more lithologically diverse than the phyllites and mica schists of the remainder of the mobile belt. A. N. Sarkar (1982a, 1982b) regarded the carbon phyllites as former euxinic sediments and the purple phyllites as abyssal red clays.

Various interpretations have been given of the Dalma suite, including subaerial eruption (Dunn, 1929; Dunn and Dey, 1942; S. N. Sarkar and Saha, 1962, 1977), island arc magmatism (Naha, 1961, 1965), and evolution as a greenstone belt in an intra-cratonic rift (Chakraborti, 1980; A. K. Gupta et al., 1980). Interpretation of the rocks as an ophiolite suite (A. K. Gupta et al., 1980; A. N. Sarkar, 1982b) is consistent with the petrology of the rocks and their tectonic occurrence.

SUMMARY OF EVOLUTION OF THE SINGHBUM OROGENIC BELT

Although neither sedimentation, volcanism, nor tectonism in the Singhbum mobile belt are well dated, it seems likely that the area represents a Proterozoic orogen. The best evidence for the age of movement on the thrust belt is from Pb isotopic dates of the associated uranium ores, which are consistent at about 1,600 m.y. ago (Krishna

Rao et al., 1979). Presumably the faulting was approximately synchronous with deformation in the mobile belt. The type of orogenic activity shown by the mobile belt, including the presence of komatiites and ophiolites, is particularly significant in rocks of this age. Most Proterozoic assemblages, in India and elsewhere, do not contain suites formed on oceanic crust and later involved in a collisional orogeny.

The Singhbum Group is characterized by thick accumulation of pelitic sediments; structurally emplaced mafic/ultramafic igneous rocks; a major thrust zone and related smaller thrusts; metamorphism to high-amphibolite facies; granites, presumably many of which are of anatectic origin; and folded structures consistent with the same north-south stresses that produced the thrusts. The belt is, therefore, a typical orogenic area. The southward vergence of the thrusts places the Singhbum Granite massif as the foreland of the deformed belt, with either the Chotanagpur area or possibly the Bundelkhand area as the hinterland. It is highly likely that the belt was formed by closure of a basin containing at least some oceanic crust about 1,600 m.y. ago. A. N. Sarkar (1982b) regarded the orogenic event as the result of collision between the Singhbum and Chotanagpur blocks (Fig. 7.6).

In Figure 7.6, the principal subdivision of the sedimentary suites is into molasse and flysch facies. Molasse sediments occur along the northern border of the Singhbum Granite and lie unconformably on the granite. This molasse sequence has a lower polymictic conglomerate that grades upward into cross-bedded arkoses and varved slates and phyllites. All the pebbles in the conglomerate could have been derived from the oro-

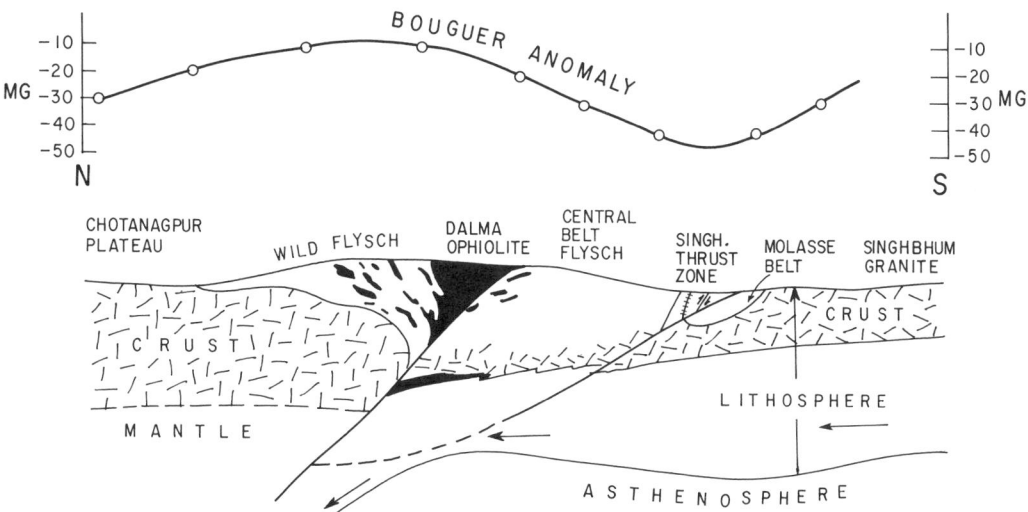

Fig. 7.6. Schematic cross-section of Singhbum mobile belt (A. N. Sarkar, 1982b). Reproduced with permission from Tectonophysics, v. 86, p. 387.

genic belt to the north, and some associated feldspathic material may have come from the Singhbhum Granite. A. N. Sarkar (1982a) also regarded the Kolhan Group (Fig. 7.2) as part of the molasse sequence.

The metaflysch (wildflysch) sediments of the central part of the mobile belt, north of the Singhbhum thrust, comprise a thick sequence of phyllites and mica schists with intercalations of quartzites and sandy conglomerates (Fig. 7.6). Sedimentary structures indicate directional currents from the south and south-southwest (Naha, 1961), consistent with a northward paleoslope. The flysch sequence is lithologically uniform and unbroken by recognizable unconformities.

Metamorphic grade in the mobile belt north of the thrust indicates maximum depths of 10 to 15 km during metamorphism. The proportions of uplift caused by thrusting and by other processes are uncertain. The occurrence of rocks of comparable metamorphic grade on both sides of the thrust throughout much of the belt indicates relatively small movement on the thrust itself. Movement of the thrust belt southward over its own molasse is consistent with sedimentary models for the evolution of the belt discussed above, although the thrust may simply have moved over an unrelated suite.

The probable occurrence of dismembered ophiolites in the mobile belt suggests the closure of an oceanic basin. The metasedimentary assemblage is also thicker and more mafic than would be expected in an intracratonic basin, and the width of the metasedimentary assemblage also indicates a large former sedimentary basin. Exposure of metamorphic rocks at high amphibolite grade, anatectic granites, and highly compressed structures indicate exposure of the metamorphic belt at a deep erosional level. The belt, however, is not comparable to the very deep levels shown in orogens that expose granulite-facies rocks.

GANGPUR GROUP

The Gangpur Group occurs within the Singhbhum-Dhalbhum mobile belt (Fig. 7.4) and differs from the supracrustal rocks of the mobile belt and possible correlatives. It appears, therefore, to be a more platformal facies than other metasediments in the area. The Gangpur Group is surrounded by low-grade metasedimentary rocks of the Dhalbhum Formation of the Singhbhum Group of S. N. Sarkar and Saha (1983). The Gangpur Group has been correlated with the Singhbhum Group (A. N. Sarkar, 1982a) and the Kolhan Group (Iyengar and Murthy, 1982), and several workers have regarded it as younger than the Singhbhum rocks (Kanungo and Mahalik, 1972, 1975). Rocks include garnet-grade mica schists, phyllites, gondites, carbonaceous quartzites and phyllites, polymictic conglomerates, calcitic and dolomitic marbles, and some mafic metavolcanic rocks (epidiorites).

The original interpretation of the Gangpur Group as an anticlinorium was challenged by Kanungo and Mahalik (1972, 1975), who showed inversion of current bedding and regarded the anticlinorium as a synclinorium. With this interpretation, the Gangpur Group lies unconformably above adjacent supracrustal rocks. Intrusions in the Gangpur Group are mafic sills and granites, which have been correlated with rocks in the Chotanagpur terrain (Mazumdar, 1978: Prasad and Mahto, 1979).

Kanungo and Mahalik (1972) explained mesoscopic structures in the Gangpur Group by one period of folding that formed the major synclinorium. Based on current-bedding orientations, the southern limb of the synclinorium appears to be overturned to the north; the northern limb dips steeply, either to the south or overturned to the north. Three-phase deformation of the Gangpur suite was proposed by Chaudhuri and Pal (1983a, 1983b). They reinterpreted the Gangpur suite as occurring in an anticlinorium plunging eastward and did not find an angular discordance with adjacent rocks.

DHANJORI GROUP AND SIMLIPAL COMPLEX

The Dhanjori and Simlipal suites are discussed together because of their geographic proximity and the tendency of some geologists to consider them as similar assemblages. For reasons summarized in this discussion, however, we do not regard them as correlative. The volcanic rocks in the Dhanjori basin (Fig. 7.2) appear to be a basinal or platformal sequence, whereas the Simlipal complex shows characteristics of a cryptovolcanic structure, possibly an astrobleme. Although Iyengar and Murthy (1982) regarded the Dhanjori Group as a thick volcanosedimentary assemblage overlying their Badampahar Group, we identify the Dhanjori Group as simply the suite in the Dhanjori basin.

Volcanic rocks of the Dhanjori basin lie on discontinuous quartzites. Two types of mafic volcanic rocks are present (A. K. Ghosh and Banerji, 1970): massive basalts with relic porphyritic and intergranular texture, and schistose, amygdaloidal basalts. The rocks are low-Al, quartz-normative tholeiites. Feeder dikes and plugs of gabbro are present, and the Dhanjori volcanic sequence has been folded into a first-order syncline with minor second-order folds superimposed (P. K. Banerjee, 1982).

The lowest sequence in the Simlipal complex (Fig. 7.2) is a thick suite of tuffs and lavas lying on a basement of quartzites and phyllites (Iyengar and Banerjee, 1964). Pyroclastic rocks and sediments alternate upward in the sequence, with

TABLE 7.8. *Compositions of rocks from Simlipal complex*

	1	2	3	4	5
SiO_2	51.0	50.3	47.4	70.0	51.9
TiO_2	0.71	0.78	0.50	0.30	0.80
Al_2O_3	17.0	14.3	14.6	11.1	15.6
Fe_2O_3	2.0	2.0	1.2	4.7	1.3
FeO	10.1	12.0	14.8	2.6	9.0
MnO	0.13	0.14	0.32	0.25	0.31
MgO	7.2	7.5	10.6	0.30	7.5
CaO	8.3	7.4	5.6	2.5	10.9
Na_2O	3.5	2.9	2.1	3.3	2.2
K_2O	0.40	0.40	0.30	3.4	0.20

All abundances of major elements are in weight percent of oxides. Where no entry is made in the FeO position, the value at Fe_2O_3 is total iron calculated as ferric oxide. All abundances of trace elements are in parts per million by weight. Values are taken directly from quoted sources and not recalculated to 100%.

Table 7.8. Column References
1. Volcanic rock (Iyengar and Banerjee, 1964).
2. "Spilite" (Iyengar et al., 1981b).
3. Gabbro (Iyengar and Banerjee, 1964).
4. Pyroxene granite (Iyengar et al., 1981b).
5. Amjori dolerite sill (Iyengar et al., 1981b).

minor flow rocks in the tuffaceous units. The sedimentary layers consist of quartzites, arkoses, and minor ferruginous shales. The center of the complex is occupied by the Amjori sill (Iyengar et al., 1981b), and the margins are intruded by alkaline and related rocks, including gabbroic anorthosites, granophyres, riebeckite granites, and pyroxene granites. The Amjori sill is highly differentiated from dunite at the base to quartz dolerite at the top (Iyengar et al., 1981b). Channakasavulu and Sahu (1980) showed that the sill is a tholeiitic magma that formed a layered body by fractional crystallization. The age of the Simlipal complex is 2,085 m.y., with an initial $^{87}Sr/^{86}Sr$ ratio of 0.745 (Iyengar et al., 1981a). This extremely high initial ratio may represent partial melting of near-surface rocks.

Typical compositions of the igneous rocks in the Simlipal complex are shown in Table 7.8. Volcanic rocks in the complex have been regarded as spilitic, although the sodium contents are not particularly high. The unusual composition of the rock suite is shown by the REE pattern, which shows a rectilineal pattern with an inflection at Gd (Iyengar et al., 1981b).

The major characteristics of the Simlipal complex are its highly explosive initial eruption, dips of all units toward the center of the basin, and alkaline nature of the rock suite. These features are very characteristic of cryptovolcanic structures (summary in McCall, 1979), and it is possible that the Simlipal complex has formed by a similar process. The presence of several layers of apparently subaqueous sediment means that the suite cannot have formed by one explosive event but must have required a series of eruptions in the same area. It is possible that the initial event was meteorite impact, which has been shown to be responsible for a large number of cryptovolcanic structures, but no impact features have been reported in the vicinity of the Simlipal basin. The extremely high initial Sr ratio is consistent with an impact origin.

KOLHAN GROUP

The Kolhan Group is the youngest Precambrian supracrustal sequence in the Singhbhum Granite nucleus (Dunn, 1940; Fig. 7.2). The sequence starts with a basal oligomictic conglomerate that separates the Kolhan Group from underlying folded IOG or Singhbhum Granite. The conglomerate is overlain by a repetitious sequence of sandstone/conglomerate that grades upward into shales with limestone intercalations. The entire sequence is about 75 m thick and is clearly intracratonic. The Kolhan basin generally slopes westward and shows a synclinal structure. The eastern part of the basin has been deformed into elongate gentle domes and basins and dome-in-dome structures. The Kolhan Group is less deformed and more compositionally mature than any other group and thus has no comparable suite in the area.

MAFIC/ULTRAMAFIC INTRUSIVE ROCKS

A variety of mafic and ultramafic bodies have intruded the supracrustal and granitic rocks of the Singhbhum craton. We have chosen to discuss only the following suites: ultramafic bodies emplaced synkinematically to late kinematically in the schistose supracrustal rocks (principally Iron Ore Group), ultramafic bodies in the Singhbhum Granite complex, gabbro-anorthosite-granophyre complexes along the eastern side of the Singhbhum Granite, and the Newer Dolerite dike swarm.

Ultramafic bodies in supracrustal rocks

The major rock types of this suite are picrites and pyroxene-rich peridotites that have largely been metamorphosed to schistose rocks consisting of talc, tremolite, chlorite, and similar minerals. Chromite is a significant constituent in many bodies. The major occurrences are along the Sukinda thrust, at the southern edge of the craton (Fig. 7.2). Small occurrences are in the Nausahi (Nausahi-Nilgiri) basin and in the Roro area.

The Iron Ore Group of the Nausahi basin has been invaded by two distinct types of ultramafic rocks (S. Mukherjee and Haldar, 1975; Haldar and Chatterjee, 1976). The older suite occurs as discordant, elongated layered bodies with a composition ranging from dunite to chromitite. The younger group is represented by dikes of peridotites and pyroxenites.

The suite at Roro is an arcuate body within folded metasedimentary rocks (Ballurkar and Pal, 1979; D. Mukhopadhyay and Dutta, 1983). It consists of chromitite and chromite-bearing dunites in which the olivine is completely serpentinized, poikilitic harzburgites, lherzolites, and pyroxenites. The ultramafic body and enclosing supracrustal rocks have been folded into a major easterly closing synform with axial plane cleavage dipping nearly east-west. S. N. Sarkar and Saha (1962) stated that structures with an age equivalent to Singhbhum thrusting were superimposed in this area on structures from an earlier deformation episode (Iron Ore orogeny). D. Mukhopadhyay and Dutta, however, explained the deformation of the area in terms of one episode.

The Sukinda ultramafic field is the largest cluster of ultramafic bodies in the area (P. K. Banerjee, 1972). It occurs along the Sukinda thrust belt, where greenschist-facies supracrustal rocks of the Singhbhum nucleus (commonly identified as Iron Ore Group) are thrust below granulite-facies rocks of the Eastern Ghats belt. The field extends over an area of 430 sq km and contains two generations of ultramafic rocks, consisting of chromite-bearing altered dunites and nonchromitiferous orthopyroxenites (P. K. Banerjee, 1972; K. L. Chakraborthy et al., 1980). Page et al. (1985) found that distribution of platinum-group elements in ultramafic rocks of the Sukinda area is similar to that found in modern ophiolites, and they proposed that the bodies in the Sukinda area represent dismembered ophiolites.

The Sukinda complex is the largest body in the ultramafic field and forms a southwest-plunging, asymmetric syncline (K. L. Chakraborthy, 1972, 1973; Mitra, 1973; Sahoo and Van der Kaaden, 1976; K. L. Chakraborthy and Baidya, 1978). It is a rhythmically layered intrusion, with the principal layering consisting of altered olivine and chromite cumulates. The Cr- and Ni-rich phases are completely altered to serpentine, tremolite schist, talc schist, and chlorite schist. P. K. Banerjee (1972) concluded that the complex represents a deep-seated, early cumulate that was later emplaced in two pulses in a semisolid/solid state. The original primary silicates were altered to the present mineralogy during the solid-state transport.

Ultramafic bodies in the Singhbhum Granite complex

Post-tectonic ultramafic suites occur as independent bodies along a linear belt from Champua to Chaibasa (Fig. 7.2). The largest body is at Keshargaria, 10 miles south of Chaibasa. It has a northeast-southwest trend, a width of 150 m, a length of 2.5 km, and consists of enstatitic pyroxenites and harzburgitic peridotites (Bose and Goles, 1970; Saha et al., 1972). Peridotites in which pyroxene is dominant contain platy enstatite, but rocks with subordinate pyroxene contain poikilitic enstatite.

Gabbro-anorthosite suites on the eastern side of the Singhbhum Granite

A number of gabbro/anorthosite bodies occur along the eastern side of the Singhbhum Granite. They extend across the Singhbhum thrust zone (Mazumdar, 1978), form the outer periphery of the Simlipal complex (Iyengar et al., 1981a), and occur within outcrop areas of the Mayurbhanj Granite and supracrustal suites (Badampahar, Iron Ore, or other groups). They have been associated with vanadium-bearing titaniferous iron ores in some areas (Das Gupta, 1969).

Many workers regarded the gabbro/anorthosite bodies as comagmatic with the Mayurbhanj and other leucogranitic rocks (Dunn and Dey, 1942; Iyengar and Alwar, 1965; R. C. Sinha and Ghose, 1975; Srivastava, 1975; S. C. Chatterjee, 1976; K. L. Chakraborthy et al., 1981). Saha (1975) and Saha et al. (1977) demonstrated a major compositional gap between the gabbro/anorthosite rocks and the leucogranites and proposed that this marked bimodality could not have resulted from fractionation of one suite from the other.

Newer Dolerites

The Newer Dolerites are a set of dikes that intrude virtually all rock types of the Singhbhum nucleus, with the notable exception of the Chak-radharpur Granite/Gneiss and the Singhbhum and Dhanjori suites (Dunn, 1940; Dunn and Dey, 1942; Guha, 1963). A variety of age determinations, mostly by K/Ar methods, has yielded ages in the range of 1,600 to 950 m.y. (S. N. Sarkar and Saha, 1977, 1983). In the Singhbhum Granite, the dikes are parallel to the joints in the granite. The dolerites have been grouped with the Amjori sill by Iyengar and Banerjee (1964) and Saha et al. (1977). Verma and Prasad (1974) proposed three intrusive episodes based on paleomagnetic studies. The dikes are mostly dolerite and consist of phenocrysts of augite and lath-shaped labradorite-bytownite in a fine-grained matrix. Both quartz- and olivine-normative varieties are present. In addition to dolerite, rock types range from poikilitic harzburgite to leucogranophyre.

Saha et al. (1973a) proposed three petrogenetic groups of Newer Dolerites. One is the product of direct magmatic crystallization, which includes two different generations of dolerites and associated micropegmatites and microgranites; each suite has a separate trend of compositional variation. A second group is dominated by cumulate processes and includes peridotites, norites, and lamprophyres. The third group was formed by local partial melting and includes leucogranophyres. The entire suite contains abnormally high K/Rb ratios and Ni and Co concentrations and low Rb, Ba, and Sr.

In addition to the Newer Dolerites, other late dikes include porphyries intruding the Singhbhum Granite. They contain large xenoliths of vein quartz and granite in a fine- to medium-grained, granophyric intergrowth of plagioclase and quartz. The porphyries and Newer Dolerites intersect and grade into each other at some places and may be comagmatic.

GRANITES AND GRANITIC GNEISSES

The major sialic terrains of the Singhbhum craton are the Singhbhum Granite complex (previously described) and the Chotanagpur area (described later). In addition to these areas, a large number of other bodies are present (Figs. 7.2 and 7.4), and we discuss them in three categories: the Mayurbhanj Granite, a large body east of the Singhbhum Granite; smaller bodies along the Singhbhum thrust zone; and miscellaneous sialic suites.

Mayurbhanj Granite

The Mayurbhanj Granite is a batholithic body with an outcrop area of about 1,000 sq km (Fig. 7.2). It intrudes the Singhbhum Granite, all of the supracrustal rocks in the area, the Dhanjori Group, and the gabbro/anorthosite bodies along the eastern side of the Singhbhum Granite (Saha, 1975; Saha et al., 1977; S. N. Sarkar and Saha,

1977, 1983; Iyengar and Murthy, 1982; S. S. Sarkar, 1982).

The batholith consists of three lithologic units. One is a fine-grained granophyric granite containing biotite, hornblende, stilpnomelane, and alkali feldspar. It occurs in the western part of the batholith. The second unit is a coarse-grained granite with biotite and ferrohastingsite that ranges from nearly massive to well foliated. The third unit consists of small injections of biotitic aplogranite. According to Saha (1975) and Saha et al. (1977), the first unit was intruded before the second phase of deformation of the Singhbhum Group, the second unit was synkinematic with the second phase of deformation, and the third unit was postkinematic. There is no evidence that the two major phases of the Mayurbhanj Granite (granophyric granite and ferrohastingsite granite) are comagmatic, and their intrusion at different periods of tectonic activity suggests that they are separate suites.

Other small granite bodies occur in the Mayurbhanj area. One of them, the Notapahar Granite, is regarded as comagmatic with the Mayurbhanj Granite (S. S. Sarkar, 1982). It ranges from quartz diorite through tonalite to granodiorite, granite, and alkali feldspar-bearing granite.

Granites associated with the Singhbhum thrust zone

Several granitic bodies occur along the Singhbhum thrust or within the mobile belt immediately north of the thrust (Fig. 7.4). We describe here the Chakradharpur Granite/Gneiss, the Arkasani Granophyre, and the Tatanagar and Surda soda granites.

The Chakradharpur Granite/Gneiss complex (CKPG) is an elliptical body elongated east-west with an outcrop area of about 225 sq km. It is the largest and westernmost of a suite of igneous bodies within the flysch sediments of the Singhbhum-Dhalbhum mobile belt (Fig. 7.7). Bandyopadhyay (1981) identified two distinct lithologic units within the CKPG complex: an old bimodal suite of tonalitic gneiss and amphibolite that covers about 75% of the outcrop, and a younger suite of granodiorite to granite. S. Sengupta et al. (1983) showed that metasediments of the surrounding Singhbhum Group rest unconformably over the older, bimodal part of the CKPG. This age relationship is consistent with earlier studies of D. S. Bhattacharyya (1966) that showed structural conformity of the CKPG and the Singhbhum Group, implying deformation of the two suites together.

Geochemistry of the CKPG has been studied by S. Sengupta et al. (1983). The tonalitic gneisses contain both high-Al and low-Al trondhjemites/tonalites, and the interlayered amphibolites contain both high-Mg and low-Mg varieties. The tonalitic suite exhibits highly fractionated REE pat-

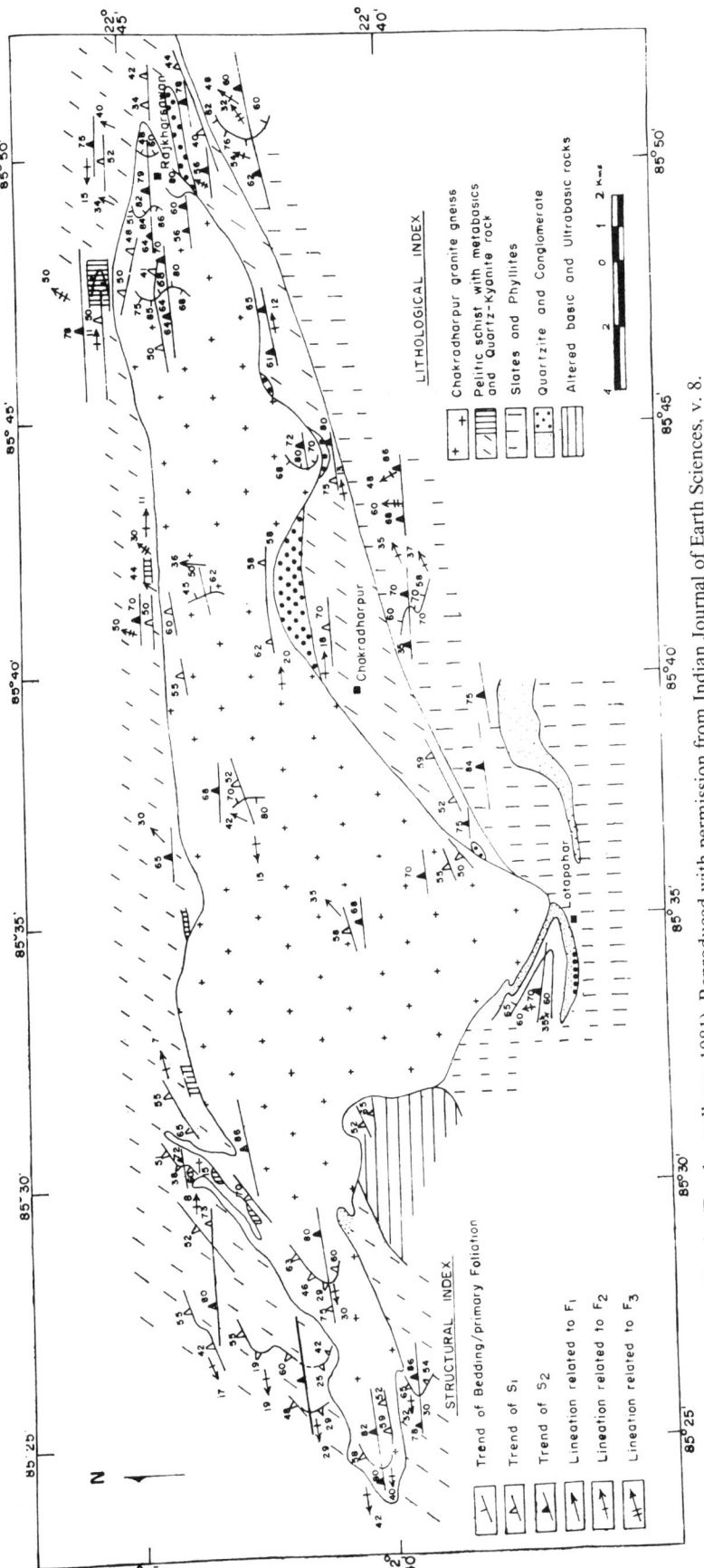

Fig. 7.7. Chakradharpur Granite (Bandyopadhyay, 1981). Reproduced with permission from Indian Journal of Earth Sciences, v. 8.

terns with distinct positive Eu anomalies. The younger suite consists of separate intrusive bodies ranging from granodiorite to quartz monzonite and granite. The REE patterns of the younger rocks are less fractionated than those of the older suite and have negative Eu anomalies.

Granophyres occur at many places in the Singhbhum craton (Bose and Sinha, 1977). The Arkasani suite consists of a number of lenticular bodies that occur along the northern part of the western extension of the Singhbhum thrust zone (A. K. Banerji, 1972; A. K. Banerji et al., 1978). The bodies have massive cores and gneissic margins that become schistose near their contacts with surrounding mica schists. Most of the bodies trend east-west and are concordant with the S1 surface of the schists, indicating syntectonic emplacement with respect to the first phase of deformation in the thrust zone.

The massive cores of the Arkasani granophyres contain oval phenocrysts of white plagioclase (an_{20-30}) in a medium- to fine-grained groundmass of darker material. Massive types range in composition from monzo- to syenogranite, and gneissic and schistose margins range from granodiorite to tonalite. S. Sengupta et al. (1984) suggested that the Arkasani magmas were generated by partial melting of pelitic rocks in the thrust zone during the first phase of deformation and at pressures of about 2 kb. The oval phenocrysts were then formed by melting at pressures greater than 5 kb during further deformation.

The Tatanagar and Surda soda granites occur along the Singhbhum thrust zone near Jamshedpur (Fig. 7–4) as areas of development of sodic plagioclase (an_{05-20}) and quartz in folded schists (Talapatra, 1969). The massive feldspathic cores grade outward into schistose rocks with muscovite, biotite, and chlorite. The bodies are particularly well developed along fold noses. Potassium feldspar is apparently absent. The bodies could have formed by local anatexis or by metasomatism caused by externally derived fluids.

Other sialic rocks

Gneissic and granitic rocks have been mapped at various places within the Singhbhum nucleus (Fig. 7.1); many of these areas are virtually unstudied. A considerable area in the southern part of the craton consists of granites and gneisses regarded as basement by Iyengar and Murthy (1982) and as unclassified gneiss by Mazumdar (1978). The Bonai and Nilgiri Granites (Fig. 7.2) have been variously correlated with the Singhbhum and/or Mayurbhanj Granites, but specific relationships are unknown.

CHOTANAGPUR TERRAIN

The Chotanagpur terrain is a vast tract of mostly gneisses and granites north of the Singhbhum-Dhalbhum orogenic belt (Fig. 7.1). It is about 200 km wide and 500 km long, extending in an arcuate east-west belt. The complex is divided into northern and southern areas by the Damodar valley, a rift that contains Gondwana sediments. The western part of the terrain is also covered by Gondwana sediments. The northwestern margin is in contact with Vindhyan sedimentary rocks along the Son lineament. The southern margin is the Singhbhum-Dhalbhum mobile belt. The northern and part of the eastern margins are covered by alluvium of the Ganges valley. The northeastern margin is covered by Jurassic Rajmahal basalts, and the southeastern margin by Tertiary sediments.

The Satpura orogenic trend (Ch. 6) of the Sausar Group of the Bhandara craton has been proposed to continue into the Shillong plateau of Meghalaya (Ch. 8) through the Chotanagpur belt (Krishnan, 1953) or Singhbhum/Dhahlbhum mobile belt. The occurrence of charnockites and other high-grade rocks in the eastern part of the Chotanagpur belt suggests a correlation with similar rocks in the Eastern Ghats (Ch. 5), leading to the possibility that the Proterozoic Eastern Ghats belt and the Satpura belt are connected in the Shillong plateau. This mobile belt may have encircled an entire Archean Dharwar-Singhbhum craton (Radhakrishna and Naqvi, 1986).

The Chotanagpur belt consists of gneisses, metasedimentary rocks, granulites (including charnockites), subordinate mafic/ultramafic schists, and minor anorthosites. Gondites have been reported locally (S. Mukherjee and Bandyopadhyay, 1975). No isochron ages have been established. A. N. Sarkar (1982b) reported K/Ar dates that suggest that rocks of the belt range from 1,500 to 800 m.y. old. In this section, we discuss rocks that appear to be older than 600 m.y.

Stratigraphy and structure

The oldest rocks of the Chotanagpur belt are metamorphosed sandy, shaly, and calcareous rocks designated as Older Metasediments by Ghose (1983; Table 7.9). The Older Metasediments occur as enclaves of various sizes in the gneisses (Ulabhaje, 1973; S. R. Chatterji and Sen Gupta, 1980) and consist of metamorphosed pelites, carbonates, and quartzites at various facies up to high amphibolite. The pelitic schists (Table 7.10) are generally medium grained and contain muscovite, chlorite, biotite, quartz, plagioclase, garnet and sillimanite $\pm$ staurolite, $\pm$ kyanite, $\pm$ cordierite, $\pm$ graphite, $\pm$ andalusite (Ghose, 1983). Calc-magnesian metasediments consist of dolomitic marble, calc-silicate rocks and skarns occurring as discontinuous bands and lenticles in the metapelites and gneisses. Major minerals in the carbonate rocks include calcite, dolomite, talc, tremolite, phlogopite, forsterite, serpentine, diopside, plagioclase, and grossularite (S. R. Chatterji

TABLE 7.9. *History of Chotanagpur block (largely from Ghose, 1983)*

Paleozoic	Various intrusive rocks
Late Proterozoic	Biotite granodiorite, tonalite gneiss, orthoamphibolite
"Satpura Orogeny"	Pegmatites and veins
Syn- to late-tectonic intrusives	Massive to foliated granite, granodiorite and tonalite
Syntectonic basic intrusives	Anorthosite, orthoamphibolite, and metamorphosed mafic and ultramafic rocks
	Quartzite
	Metamorphosed ultramafic rocks (schists with talc, tremolite, actinolite, etc.) and layered anorthosites
Older Metasediments	Limestone, dolomite, calc-silicates; pelitic schists, gneisses and migmatites
Crystalline Basement	Tonalite gneiss, charnockite, khondalite, granulite and leptynite

TABLE 7.10. *Compositions of rocks in Chotanagpur block (Ghose, 1983)*

	1	2	3
SiO_2	63.5	50.2	70.0
TiO_2	0.88	1.3	0.45
Al_2O_3	14.3	14.9	13.3
Fe_2O_3	4.6	2.7	1.8
FeO	1.8	9.1	2.4
MnO	0.14	0.26	0.10
MgO	3.6	7.0	1.8
CaO	1.0	9.8	2.1
Na_2O	1.0	2.7	3.1
K_2O	4.2	0.95	4.0

All abundances of major elements are in weight percent of oxides. Where no entry is made in the FeO position, the value at Fe_2O_3 is total iron calculated as ferric oxide. All abundances of trace elements are in parts per million by weight. Values are taken directly from quoted sources and not recalculated to 100%.

Table 7.10. Column References
1. Average pelitic schist.
2. Average metabasic rock (mostly amphibolite).
3. Average gneiss.

and Sen Gupta, 1980; Ghose, 1983). Cross-bedded quartzites occur as small bands in the schists and gneisses. Amphibolites associated with the Older Measediments have a uniform mineralogy consisting largely of hornblende, actinolite and plagioclase (S. R. Chatterji and Sen Gupta, 1980; Ghose, 1983); compositions generally indicate a tholeiitic parent prior to metamorphism (Table 7.10).

The Older Metasediments were deformed in what has been regarded as the Satpura orogeny. Ultramafic bodies in the supracrustal suites were probably emplaced before deformation, and mafic bodies appear to be syntectonic. The mafic rocks now occur as amphibolites, metadolerites, metanorites, metagabbros, and pyroxene granulites (Ghose, 1983). The regional metamorphism was accompanied by large-scale syn- to late-tectonic granite emplacement and migmatization. Ghose (1983) considered the Chotanagpur belt to be a more granitoid part of a mobile belt that included the Singhbhum-Dhalbhum belt and extended across the Dharwar, Singhbhum, Aravalli, and Bundelkhand cratonic areas.

The northernmost part of the Chotanagpur area is occupied by the Gaya-Monghyr (Rajghyr) belt (Fig. 7.1). This belt contains thick sequences of quartzose and pelitic metasediments with interspersed gneissic terrain (Hassan and Sarkar, 1968; Ghose, 1983). The area has been complexly deformed. Age relationships of rocks in this belt to the Older Metasediments or other suites are unknown.

Detailed structural studies have shown three phases of deformation in the Chotanagpur terrain (B. K. Chattopadhyay, 1973; B. K. Chattopadhyay and Saha, 1974; B. P. Bhattacharyya, 1975; N. Chattopadhyay, 1975; Mehrotra and Verma, 1978; and S. R. Chatterji and Sen Gupta, 1979, 1980). The first generation of folds (F1) was developed on compositional layers (S1). They are isoclinal and accompanied by development of schistosity (S2) parallel to F1 axial planes. The second phase of folding (F2) affected both S1 and S2 and resulted in dome-and-basin structure. The F2 folds are open and disharmonic, with development of crenulation cleavage parallel to F2 axial planes in the metaphyllites; quartzites show fracture cleavage. The third phase of deformation (F3) produced large-scale warps of F2 axes.

Gneisses, granites and related rocks

The major part of the Chotanagpur region consists of gneisses (Ghose, 1970; Ghose and Chakraborty, 1978; Yadav, 1981). They engulf the Older Metasediments and associated amphibolites, and they are intruded by granitic rocks, including pegmatites of the Bihar mica belt. Most of the gneisses are granitic, but granodioritic and tonalitic varieties are present (Table 7.10). Some amphibolitic varieties contain both amphibole and biotite. Ghose regarded the composition of the rocks as indicative of a parent assemblage of pelites and graywackes.

Migmatization has occurred extensively in the gneissic terrain (Akella, 1970). Both trondhjemitic and granitic migmatites have been proposed to have formed by metasomatism, metamorphic differentiation, and pegmatite melt injection. In one area, injection of pegmatite fluids into paleo-somes of hornblende-biotite schist and amphibo-lite caused separation into leucosomes and mela-nosomes (Saha et al., 1976).

Charnockites and other granulitic rocks occur primarily in the eastern part of the belt, but some occurrences have been described to the west (Sen, 1967; Manna and Sen, 1974; B. P. Bhattacharyya, 1975, 1976; Sen and Manna, 1976; Ghose, 1983). They are commonly associated with amphibo-lites, norites, khondalites, leptynites, calc-granu-lites, and anorthosites. No pattern of relation-ships to tectonic features has been established, but little general mapping has been done.

Anorthosites and related igneous suites are broadly associated with granulite-facies rocks (J. K. Sinha and Bose, 1977; Mazumdar, 1978). An-orthosites occur both as independent bodies and in layered mafic/ultramafic complexes (P. K. Mu-kherjee, 1955; Ghose, 1983).

The best-studied anorthosite in the Chotanag-pur terrain is the Bengal anorthosite, which oc-cupies an area of 200 sq km at the eastern end of the belt. It consists of anorthosites, gabbroic an-orthosites and anorthositic gabbros emplaced in the Older Metasediments (S. C. Chatterjee, 1959, 1976; A. K. Roy, 1977). The main body forms an elongated dome overturned toward the south (Verma et al., 1975). Plagioclase in megacrysts and groundmass is an_{60-70}. The main body and wall rocks contain a large number of small bands of dioritic anorthosites, quartz-hornblende nor-ites, and hornblende syenites. The anorthosites are synkinematic with the major deformation (Ray, 1976), and S. C. Chatterjee (1959) regarded them as having formed as plagioclase cumulates from an original magma of anorthositic gabbro. Based on trace-element data, however, K. R. Roy and Saha (1975) stated that the main anorthosite body was not comagmatic with associated norites.

Silicic intrusive rocks, primarily adamellites, are widespread in the Chotanagpur terrain (Dunn, 1929; Ghose, 1983). Most of the bodies were em-placed along the trends of major fold axes during and after the major orogeny in the belt and have not been affected by later recrystallization (Sen, 1956; P. Mukhopadhyay, 1978). Two types of granites have been recognized in the Bihar mica belt (Mahadevan and Maithani, 1967): phacol-ithic sheets along the noses and limbs of folds that consist of gneissic, mostly nonporphyritic rocks; and subelliptical bodies, foliated only at the mar-gins and consisting of equigranular granite. D. K. Banerjee (1975) proposed formation of granodio-rites and adamellites by granitization of Older Metasediments. One pluton was described by B. K. Chattopadhyay and Saha (1974) as related to the Singhbhum Granite. Rapakivi granites occur at numerous places (S. Bhattacharyya, 1976; U. Bhattacharya, 1980).

Extensive pegmatite development in the north-ern part of the Chotanagpur terrain has resulted in the formation of large, economic deposits of muscovite, commonly referred to as the Bihar Mica Belt (Fig. 7.1; Holland, 1902; Das, 1965, 1971; Mahadevan and Maithani, 1967; and Deb and Kutty, 1978). The belt covers an area of about 2,400 sq km. The larger bodies follow axial plane schistosity in the country rocks. Both albite- and K-feldspar-rich pegmatites are present. The albi-tic varieties contain commercial-grade muscovite. Pegmatites with predominantly K-feldspar have little muscovite but locally contain columbite, tantalite, and beryl. One suite of pegmatites con-tains rare earth and radioactive minerals, Li-bear-ing minerals, and clevelandite (Saha and Ghosh, 1973). An evolutionary sequence has been pro-posed from K-feldspar pegmatites through albite pegmatites to the rare metal pegmatites (B. K. Chattopadhyay, 1973).

ECONOMIC MINERAL DEPOSITS

The Singhbhum craton contains copper, thorium and uranium, manganese, chromium, iron, tung-sten, apatite, mica, kyanite, graphite and other materials (summaries by Dunn, 1937; P. R. Sen Gupta, 1972). We discuss here only the copper and uranium deposits of the Singhbhum thrust zone.

Copper deposits

Copper ores occur along 50 km of the eastern part of the Singhbhum thrust zone (P. R. Sen Gupta, 1972; S. C. Sarkar and Deb, 1974; and Deb and Sarkar (1975). The ore bodies occur in a variety of rock types, including chlorite-quartz schists, magnetite-bearing chlorite schists, biotite-quartz schists, soda granites, and feldspathic schists. General amphibolite-facies metamorphism is characteristic of the area, but country rocks around the ore deposits are commonly in greenschist facies, possibly because of retrogres-sion. The ore bodies are tabular, sheet like, len-soid, and peneconcordant at most mines, al-though branching veins occur locally. Mineralization occurred about 1,600 m.y. ago. (Krishna Rao et al., 1979).

The copper ores contain both hypogene and su-pergene minerals. Hypogene minerals include chalcopyrite, pyrite, pyrrhotite, pentlandite, mil-lerite, ilmenite, molybdenite, cubanite, valleriite, uraninite, and rutile. Supergene minerals include marcasite, pyrite, cuprite, native copper, chalco-cite, and covellite (Dunn, 1937; P. R. Sen Gupta, 1972; S. C. Sarkar and Deb, 1974). Gangue min-erals include quartz, chlorite, biotite, sericite, pyr-ite, and magnetite. Dynamothermal metamor-phism of the ores is shown by a variety of textures, including polygonization of equigranular triple-junction points of grains in polycrystalline aggregates, annealing of twins, and poikiloblastic growth.

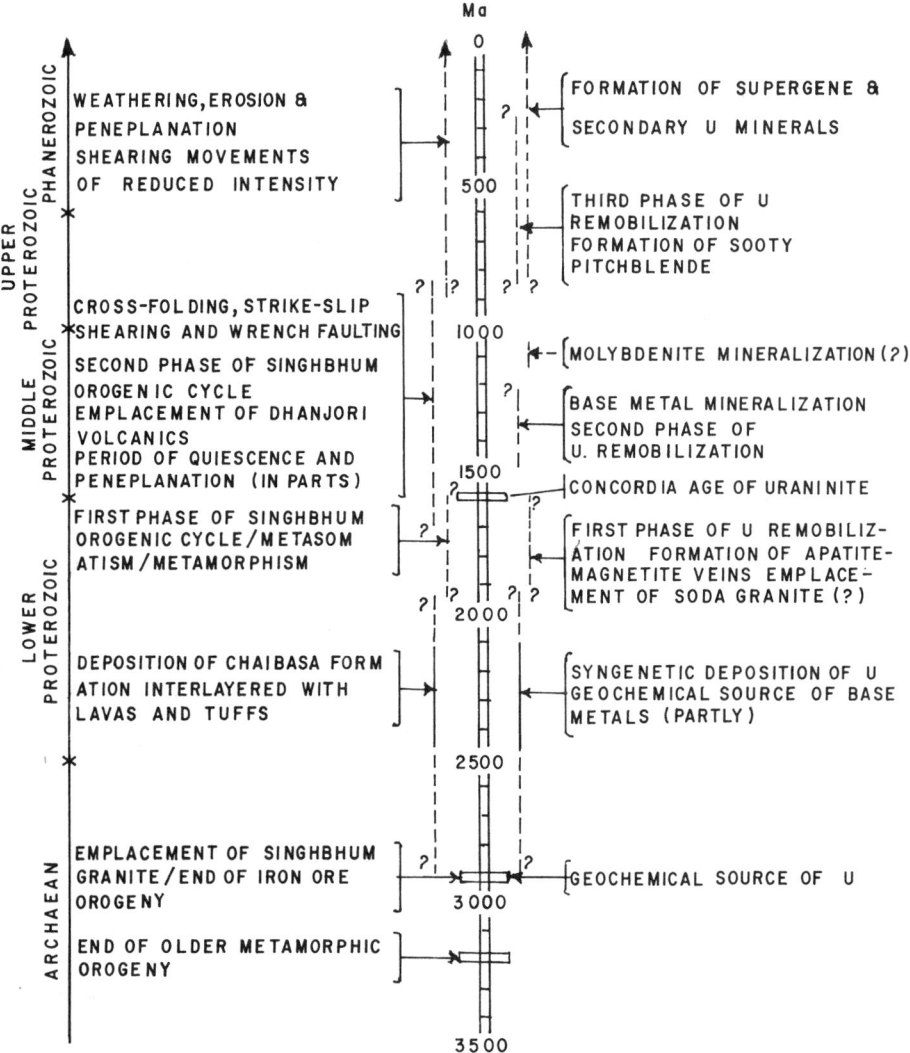

Fig. 7.8. Depositional history of uranium minerals in Singhbhum thrust zone (Krishna Rao and Rao, 1983d). Reproduced with permission from the Journal of the Geological Society of India, v. 24, p. 625.

The origin of the copper ores has been extensively debated. Dunn (1937) proposed that the ores were derived from the soda granites of the thrust zone. A combined metasomatic/hydrothermal model has been proposed by A. K. Banerji et al. (1972), A. K. Ghosh (1972), and A. K. Banerji (1981a). S. C. Sarkar et al. (1971) classified the deposits as exhalative volcanogenic, formed by metamorphic redistribution of copper originally contained in volcanic rocks along the thrust zone. Ray (1972) proposed that mineralization occurred from late hydrothermal solutions produced by metamorphism and partial melting of rocks of the Singhbhum-Dhalbhum mobile belt. Ray's model is consistent with the work of P. R. Sen Gupta (1972), who concluded that mineralization started when deformation and metamor-

phism were well advanced, with high-temperature mineralization caused by intermittent influx of hydrothermal fluids.

Uranium

The occurrence of uranium deposits in the Singhbhum thrust zone was first reported in 1921 (Bhola et al., 1966). The uranium deposits have recently been studied in a series of papers by Krishna Rao and Rao (1983a, 1983b, 1983c, 1983d). Earlier work was summarized by Krishna Rao and Rao (1983a).

The uranium in the thrust zone occurs in areas that have been extensively affected by chloritization, biotitization, sericitization, and feldspathization, with the introduction of such minerals as

157

tourmaline, apatite, allanite, xenotime and fluorite. This alteration has affected all rocks along the thrust belt except the mafic varieties. Localization of mineralization appears to have been controlled predominantly by small-scale structures. Ore mineralization is generally confined to the lower part of the Chaibasa Formation. Minerals are subparallel to bedding planes, schistosity, or shear zones, most of which features are also parallel to each other. Uranium lodes have been folded in the second generation of folding in the thrust belt.

Uranium occurs primarily in uraninite, which has three textural forms resulting from at least three episodes of mineralization (Krishna Rao and Rao, 1983d). Other uranium-bearing minerals include sooty pitchblende, brannerite, allanite, sphene, davidite, xenotime, monazite, uraniferous iron oxides, and secondary minerals (Krishna Rao and Rao, 1983a). Apparently, mineralization occurred over a broad range of temperatures in order to form this diverse assemblage.

The sequence of mineralization and other events in the uraniferous areas of the Singhbhum thrust zone are shown in Figure 7.8. According to this model, the uranium was originally derived by erosion of the Singhbhum Granite, which placed uranium in the Chaibasa sediments of the Singhbhum Group. Most of the deposition was in the form of detrital uraninite in conglomerates at the base of the section. The sedimentary uranium was then remobilized during deformation and locally enriched along the thrust zone. Later exogenic mobilization formed many of the low-temperature minerals.

REFERENCES

Acharya, S. (1964). Stratigraphy of the banded iron formations of Daiteri-Tomka area, Cuttack District, Orissa, India. 22nd Internat. Geol. Cong. Rept., Delhi, Sect. 10, 13–21.

Akella, J. (1970). Petrological studies on the rocks around Sibsagar, Bihar, India, and the origin of migmatites in the light of present experimental data. Ind. Mineralogist, v. 11, 11–23.

Baksi, A. K., Sarkar, S. N., and Saha, A. K. (1984). ^{40}Ar/^{39}Ar incremental heating studies on mineral separates from the oldest rocks of Singhbhum craton, eastern India—Their bearing on the tectonothermal history of the terrane (Abstract). Program Internat. Assoc. Seis. Phys. Earth's Interior, 31 Oct to 7 Nov., 1984, Natl. Geophys. Res. Inst., Hyderabad, p. 263.

Ballurkar, A., and Pal, P. C. (1979). Controls of asbestos mineralization in Roro ultramafics, Singhbhum District, Bihar. Geol. Soc. India J., v. 20, 158–169.

Bandyopadhyay, P. K. (1981). Chakradharpur granite gneiss—A composite batholith in the western part of the Singhbhum shear zone, Bihar. Ind. J. Earth Sci., v. 8, 109–118.

Banerjee, D. K. (1975). A tentative classification of the granitic rocks around Purulia Town, West Bengal, with a note on their mode of emplacement. Geol. Surv. India Misc. Publ. 23, Part 1, 139–147.

Banerjee, P. K. (1972). Geology and geochemistry of the Sukinda ultramafic field, Cuttack District, Orissa. Geol. Surv. India Mem. 103, 171 p.

—— (1982). Stratigraphy, petrology and geochemistry of some basic volcanic and associated rocks of Singhbhum District, Bihar, and Mayurbhanj and Keonjhar Districts, Orissa. Geol. Surv. India Mem. 111, 58 p.

Banerji, A. K. (1972). On the sequence of granitic activity and tectonism in the western part of the Singhbhum shear zone, Bihar. Geol. Min. Metall. Soc. India Quart. J., v. 44, 217–220.

—— (1975). On the evolution of the Singhbhum nucleus, eastern India. Geol., Min. Metall. Soc. India Quart. J., v. 47, 51–60.

—— (1977). On the Precambrian banded iron formations and the manganese ores of the Singhbhum region, eastern India. Econ. Geol., v. 72, 90–98.

—— (1981a). Ore genesis and its relationship to volcanism, tectonism, granitic activity, and metasomatism along the Singhbhum shear zone, eastern India. Econ. Geol., v. 76, 905–912.

—— (1981b). On the genesis of Lapsa Buru kyanite, Singhbhum District, Bihar. Geol. Soc. India J., v. 22, 496–501.

——, Talapatra, A. K., Sankaran, A. V., and Bhattacharyya, T. K. (1972). Ore genetic significance of geochemical trends during progressive migmatisation within part of the Singhbhum shear zone, Bihar. Geol. Soc. India J., v. 13, 39–50.

——, Bhattacharyya, M. C., and Chattopadhyay, B. (1978). Arkasani granophyres from the Singhbhum shear zone, Bihar. Geol. Soc. India J., v. 19, 350–358.

Basu, A., Gupta, A., and Gupta, P. K. (1979). Surface and subsurface structural framework of the area around Turamdils, Singhbhum District, Bihar. Ind. J. Earth Sci., v. 6, 52–66.

Basu, A. R., Ray, S. L., Sahu, A. K., and Sarkar, S. N. (1981). Eastern Indian 3800-million-year-old crust and early mantle differentiation. Science, v. 212, 1502–1506.

Bhattacharya, U. (1980). The Gariadih rapakivi granite of the Bihar mica belt, Hazaribagh District, Bihar. Geol. Soc. India J., v. 21, 21–29.

Bhattacharyya, B. P. (1975). Structural evolution in central part of Santhal Parganas District, Bihar. Geol., Min. Metall. Soc. India Bull., v. 48, 40–47.

—— (1976). Metamorphism of the Precambrian rocks of the central part of Santhal Parganas Distict, Bihar. Geol., Min. Metall. Soc. India Quart. J., v. 48, 183–196.

Bhattacharyya, D. S. (1966). Structure of the rocks around Chakradharpur, Bihar. Geol., Min. Metall. Soc. India Quart. J., v. 38, 143–151.

——— (1983). Tectonic evolution of northwest Singhbhum. Recent Res. Geol., v. 10, 72–80.

———, and Bhattacharyya, T. K. (1970). Geological and geophysical investigation of a basaltic layer in Archean terrain of eastern India. Geol. Soc. Amer. Bull., v. 81, 3073–3078.

———, Pasayat, S., and Sarkar, A. N. (1976). Zones of metamorphism in western Singhbhum, eastern India. Ind. J. Earth Sci., v. 3, 26–36.

Bhattacharyya, S. (1976). Emplacement of the Kohbarwa pluton in Bihar mica belt. Ind. Geol. Assoc. Bull., v. 9, 59–75.

Bhola, K. L., Rama Rao, Y. N., Sastry, C. S., and Mehta, N. R. (1966). Uranium mineralization in Singhbhum thrust belt, Bihar, India. Econ. Geol., v. 61, 162–173.

Bose, M. K. (1982). Precambrian picritic pillow lavas from Nomira, Keonjhar, eastern India. Curr. Sci., v. 51, 677–684.

———, and Chakraborty, M. K. (1981). Fossil marginal basin from the Indian shield: A model for the geology of Singhbhum Precambrian belt, eastern India. Geol. Rundschau, v. 70, 504–518.

———, and Goles, G. G. (1970). Chemical petrology of the ultramafic minor intrusions of Singhbhum, Bihar. Second Symp. Upper Mantle Proj. Proc., Natl. Geophys. Res. Inst., Hyderabad, 305–326.

———, and Sinha, J. K. (1977). Observations on Indian granophyres with special reference to granophyres of Barabar Hills, Gaya, Bihar. Ind. J. Earth Sci., S. Ray Volume, 225–250.

Chakraborthy, K. L. (1972). Some primary structures in the chromitites of Orissa, India. Mineralium Deposita, v. 7, 280–284.

——— (1973). Some characters of the bedded chromite deposits of Kalrangi, Cuttack District, Orissa, India. Mineralium Deposita, v. 8, 73–80.

———, and Baidya, T. (1978). Geological setting and mineralogy of of the chromite and some associated minerals of Kathapal, Dhenkanal District, Orissa. Geol. Soc. India J., v. 19, 303–309.

———, and Sen, S. K. (1967). Regional metamorphism of pelitic rocks around Kandra, Singhbhum, Bihar. Contrib. Mineral. Petrol., v. 16, 210–232.

———, Chakraborthy, T. L., and Majumder, T. (1980). Stratigraphy and structure of the Precambrian banded iron formation and chromite-bearing ultramafic rocks of Sukinda valley, Orissa. Geol. Soc. India J., v. 21, 398–404.

———, Bhattacharyya, A., and Roy, B. (1981). Gabbro-anorthosite-granophyre-leucogranite relationship around Gorumahisani Hill, Mayurbhanj District, Orissa. Geol. Soc. India J., v. 22, 336–341.

Chakraborti, M. K. (1980). On the pyroclastic rocks of Dalma volcanic sequence, Singhbhum, Bihar. Ind. J. Earth Sci., v. 7, 216–222.

Channakasavulu, N., and Sahu, K. C. (1980). Olivine and pyroxene of Amjori sill, Simlipal complex, Mayurbhanj. Geol. Soc. India J., v. 21, 211–231.

Chatterjee, A. K., and Mukherji, P. (1981). The structural set up of a part of the Malangatoli iron ore deposit, Orissa. Geol. Soc. India J., v. 22, 121–130.

Chatterjee, S. C. (1959). The problem of the anorthosites with special reference to the anorthosites of Bengal. 46th Ind. Sci. Cong. Proc., Part 2, 75–95.

——— (1976). Anorthosites of West Bengal and their metamorphism. Geol. Surv. India Misc. Publ. 23, Part 2, 245–254.

Chatterji, S. R., and Sen Gupta, D. K. (1979). A study of geometry and mechanics of folds in the Precambrian complex around Banka, Bhagalpur District, Bihar, India. Modern Geol., v. 7, 1–10.

———, and ——— (1980). Structural and petrological evolution of the rocks around Jamua-Kakwara-Bhitia of Satpura orogeny, Bhagalpur District, Bihar. Geol. Soc. India J., v 21, 171–183.

Chattopadhyay, B. K. (1973). On the migmatites around Inderwa, Hazaribagh District, Bihar. Ind. Acad. Geosci. J., v. 16, 61–80.

——— (1983). On the relationship between two granitic units of the Singhbhum Granite near Karanjia, Mayurbhanj District, Orissa. Geol. Soc. India J., v. 24, 639–654.

———, and Saha, A. K. (1974). The Neropahar pluton in eastern India: A model of Precambrian diapiric intrusion. Neues Jahrb. Mineral. Abh., v. 121, 103–126.

Chattopadhyay, N. (1975). Emplacement and evolutionary history of two granitic plutons from Bihar mica belt, Hazaribagh District, Bihar. Ind. Natl. Sci. Acad. Proc., v. 44A, 600–612.

Chaudhuri, A. K., and Pal, A. B. (1983a). Structural history as an aid in Precambrian correlation: An example from the Gangpur Group in eastern India. Geol. Soc. India J., v. 24, 522–532.

———, and ——— (1983b). Three phases of folding with macroscopic axial plane folding in the Gangpur Group in eastern India. Recent Res. Geol., v. 10, 110–119.

Das, B. (1965). On the origin and mechanics of emplacement of pegmatites and related bodies, Gurpa, Gaya District, Bihar (India). Geol. Soc. India J., v. 6, 94–107.

——— (1971). Tectonics of the western part of the mica belt around Bagai, Gurpalabni Marrni area, Bihar (India). In Studies in Earth Sciences (ed. T. V. V. G. R. K. Murty and S. S. Rao), Today and Tomorrow Printers, New Delhi, 161–180.

Das Gupta, H. C. (1969). Fe-Ni-V oxide ores associated with gabbro rocks around Dubabera, Bihar. Geol., Min. Metall. Soc. India Quart. J., v. 41, 51–64.

Deb, A., and Kutty, D. P. (1978). A study of the pegmatites and associated rocks of Saphitola and Jorasemar area, District Hazaribagh, Bihar. Recent Res. Geol., v. 4, 267–285.

———, and Sarkar, S. C. (1975). Sulphide orebodies and their relation to structure at Roam-Rakha Mines, Tamapahar section copper belt, Bihar. Recent Res. Geol, v. 1, 247–264.

Dunn, J. A. (1929). Geology of northern Singhbhum, including parts of Ranchi and Manbhum Districts, Geol. Surv. India Mem. 54, 1–66.

——— (1937). Mineral deposits of eastern Singhbhum and surrounding areas. Geol. Surv. India Mem. 63, Part 1, 280 p.

——— (1940). The stratigraphy of south Singhbhum. Geol. Surv. India Mem. 63, Part 3, 303–369.

———, and Dey, A. K. (1942). The geology and petrology of eastern Singhbhum and surrounding areas. Geol. Surv. India Mem. 69, part 2, 281–456.

Ghose, N. C. (1970). Geochemistry of thermal metamorphic and granitization processes in the aureole rocks around Richuguta, Dt. Palamau, Bihar, India. Geol. Rundschau, v. 59, 686–724.

——— (1983). Geology, tectonics and evolution of the Chotanagpur granite-gneiss complex, eastern India. Recent Res. Geol., v. 10, 211–247.

———, and Chakraborty, S. K. (1978). Petrology of the McCluskieganj granites, District Palamau, Bihar. Recent Res. Geol., v. 7, 33–52.

Ghosh, A. K. (1972). Trace element geochemistry and genesis of the copper ore deposits of the Singhbhum shear zone, eastern India. Mineralium Deposita, v. 7, 292–313.

———, and Banerji, A. K. (1970). On the nature and petrogenesis of Dhanjori lava near Rakha Mines, Singhbhum, Bihar. Geol. Soc. India. J., v. 11, 77–81.

Ghosh, S. K. (1963). Structural, metamorphic, and migmatitic history of the area around Kuilapal, eastern India. Geol., Min. Metall. Soc. India Quart. J., v. 35, 211–234.

Guha, P. K. (1963). Structural relations of Newer Dolerite intrusions of Bihar and Orissa. Geol., Min. Metall. Soc. India Quart. J., v. 41, no. 4, 205–210.

Gupta, A. K. , Basu, A., and Ghosh, P. K. (1980). The Proterozoic ultramafic and mafic lavas and tuffs of the Dalma greenstone belt, Singhbhum, eastern India. Can. J. Earth Sci., v. 17, 210–231.

———, ———, and Srivastava, D. (1981). Mafic and ultramafic volcanism of Ongarbira greenstone belt, Singhbhum, Bihar. Geol. Soc. India J., v. 22, 593–596.

Gupta, M. L. (1982). Heat flow in the Indian peninsula and its geological and geophysical implications. Tectonophysics, v. 83, 71–90.

Haldar, D., and Chatterjee, P. K. (1976). Interrelationship of the ultramafic and mafic-granophyric suites of rocks in the Nausahi-Nilgiri belt, Orissa. Geol. Surv. India Misc. Publ. 23, Part 2, 299–309.

Hassan, Z., and Sarkar, S. N. (1968). Structural analysis of the Monghyr area, India. Norsk Geol. Tiddskr., v. 48, 101–116.

Holland, T. H. (1902). The mica deposits of India. Geol. Surv. India Mem. 34, Part 2, 11–111.

Holmes, A. (1955). Dating the Precambrians of peninsular India and Ceylon. Geol. Assoc. Canada Proc., v. 7, 81–106.

Iyengar, S. V. P., and Alwar, M. A. (1965). The Dhanjori eugeosyncline and its bearing on the stratigraphy of the Singhbhum, Keonjhar, and Mayurbhanj Districts. In D. N. Wadia Commemorative Volume, Geol., Min. Metall. Inst. India, Calcutta, 138–162.

———, and Banerjee, S. (1964). Magmatic phases associated with the Precambrian tectonics of Mayurbhanj District, Orissa, India. 22nd Internat. Geol. Cong. Rept., Delhi, Sect. 10, 515–538.

———, and Murthy, Y. G. K. (1982). The evolution of the Archaean-Proterozoic crust in parts of Bihar and Orissa, eastern India. Geol. Surv. India Records, v. 112, Part 3, 1–5.

———, Chandy, K. C., and Narayanswamy, R. (1981a). Geochronology and Rb-Sr systematics of the igneous rocks of the Simlipal Complex, Orissa. Ind. J. Earth Sci., v. 8, 61–65.

———, Kresten, P., Paul, D. K., and Brunfelt, A. O. (1981b). Geochemistry of Precambrian magmatic rocks of Mayurbhanj District, Orissa. Geol. Soc. India J., v. 22, 305–315.

Kanungo, D. N., and Mahalik, N. K. (1972). Metamorphism in the eastern part of Gangpur series. Geol. Soc. India J., v. 13, 112–130.

———, and ——— (1975). A revision of the stratigraphy and structure of Gangpur series in Sundergarh District, Orissa, and their tectonic history. Geol. Surv. India Misc. Publ. 23, Part 1, 130–138.

Krishna Rao, N., and Rao, G. V. U. (1983a). Uranium mineralization in Singhbhum shear zone, Bihar: I, Ore mineralogy and petrography. Geol. Soc. India J., v. 24, 437–453.

———, and ——— (1983b). Uranium mineralization in Singhbhum shear zone, Bihar: II, Occurrence of brannerite. Geol. Soc. India J., v. 24, 489–501.

———, and ——— (1983c). Uranium mineralization in Singhbhum shear zone, Bihar: III, Nature of occurrence of uranium in apatite-magnetite rocks. Geol. Soc. India J., v. 24, 555–561.

——, and —— (1983d). Uranium mineralization in Singhbhum shear zone, Bihar. IV, Origin and geological time frame. Geol. Soc. India J., v. 24, 615–627.

——, Agarwal, S. K., and Rao, G. V. U. (1979). Lead isotopic ratios of uraninites and the age of uranium mineralization in Singhbhum shear zone, Bihar. Geol. Soc. India J., v. 20, 124–127.

Krishnan, M. S. (1953). Structural and tectonic history of India. Geol. Surv. India Mem. 81, 9–10.

Lal, R. K., and Ackermand, D. (1979). Coexisting chloritoid-staurolite from the sillimanite (fibrolite) zone, Sini, District Singhbhum, India. Lithos, v. 12, 133–142.

——, and Singh, J. B. (1977). Prograde regional metamorphism of pelitic rocks in Sini, District Singhbhum. Recent Res. Geol., v. 3, 217–232.

—— and —— (1978). Prograde polyphase regional metamorphism and metamorphic reactions in pelitic schists at Sini, District Singhbhum, India. Neues Jahrb. Mineral. Abh., v. 131, 304–333.

Mahadevan, T. M., and Maithani, J. B. P. (1967). Geology and petrology of the mica pegmatites in parts of the Bihar mica belt. Geol. Surv. India Mem. 93, 125 p.

Majumder, P., and Chakraborty, K. L. (1977). Primary sedimentary structures in the Banded Iron Formation of Orissa, India. Sediment. Geol., v. 19, 287–300.

——, and —— (1979). Petrography and petrology of the Precambrian banded iron formation of Orissa, India, and reformation of the bands. Sediment. Geol., v. 22, 243–265.

——, Pal, J. C., and Das, S. B. (1979). Trace element study of banded iron formation and associated rocks of Orissa. Geol. Surv. India Spec. Publ. 1, 585–597.

Manna, S. S., and Sen, S. K. (1974). Origin of garnet in the basic granulites around Saltora, W. Bengal, India. Contrib. Mineral. Petrol., v. 44, 195–218.

Mazumdar, S. K. (1978). Precambrian geology of eastern India between the Ganga and the Mahanadi: A review. Geol. Surv. India Records, v. 110, Part 2, 60–116.

McCall, G. J. R. (ed.) (1979). Astroblemes—Cryptoexplosion Structures. Dowden, Hutchinson Ross, Stroudsburg, Pennsylvania, 437 p.

Mehrotra, R. C., and Verma, S. N. (1978). On metasedimentary origin of the Precambrian quartzo-feldspathic rocks around Pipra, District Sidhi, India. Recent Res. Geol., v. 7, 339–377.

Mitra, S. (1973). Olivines from Sukinda ultramafites and the nature of the parental magma. Neues Jahrb. Mineral. Abh., v. 118, 177–189.

Mukherjee, P. K. (1955). The geology of the area around Simra, Santhal Parganas. Geol., Min. Metall. Soc. India Quart. J., v. 28, no. 2, 67–84.

Mukherjee, S., and Bandyopadhyay, B. K. (1975). The gonditic rocks of Manbazar, Purulia District, West Bengal. Geol. Soc. India J., v. 16, 485–490.

——, and Haldar, D. (1975). Sedimentary structures displayed by the ultramafic rocks of Nausahi, Keonjhar District, Orissa, India. Mineralium Deposita, v. 10, 109–119.

Mukhopadhyay, A., and Chanda, S. K. (1972). Silica diagenesis in the banded hematite jasper and bedded chert associated with the Iron Ore Group of Jamda-Koira valley, Orissa, India. Sediment. Geol., v. 8, 113–135.

Mukhopadhyay, D. (1976). Precambrian stratigraphy of Singhbhum—The problems and a prospect. Ind. J. Earth Sci., v. 3, 208–219.

——, and Dutta, D. R. (1983). Structures in the Roro ultramafics and their country rocks, Singhbhum District, Bihar. Recent Res. Geol., v. 10, 98–109.

——, Ghosh, A. K., and Bhattacharya, S. (1975). A reassessment of structures in the Singhbhum shear zone. Geol., Min. Metall. Soc. India Bull. 48, 49–67.

——, Dasgupta, S., and Dhar, K. (1980). Basement slices within the cover sediments near Hakegora and Jaikan, south of Tatanagar, Singhbhum District, Bihar. Geol. Soc. India J., v. 21, 286–294.

Mukhopadhyay, P. (1978). Mode of emplacement of two catazonal granite plutons in the Bihar mica belt. Ind. Natl. Sci. Acad. Proc., Part A, v. 44, 229–235.

Murthy, V. N. (1975a). Note on the stratigraphy of the Iron Ore Group rocks in the southern part of the Bonai-Keonjhar belt, Orissa. Geol. Surv. India Misc. Publ. 23, Part 1, 115–128.

—— (1975b). Paleosalinity studies on some Precambrian shales of the Bonai-Keonjhar belt, Orissa: A preliminary appraisal. Geol. Soc. India J., v. 16, 230–234.

Naha, K. (1961). Precambrian sedimentation around Ghatsila in east Singhbhum in eastern India. Ind. Natl. Inst. Sci. Proc., Part A, v. 27, 361–372.

—— (1965). Metamorphism in relation to stratigraphy, structure, and movements in part of east Singhbum, eastern India. Geol., Min. Metall. Soc. India Quart. J., v. 37, 41–88.

Page, N. J., Banerji, P. K., and Haffty, J. (1985). Characteristics of the Sukinda and Nausahi ultramafic complexes, Orissa, India by platinum-group element geochemistry. Precamb. Res., v. 30, 27–42.

Pandey, N., and Chatterjee, D. K. (1984). Petrology of the Precambrian banded iron formation of Gandhamardhan Hill, west of Keonjhargarh, Orissa. Geol. Soc. India J., v. 25, 286–294.

Prasad, A. K., and Mahto, R. A. (1979). Petrochemistry of the Precambrian metabasites from Sundargarh District, Orissa, India. Ind. Geol. Assoc. Bull., v. 12, 91–110.

Prasada Rao, G. H. S. V., Murthy, Y. G. K., and Deekshitulu, M. N. (1964). Precambrian structures and igneous activity in parts of Balasore, Keonjhar, Cuttack, Dhenkanal, and Sundergarh Districts, Orissa. 22nd Internat. Geol. Cong. Rept. Delhi, Sect. 4, 390–401.

Radhakrishna, B. P., and Naqvi, S. M. (1986). Precambrian continental crust of India and its evolution. J. Geol., v. 94, 145–166.

Rai, K. L., Sarkar, S. N., and Paul, P. R. (1980). Primary depositional and diagenetic features in the banded iron formation and associated iron ore deposits of Noamundi, Singhbhum District, Bihar, India. Mineralium Deposita, v. 15, 189–200.

Rao, R. U. M., and Rao, G. V. (1974). Results of some geothermal studies in Singhbhum thrust belt, India. Geothermics, v. 3, 153–161.

——, Rao, G. V., and Narain, H. (1976). Radioactive heat generation and heat flow in the Indian shield. Earth Planet. Sci. Lett., v. 30, 57–64.

Ray, S. (1972). An integrated model for ore mineralization in the Singhbhum belt. Geol., Min. Metall. Soc. India Quart. J., v 44, 171–184.

—— (1976). Metamorphic history of the Singhbhum Precambrian belt—A preliminary synthesis. Geol. Surv. India Misc. Publ. 23, Part 2, 291–298.

——, Ghosh, S., and Saha, A. K. (1980). Biotites from the Older Metamorphic tonalite-gneiss and the Singhbhum Granite, eastern India. Ind. Natl. Sci. Acad. Proc., Part A, v. 46, 410–422.

Roonwal, G. S. (1972). Trace elements in the Singhbhum Granite, India. Geol. Rundschau, v. 61, 282–301.

Roy, A. K. (1977). Structural and metamorphic evolution of the Bengal anorthosites and associated rocks. Geol. Soc. India J., v. 18, 203–223.

Roy, K. R., and Saha, A. K. (1975). Trace element geochemistry of the Bengal anorthosites and associated rocks. Neues Jahrb. Mineral. Abh., v. 125, 297–314.

Saha, A. K. (1972). Petrogenetic and structural evolution of the Singhbhum granitic complex, eastern India. 24th Internat. Geol. Cong. Rept., Montreal, No. 24, 147–155.

—— (1975). The Mayurbhanj Granite; A Precambrian batholith in eastern India. Geol. Soc. India J., v. 16, 37–43.

—— (1979). Geochemistry of the Archaean granites of the Indian shield—A review. Geol. Soc. India J., v. 20, 375–391.

——, and Ghosh, K. P. (1973). Mineralogical studies on some rare metal pegmatite bodies of Bihar mica belt. Ind. Mineralogist, v. 14, 50–59.

——, and Rao, S. V. L. N. (1971). Quantitative discrimination between magmatic units of the Singhbhum Granite. Internat. Assoc. Math. Geol. J., v. 3, 123–133.

——, and Ray, S. L. (1984). The structural and geochemical evolution of the Singhbhum granite batholithic complex, India. Tectonophysics, v. 105, 163–176.

——, Bose, M. K., Sankaran, A. V., and Bhattacharya, T. K. (1972). Petrology and geochemistry of the ultramafic intrusion of Keshargaria, Singhbhum, Bihar. Geol. Soc. India J., v. 13, 118–121.

——, Sankaran, A. V., and Bhattacharyya, T. K. (1973a). Geochemistry of the Newer Dolerite suite of intrusions within the Singhbhum Granite; A preliminary study. Geol. Soc. India J., v. 14, 329–346.

——, Chakrabarti, S., and Sankaran, A. V. (1973b). A statistical study of the trace-element distribution in the magmatic members of Singhbhum Granite. Ind. Natl. Acad. Sci. Proc., Part A, v. 39, 171–184.

——, Bhattacharyya, C., and Lakshmipathy, S. (1976). Quantitative studies of migmatites of Tulin, Purulia District, West Bengal. Ind. J. Earth Sci., v. 3, 44–54.

——, Bose, R., Ghosh, S. N., and Roy, A. (1977). Petrology and emplacement of the Mayurbhanj granite batholith, eastern India. Geol., Min. Metall. Soc. India Bull. 49, 1–34.

Sahoo, R. K., and Van der Kaaden, C. (1976). Chemistry of the Sukinda chromites, Orissa, India and its petrogenetic significance. Neues Jahrb. Mineral. Monats., no. 11, 484–494.

Sarkar, A. N. (1977). Structure and metamorphism of the Baraganda-Dondlo area, Hazaribagh District, Bihar, and their bearing on the base metal mineralisation. Geol. Surv. India Records, v. 109, Part 2, 170–202.

—— (1982a). Structural and petrological evolution of the Precambrian rocks in western Singhbhum, Bihar. Geol. Surv. India Mem. 113, 92 p.

—— (1982b). Precambrian tectonic evolution of eastern India; A model of converging microplates. Tectonophysics, v. 86, 363–397.

——, and Bhattacharyya, D. S. (1978a). Geometry of superposed deformation in Hesadi-Tebo area, western Singhbhum, Bihar. Geol. Soc. India J., v. 19, 39–45.

——, and —— (1978b). Deformation and metamorphism in the western part of Singhbhum shear belt around Lapsa Buru and Kharsawan, Bihar. Geol. Soc. India J., v. 19, 310–320.

——, and —— (1978c). Deformation cycles and intersecting isograds; The Hesadi antiform of the Singhbhum orogenic belt. Contrib. Mineral. Petrol., v. 66, 333–340.

Sarkar, S. C., and Deb, M. (1974). Metamorphism of sulfides of the Singhbhum copper belt, India;

The evidence from the ore fabric. Econ. Geol., v. 69, 1282–1293.

——, ——, and Roy Chowdhury, K. (1971). Sulfide ore mineralization along Singhbhum shear zone, Bihar, India. In Geochemistry and Crystallography of Sulphide Minerals in Hydrothermal Deposits, Internat. Mineral. Assoc. and Internat. Assoc. Genesis Ore Deposits Joint Symp. on Volcanology, Soc. Mineral. Geol. Japan, Spec. Issue 3, 226–234.

Sarkar, S. N., and Saha, A. K. (1962). A revision of the Pre-Cambrian stratigraphy and tectonics of Singhbhum and adjacent regions. Geol., Min. Metall. Soc. India Quart. J., v. 34, 97–136.

——, and —— (1977). The present status of the Precambrian stratigraphy, tectonics and geochronology of Singhbhum-Keonjhar-Mayurbhanj region, eastern India. Ind. J. Earth Sci, S. Ray Volume, 37–65.

——, and —— (1983). Structure and tectonics of the Singhbhum-Orissa-Iron Ore craton, eastern India. Recent Res. Geol., v. 10, 1–25.

——, ——, Boelrijk, A. I. M., and Hebeda, E. H. (1979). New data on the geochronology of the Older Metamorphic Group and the Singhbhum Granite of Singhbhum-Keonjhar-Mayurbhanj region, eastern India. Ind. J. Earth Sci., v. 6, 32–51.

Sarkar, S. S. (1982). Notopahar granitic complex; An example of magmatic differentiation from Mayurbhanj District, Orissa. Geol. Soc. India J., v. 23, 53–66.

Sen, S. (1956). Structures of the porphyritic granites and associated metamorphic rocks of east Manbhum, Bihar, India. Geol. Soc. Amer. Bull., v. 67, 647–670.

—— (1967). Charnockites of Manbhum and the charnockite problem. Geol. Soc. India J., v. 8, 8–17.

——, and Manna S. S. (1976). Pattern of cation fractionation among pyroxene, hornblende and garnet in the basic granulites of Saltora, W. Bengal. Ind. J. Earth Sci., v. 5, 117–128.

Sen Gupta, P. R. (1972). Studies on mineralization in the southeastern part of the Singhbhum copper belt, Bihar, India. Geol. Surv. India Mem. 101, 82 p.

Sengupta, S., van den Hul, H. J., and Bandyopadhyay, P. K. (1983). Geochemistry of the Chakradharpur granite-gneiss complex—A Precambrian trondhjemite body from West Singhbhum, eastern India. Precamb. Res., v. 23, 57–78.

——, Bandyopadhyay, P. K. , van den Hul, H. J., and Chattopadhyay, B. (1984). Arkasani granophyre: Proterozoic intraplate acid magmatic activity in the Singhbhum craton, eastern India. Neues Jahrb. Mineral. Abh., v. 148, 328–343.

Sinha, J. K., and Bose, M. K. (1977). Petrology of the anorthosite suite, east of Bala, District Gaya, Bihar. Geol. Soc. India J., v. 18, 129–138.

Sinha, R. C., and Ghose, N. C. (1975). Petrochemistry of differentiated pre-Cambrian gabbroic rocks around Dalki, Mayurbhanj District, Orissa, India. Neues Jahrb. Mineral. Abh., v 124, 326–348.

Srivastava, R. K. (1975). Petrochemistry of differentiated pre-Cambrian gabbroic rocks around Dalki, Mayurbhanj District, Orissa, India. Neues Jahrb. Mineral. Abh., v. 124, 326–348.

Talapatra, A. K. (1969). Study of feldspathic schists and soda granites from the southeastern part of the Singhbhum shear zone, eastern India. Geol., Min. Metall. Soc. India Quart. J., v. 41, 65–81.

Ulabhaje, A. V. (1973). Petrochemistry and study of the thermal metamorphic processes in the calcareous rocks around Kodag, District Palamau, Bihar. Geol. Soc. India J., v. 14, 423–428.

Verma, R. K., and Prasad, S. N. (1974). Paleomagnetic study and chemistry of Newer Dolerites from Singhbhum, Bihar, India. Can. J. Earth Sci., v. 11, 1043–1054.

——, Ghosh, D., Roy, S. K., and Ghosh, A. (1975). Gravity survey over Bankura anorthosite complex, West Bengal. Geol. Soc. India J., v. 16, 361–367.

——, Sarma, A. U. S., and Mukhopadhyay, M. (1984). Gravity field over Singhbhum; Its relationship to geology and tectonic history. Tectonophysics, v. 106, 87–107.

Yadav, R. P. (1981). The paragneisses of Balumath, District Palamau, Bihar. Ind. Acad. Geosci. Bull., v. 24, 29–40.

8. MEGHALAYA

The Shillong plateau covers parts of the states of Assam and Meghalaya (Figs. 1.1 and 8.1). The southern edge of the plateau is the Dawki lineament (fault), and the northern edge is covered by alluvium of the Brahmaputra River. Precambrian rocks crop out throughout the plateau and in isolated areas in surrounding terrains. They are separated from the Chotanagpur area of the Singhbhum craton (Ch. 7) by the north-south-trending Garo-Rajmahal gap. Along the eastern boundary of the plateau the Precambrian rocks disappear beneath folded sediments of the Assam area.

The Meghalayan suite is largely surrounded by complexly deformed Phanerozoic rocks of the Himalayas and associated mountain belts. The structural pattern is known as the Assam syntaxis, a term implying a pivot around which wrapping of the orogenic belts surrounded the Meghalayan suite. Nandy (1976), however, showed that no simple pivot exists.

Deep drilling and geophysical data show that Precambrian rocks of the Shillong plateau extend for many tens of kilometers northeast of the plateau, beneath the Brahmaputra alluvium, and also westward to the Chotanagpur area (Evans, 1964; Desikachar, 1974). The connection between the Chotanagpur and Meghalaya Precambrian suites has been designated the Malda-Dinajpur ridge, which at some places is covered by less than 300 m of alluvium. The presence of this continuous ridge contradicts the proposal of Evans (1964) of separation of Meghalaya from the remainder of peninsular India by up to 200 km of movement along the Dawki fault (Fig. 8.1; Auden, 1974; Desikachar, 1974).

STRATIGRAPHY AND STRUCTURE

Recent information on the Shillong plateau has been summarized by Mazumdar (1976) and Murthy et al. (1976a, 1976b). Mazumdar (Table 8.1) subdivided the Precambrian rocks of the Khasi Hills part of the plateau into the Gneissic Complex, nonporphyritic granitic rocks, the Shillong Group of supracrustal rocks, the Khasi Greenstone, and porphyritic plutons. The Gneissic Complex and the nonporphyritic granitic rocks form the basement for the Shillong Group. An unconformity at the base of the Shillong Group is marked by a quartz pebble conglomerate. The Khasi Greenstone was emplaced into the Shillong Group as sills and dikes prior to metamorphism. The youngest Precambrian rocks of the area are the porphyritic granites, which intrude all other formations. Mazumdar's classification is similar to that of Choudhury and Narayana Rao (1975) except that Choudhury and Narayana Rao grouped the Gneissic Complex and nonporphyritic granites into an older metamorphic group and granitic gneiss and separated a Mylliem Granite from other porphyritic granite plutons.

Both the gneisses and supracrustal rocks of the Meghalaya area show polyphase deformation, but the relationships between the deformational episodes in the two suites are uncertain (Barooah, 1976; Mazumdar, 1976; Maswood, 1982). Polyphase deformation of the gneisses is shown as early foliation defined by mineral elongation and compositional banding parallel to isoclinal fold axes and reorientation of this foliation by refolding during later deformational events. A late phase of open folding is well preserved.

Polyphase deformation in the weakly metamorphosed Shillong Group is better known than in the gneisses (Mazumdar, 1976). Primary sedimentary structures, such as current bedding and ripple marks, have been preserved. Mazumdar proposed that the principal folding style was "crest-like," consisting of sharp anticlinal crests between areas of flat dips. A different interpretation of the structures of the Shillong Group was proposed by Barooah (1976), who recognized three phases of deformation. The earliest phase caused development of foliation, folding of foliation, and the major metamorphism. The second phase of deformation produced asymmetrical to isoclinal folds of various dimensions, resulting in the development of slaty cleavage and dispersion of the earlier fold axes. The map pattern of the area is largely controlled by the second folds, and their planes localized intrusion of the porphyritic granites and hydrothermal solutions. Folds of the third phase are open, steeply plunging and accompanied by axial plane jointing. The major deformation of the area presumably occurred during the Precambrian. Phanerozoic deformation resulted in extensive fracturing and uplift of the region.

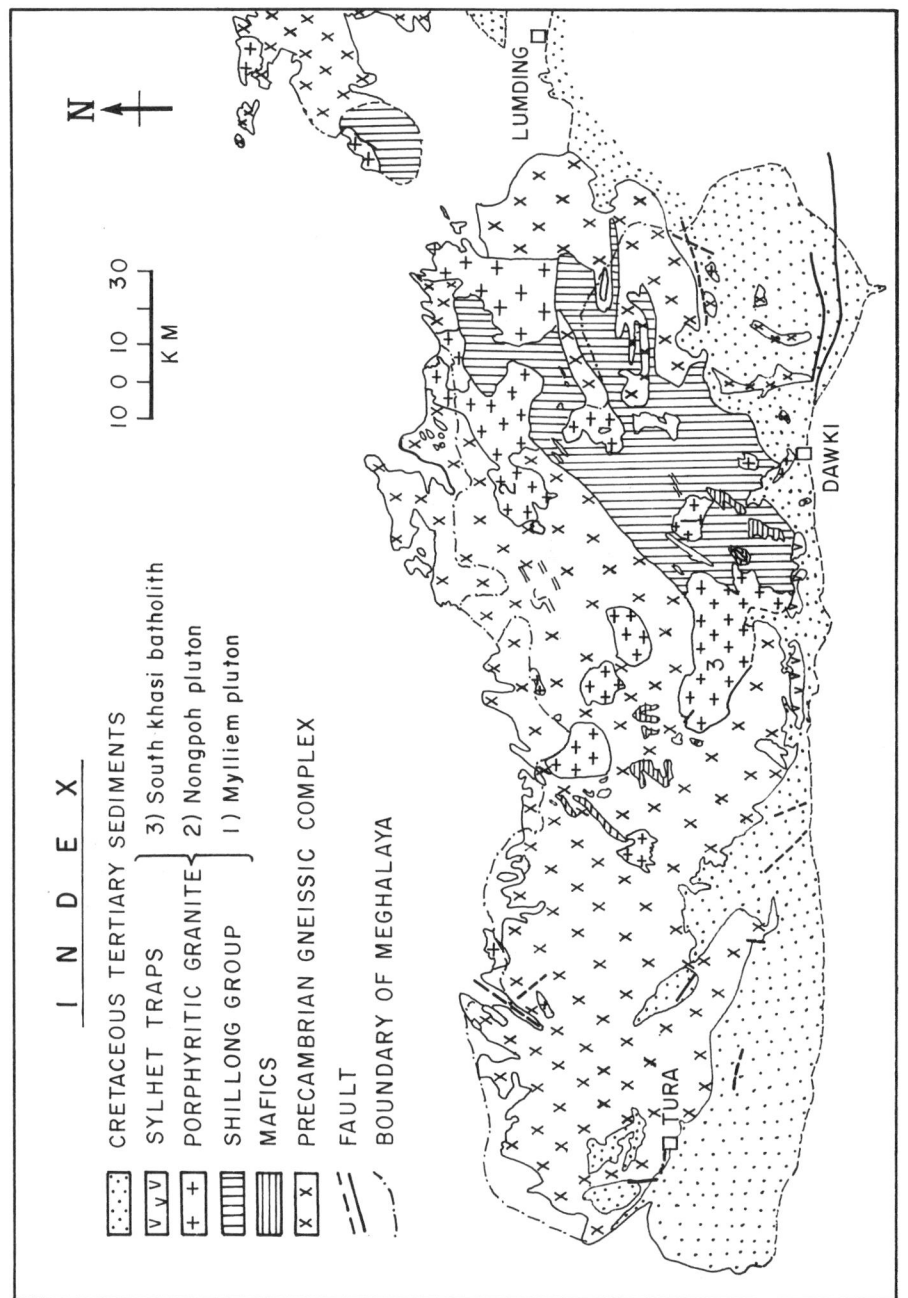

I N D E X

CRETACEOUS TERTIARY SEDIMENTS

SYLHET TRAPS

PORPHYRITIC GRANITE ⎱ 3) South khasi batholith
⎰ 2) Nongpoh pluton
1) Mylliem pluton

SHILLONG GROUP

MAFICS

PRECAMBRIAN GNEISSIC COMPLEX

FAULT

BOUNDARY OF MEGHALAYA

10 0 10   30
K M

N

LUMDING

DAWKI

TURA

Fig. 8.1. Meghalaya (Mazumdar, 1976).

164

TABLE 8.1. *General stratigraphy of Meghalaya (Mazumdar, 1976)*

Porphyritic Granite	Plutons throughout Meghalaya and Assam, intrusive into all older suites; posttectonic intrusions at the end of a major orogeny; extensively recrystallized
--Intrusive Contact --	
Khasi Greenstone	Mafic sills and dikes, mostly within Shillong Group; emplaced before metamorphism of Shillong Group but of uncertain relationship to major folding
Shillong Group	Metamorphosed conglomerates, sandstone, siltstones, and shales; metamorphism increases near Porphyritic Granite plutons; persistent NE strikes, with open folds alternating with areas of steep dips
--Unconformity ---	
Nonporphyritic Migmatitic Granitoids	Amphibolite facies rocks within the Gneissic Complex; mobile, syn- to late tectonic gneiss domes formed during the youngest phase of regional deformation
--Diffuse Contact---	
Gneissic Complex	Various gneisses that represent a major deformational event; enclaves of augen gneisses may represent relics of an earlier orogeny

PETROLOGY

The Gneissic Complex consists of both paragneisses (Gogoi, 1975) and orthogneisses. The paragneisses apparently formed from a variety of pelitic, mafic, impure calcareous, and banded ferruginous rocks. These rocks now occur as hornblende ($\pm$ biotite) schists, amphibolites, quartz-magnetite schists, tremolite schists, cordierite-anthophyllite schists, garnet-mica schists, and a variety of granulites containing such index minerals as sapphirine, cordierite, sillimanite, and corundum (Lal et al., 1980). The orthogneisses are typically biotitic granite gneisses that grade into migmatized and augen-bearing varieties (Choudhury and Narayana Rao, 1975; Maswood, 1981).

The Shillong Group consists mainly of sandy and clayey rocks that have undergone low-grade metamorphism (Ahmed, 1983). The dominant rock types are quartzites, consisting of quartz grains in a sericitic matrix. The Khasi Greenstone is an amphibolitic rock consisting of hornblende and plagioclase (Rahman, 1981). The young porphyritic granites consist of large K-feldspar phenocrysts in a matrix of quartz, plagioclase, biotite, and minor hornblende; minor minerals include apatite, zircon, sphene, and opaque ores (Maswood, 1979).

REFERENCES

Ahmed, M. (1983). Depositional environment of the basal conglomerate of the Barapani Formation, Shillong Group, Khasi Hills, Meghalaya. Geol., Min. Metall. Soc. India Quart. J., v. 55, 62–68.

Auden, J. B. (1974). Review of "Himalayan Geology, v. 2, edited by A. G. Jhingran, K. S. Valdiya, and A. K. Jain, Wadia Inst. Himalayan Geol., Dehra Dun." Geol. Soc. India J., v. 15, 216–218.

Barooah, B. C. (1976). Tectonic pattern of the Precambrian rocks around Tyresad, Meghalaya. Geol. Surv. India Misc. Publ. 23, Part 2, 485–496.

Choudhury, J. M., and Narayana Rao, M. (1975). A review of the Precambrian stratigraphy of the Assam-Meghalaya plateau. Geol. Surv. India Misc. Publ. 23, Part 1, 27–36.

Desikachar, S. V. (1974). A review of the tectonic and geologic history of eastern India. Geol. Soc. India J., v. 15, 137–149.

Evans, P. (1964). The tectonic framework of Assam. Geol. Soc. India J., v. 5, 80–96.

Gogoi, K. (1975). The geology of the Precambrian rocks in the northwestern part of the Khasi and Jaintia Hills, Meghalaya. Geol. Surv. India Misc. Publ. 23, Part 1, 37–48.

Lal, R. K., Ackermand, D., Seifert, F., and Halder, S. K. (1980). Chemographic relationships in sapphirine-bearing rocks from Sonapahar, Assam, India. Contrib. Mineral. Petrol., v. 67, 169–187.

Maswood, M. (1979). Grey porphyritic granite around Gauhati, Assam. Geol., Min. Metall. Soc. India Quart. J., v. 51, 127–132.

—— (1981). Granite gneiss around Gauhati, Assam. Geol., Min. Metall. Soc. India Quart. J., v. 53, 115–124.

—— (1982). Structural geology of the Precambrian rocks around Gauhati, Assam. Geol., Min. Metall. Soc. India Quart. J., v. 54, 33–36.

Mazumdar, S. K. (1976). A summary of the Precambrian geology of the Khasi Hills, Meghalaya. Geol. Surv. India Misc. Publ. 23, Part 2, 311–334.

Murthy, M. V. N., Nandy, D. R., and Chakrabarti, C. (1976a). A note to accompany the tectonic map of northeast India and adjoining areas. Geol. Surv. India Misc. Publ. 23, Part 2, 347–362.

——, Mazumdar, S. K., and Bhaumik, N. (1976b). Significance of tectonic trends in the geological evolution of the Meghalaya uplands since the Precambrian. Geol. Surv. India Misc. Publ. 23, Part 2, 471–484.

Nandy, D. R. (1976). The Assam syntaxis of the Himalayas—A reevaluation. Geol. Surv. India Misc. Publ. 23, Part 2, 363–367.

Rahman, S. (1981). Petrology and petrochemistry of the Khasi greenstone occurring around the Mylliem Granite, Khasi Hills, Meghalaya. Ind. Geol. Assoc. Bull., v. 14, 133–144.

9. ARAVALLI CRATON

The Aravalli craton (Fig. 1.1) is delineated by the boundary fault of the Himalayas in the north, the Cambay graben in the southwest, and the Narmada-Son lineament in the south and southeast (Fig. 9.1). In the west the craton is covered by Recent deposits and may extend to the Pakistan Himalayas. The best exposures are along the Aravalli mountain ranges and in the Bundelkhand massif and the Son valley region. The Aravalli ranges and the Bundelkhand Massif-Son valley areas are separated by a major lineament known as the Great Boundary Fault (Fig. 9.1).

The Aravalli supracrustal rocks are different from those of other parts of the Indian shield. The Dharwar supracrustal rocks (Ch. 2) are Archean and predominantly mafic/ultramafic schists (extrusive rocks), chemogenic sediments, argillites, volcaniclastics, and graywackes, with subordinate detrital orthoquartzites and carbonates. In contrast, the Aravalli supracrustal rocks are mostly Proterozoic and consist of phyllites and graywackes with abundant orthoquartzites, carbonates, and minor mafic/ultramafic schists. Another difference is shown by the distribution of iron. With the exception of the Bijawar suite and minor deposits in the Delhi Group, banded iron formations (BIF) are almost wholly absent from the Aravalli craton. The Dharwar, Singhbhum, and Bhandara cratons, however, contain major BIF deposits. Widespread development of stromatolites and phosphorites is also confined to the Aravalli craton, showing abundant availability of photosynthetic oxygen during Aravalli time. The presence of phosphorite, requiring oxygen, and the absence of BIF shows that Aravalli basins were not supplied with the iron to be oxidized and deposited, which can be attributed to negligible subaqueous volcanism and the sialic composition of the source.

Tectonic features of the Aravalli craton include the major boundaries of the craton; the Great Boundary fault; the Delhi-Haridwar ridge, a northern subcrop extension of the Arvalli-Delhi belt; and the Faizabad ridge, which appears to be the subcrop extension of the Bundelkhand massif into the Gangetic plain and Himalayas (Fig. 9.1; Narain and Kaila, 1982; Ch. 11). Also, a large number of major lineaments have been shown by multispectral LANDSAT images (R. P. Gupta and Bharktya, 1982; Bharktya and Gupta, 1983).

Stratigraphic terminology in the Aravalli craton has been highly controversial. Summaries have been presented by K. K. Basu et al. (1976), Geological Survey of India (1977), Naha (1983), Naha and Roy (1983), Sen (1983), and Naha et al. (1984). Stratigraphic successions of the Precambrian rocks as envisaged in various localities and by various workers are given in Table 9.1. Aravalli supracrustal rocks are metamorphosed to low to medium grade. Their overlying formations (Delhi) and the underlying (Banded Gneiss Complex) are generally metamorphosed to higher-grade amphibolite or granulite facies.

The free-air gravity map (Fig. 1.6) shows small positive anomalies along the Aravalli-Delhi supracrustal outcrops and a very small negative anomaly over the Bundelkhand massif. Geomagnetic induction patterns reveal a primary conductivity at right angles to the Himalayan ranges (B. J. Srivastava et al., 1984). A major conductivity is found along the Faizabad ridge. High heat flow values are now recorded from the Khetri copper belt (M. L. Gupta et al., 1967).

ROCK SUITES AND THEIR RELATIONSHIPS

The Aravalli craton can be divided into two parts, east and west of the Great Boundary Fault (Fig. 9.1). Only very tentative correlations of suites can be made across the fault, and we describe the different sides separately.

The oldest rocks west of the Great Boundary Fault are either members of the Banded Gneissic Complex (BGC) or metasedimentary/metavolcanic units engulfed in parts of the complex (Fig. 9.2). The supracrustal enclaves, some gneisses/ granites, and some more coherent schist belts have been grouped as the Bhilwara suite by several workers. The Aravalli Supergroup is a supracrustal sequence that, according to reports from various areas, unconformably overlies part of the BGC and is intruded by other parts of the BGC. Apparently, the BGC is a complex sequence of magmatic/metamorphic rocks with a wide variety of ages (3,500 to 2,000 m.y.) that overlap the period of Aravalli deposition.

Relationships between the Aravalli and overlying Delhi Supergroup are also uncertain. An old term, Raialo series, represents a suite of metase-

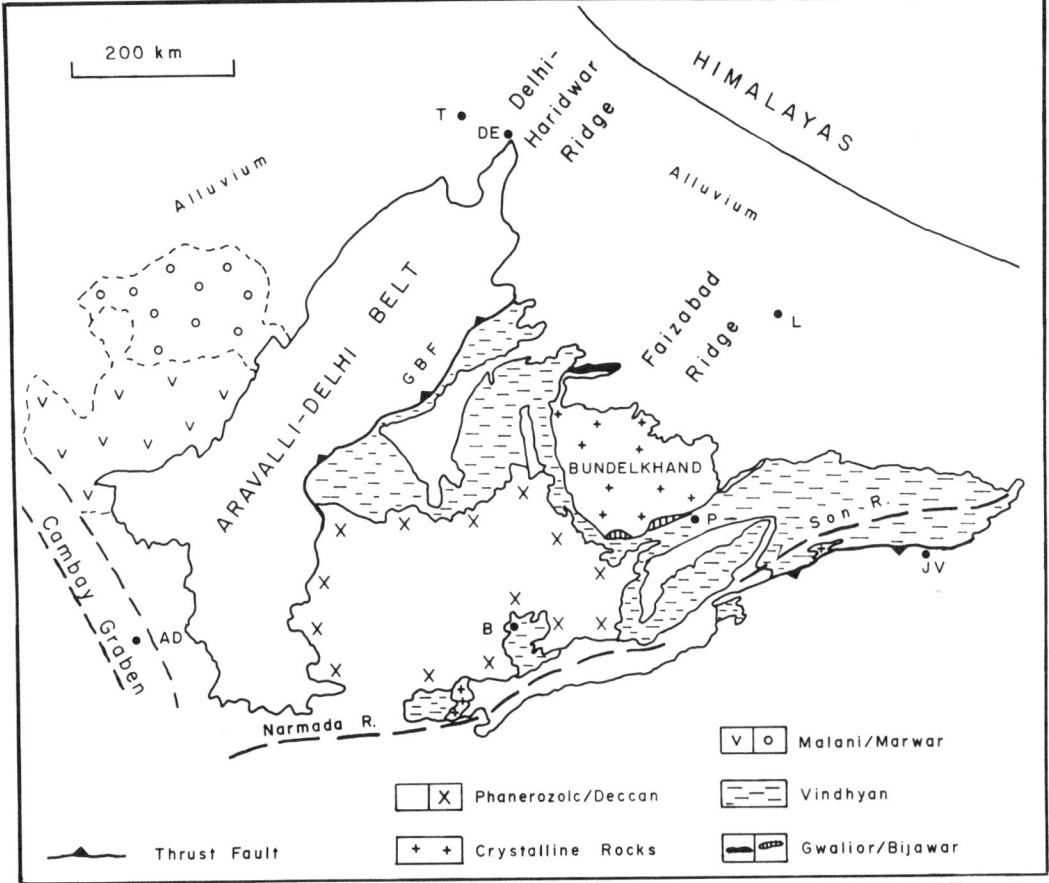

Fig. 9.1. Aravalli craton. The Delhi-Haridwar and Faizabad ridges are basement highs extending under the Ganges plain to the Himalayas. GBF designates the Great Boundary Fault. Locations include T, Tusham; D, Delhi; L, Lucknow; P, Panna; JV, Jungel Valley; B, Bhopal; AD, Ahmedabad.

diments that has been grouped with both the Aravalli and Delhi suites by various workers. The Delhi Supergroup is a metasedimentary assemblage that constitutes the main Aravalli mountain range. Part of it is at least 1,600 m.y. old, but parts may be much younger. The entire Aravalli-Raialo-Delhi sequence has clearly been deformed as a unit, probably in the Middle Proterozoic. Accumulation of the suite presumably occurred throughout much of the Early Proterozoic, and older rocks are probably Archean. The Champaner supracrustal rocks of Gujarat are apparently a southern extension of the main Aravalli Supergroup.

East of the Great Boundary Fault is the Bundelkhand region, where outcrops are scarce and radiometric age dating is almost absent. Sialic rocks, including granites and gneisses, are broadly grouped as the Bundelkhand Complex, representing rocks formed over a considerable age range. The Bijawar and Gwalior supracrustal suites

occur in separate areas and are presumably Archean to Early Proterozoic.

Separate intrusive bodies occur throughout much of the Aravalli terrain west of the Great Boundary Fault, and their ages range through much of the Precambrian. Older rocks are generally "granitic," but the youngest ones tend to be more alkaline. The young plutons are probably correlative with the Malani igneous suite (younger than 1,000 m.y.) to the west of the Aravalli range.

Late Proterozoic sedimentary sequences are generally unmetamorphosed and flat lying. The Vindhyan Supergroup is the remnant of an enormous basin east of the boundary fault. It is deformed only near the Great Boundary Fault and its southern margin, near the Satpura orogenic belt (Chapter 6; Chapter 10). At one time, the Vindhyan suite was regarded as extending to the west of the Aravalli range, but later work has shown that these western sequences lie on top of

TABLE 9.1. *General stratigraphy of Aravalli craton*

Age (in b.y.)	Aravalli-Delhi Belt and Farther West		Bundelkhand—Son Valley	
<1	Marwar Supergroup	Nagaur Group Bilara Group Jodhpur Group		
	Malani igneous rocks			
--Unconformity--				
1.4–0.9			Vindhyan Supergroup	Bhander Group Rewa Group Kaimur Group Semri Group
--Unconformity--				
1.8–1.5	Delhi Supergroup	Ajabgarh Group Alwar Group	Gwalior Group	
--Unconformity--				
2.5–2.0	Aravalli Supergroup	Jharol Group Udaipur Group	Bijawar Group	
--Unconformity--				
2.5–2.6			Bundelkhand Complex	Bundelkhand granites and gneisses Mehroni Group
	Banded Gneissic Complex/Bhilwara Group			

the Malani suite and are, therefore, younger than the Vindhyan rocks. The western suite is now designated as the Marwar sequence.

BANDED GNEISSIC COMPLEX (BGC)

The Banded Gneissic Complex (BGC) of the Aravalli craton was originally defined as a suite of pre-Aravalli metasediments, migmatites, agmatites, granites, pegmatites, aplites and metabasic rocks (Fig. 9.2; Heron, l953; Sen, 1983). Evidence for an unconformity between the BGC and overlying supracrustal suites has been widely proposed (Heron, 1953). Basal conglomerates and unconformities above the BGC have been described by Poddar (1966) and Naha and Majumder (1971).

The BGC occurs in five major regions (Fig. 9.2). The large northern area, around Karera, and the area just east of Udaipur are separated by a suite of gneissic rocks between Nathdwara and Amet. Some workers have considered these rocks to be part of the BGC, making a continuous BGC terrain, but Naha (1983) and Naha and Roy (1983) summarized evidence that these rocks are migmatized Aravalli metasediments. The large area of BGC just west of Chitorgarh contains the Berach Granite (2,600-m.y. old), and some maps show the whole BGC outcrop as Berach Granite, younger than the BGC. Scattered BGC outcrops

southeast of Udaipur are surrounded by Aravalli metasediments; a major outcrop of this group is the undated gneiss dome at Sarara. The BGC shown in the southern part of the Aravalli belt, in Gujarat, is only tentatively correlated with the type areas farther north.

Contrasting stratigraphies of the BGC are based on several lines of evidence. Crookshank (1948) stated that the BGC had invaded the Aravalli sediments and incorporated them partly as enclaves. R. S. Sharma (1983) described the BGC as a polymetamorphic basement containing older supracrustal rocks. Naha and Roy (1983) showed structural concordance between the BGC and the Aravalli-Raialo suite and concluded that at least part of the BGC formed by migmatization of metasedimentary parts of the Aravalli sequence. They suggested that the basement gneisses were involved in ductile deformation with the Aravalli rocks and that migmatization was synkinematic with the first Aravalli deformational episode. This interpretation means that the BGC, as exposed now, represents remobilized basement, and the BGC-Aravalli boundary is largely a migmatized front.

Several lines of evidence can be used to provide approximate dates for the BGC. The Untalla Granite, which intrudes the BGC, was dated at 2,950 ± 150 m.y. by Rb/Sr methods (Choudhary et al., 1984), and the Berach Granite was dated at

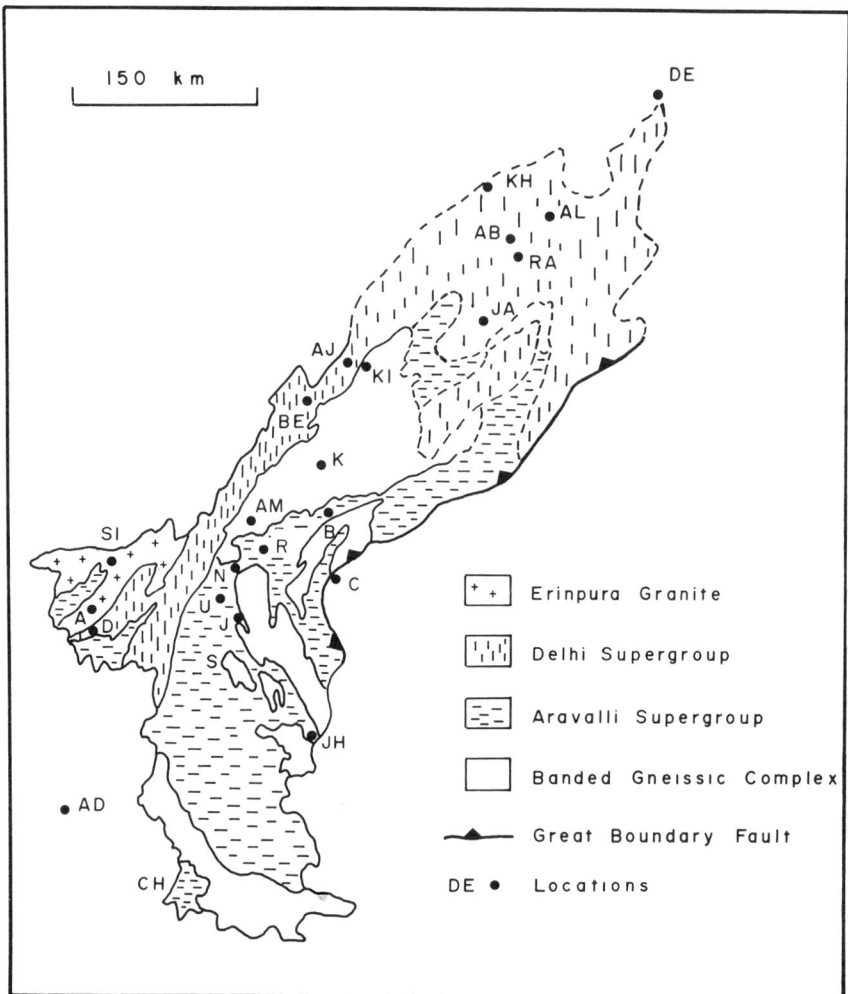

Fig. 9.2. Aravalli-Delhi belt. The northern part of the belt is mostly scattered outcrops of Delhi Supergroup through alluvium. Locations include A, Abu (Abu Roads, Mt. Abu); AB, Ajabgarh; AD, Ahmedabad; AJ, Ajmer; AL, Alwar; AM, Amet; B, Bhilwara; BE, Beawar; C, Chitorgarh; CH, Champaner area; D, Deri and Ambaji; DE, Delhi; J, Jhamarkotra; JA, Jaipur; JH, Jhabua; K, Karera; KH, Khetri; KI, Kishangarh; N, Nathdwara; R, Rajpura and Dariba; RA, Raialo; S, Sarara dome; SI, Sirohi; U, Udaipur. Locations of other places mentioned in the text include Ajitpura, Bhandanwara, Deogarh, and Sand Mata—all near Karera; Kankroli—near Amet.

2,600 m.y. by Pb/U techniques (Sivaraman and Odom, 1982). Crawford (1970) found Rb/Sr isochrons for the Berach Granite of 2,550 m.y. and the BGC of 2,000 m.y. Amphibolites and gray gneisses near Udaipur define an Sm/Nd isochron giving an age of 3,500 m.y. with an initial ratio corresponding to $e_{JUV}(T) = +3.5$ (McDougall et al., 1983). Thus, the BGC apparently includes rocks with ages from about 3,500 to 2,000 m.y.

The BGC has been affected by several periods of deformation and metamorphism. One period apparently occurred prior to deposition of most of the Aravalli Supergroup (R. S. Sharma, 1977, 1983; K. K. Sharma, 1978). These early defor-

mations are not well understood. Most deformation of the BGC also involved the Aravalli suite and is described later.

Petrology, mineralogy, geochemistry and metamorphism

The BGC consists of impure quartzites, calc-silicates, pelitic (mica) schists, biotite- and hornblende-bearing granitic gneisses, charnockites and other high-grade rocks, paragneisses, mafic and ultramafic schists, granites, and pegmatites and aplites (R. S. Sharma and Narayana, 1975a, 1975b, 1975c; R. S. Sharma and Ray, 1979, 1980;

Pandya, 1981; R. S. Sharma, 1982; R. S. Sharma and Macrae, 1981).

The major rock type of the BGC is biotite- and hornblende-bearing granitic gneisses that range from tonalite to adamellite in composition. Other rock types are mostly supracrustal enclaves. Impure quartzites contain quartz and varietal minerals such as hornblende, biotite, tourmaline, and garnet (almandine and grossularite). Impure marbles and calc-silicates contain calcite and varietal minerals, including quartz, forsterite, phlogopite, fuchsite, dolomite, muscovite, diopside, scapolite, and serpentine. Amphibolites consist of hornblende, plagioclase and quartz ± clinopyroxene, ± garnet, ± biotite and show small to moderate LREE enrichment (McDougall et al., 1983). Pegmatites are abundant. Mafic and ultramafic rocks have an uncertain relationship to other rocks in the BGC (Somayajulu, 1975, 1979). Metaperidotites contain relic olivine and varietal minerals such as anthophyllite, tremolite, phlogopite, and serpentine. Norites are also present.

Pelitic (mica) schists occur at numerous places in the BGC. Southeast of Beawar (R. S. Sharma and Narayana, 1975a, 1975b) they consist mostly of micas, quartz, and plagioclase with kyanite, staurolite, garnet, or sillimanite porphyroblasts. Garnets depict two periods of crystallization, possibly related to pre-Delhi and Delhi orogenies. The metamorphic conditions were estimated as about 540° to 650° at 5 to 5.5 kb. Metapelites near Ajmer occur in staurolite, kyanite, and sillimanite-muscovite zones of the amphibolite facies (Lal and Shukla, 1970). Sillimanite-bearing rocks also occur at Kuanthal, near Udaipur (Goel and Chaudhuri, 1979).

Paragneisses are particularly well studied in the Sand Mata Hills, near Karera (B. C. Gupta, 1934; Somayajulu, 1979). The area consists of basic garnet granulites with inclusions of aluminous gneisses, amphibolites ± garnets, and engulfing augen gneisses and migmatites. Some of the assemblages in the aluminous gneisses contain kyanite, garnet, and/or cordierite. The basic granulites contain garnet and clinopyroxene, and some have quartz coronas around orthopyroxenes where they are in contact with plagioclase. R. S. Sharma and Joshi (1984) estimated that the Sand Mata rocks were formed at temperatures of 550° to 650° and pressures of 9 to 12 kb, corresponding to a depth of 35 to 40 km.

Gedrite-cordierite-garnet-staurolite-sillimanite gneisses occur as lenses in the BGC in the Karera region (R. S. Sharma and Ray, 1979, 1980). The assemblage is uncommon because aluminous parent rocks rich in Mg and Fe should produce garnet and cordierite instead of gedrite in association with sillimanite. The paragneisses may represent the residuum left after a granitic melt separated from the parent pelitic rock. Similar gedrite-bearing paragneisses at Ajitpura, near Karera, were formed at 700° and approximately 6 kb (R. S. Sharma and Macrae, 1981; Fig. 9.3).

Basic and acid charnockites occur in the BGC in the Karera region (including Bhim, Deogarh, and Bhandanwara). Garnetiferous and nongarnetiferous enderbites, hypersthene-quartz syenites, and alaskites show gradational contacts with the charnockites (Gyani, 1971). Rocks of similar high metamorphic grade but uncertain age occur in the Abu Roads area (Fig. 9.2; Fig. 9.4) and southward (Desai et al., 1978). It is not clear that they should be considered part of the BGC or younger suites, and they are shown in the Aravalli outcrop area in Figure 9.2. Some rocks at Abu Roads are in pyroxene-granulite facies, with retrogression in the neighborhood of younger granites.

BHILWARA GROUP/SUPERGROUP

The metasediments surrounded by the granitic/ gneissic component of the BGC along an extended belt east of Karera and west of the Great Boundary Fault (Fig. 9.2) are considered by some workers to be part of a continuous sequence. This group of rocks was designated the Bhilwara Supergroup by Raja Rao et al. (1971) and Raja Rao (1976). The base of the sequence is not exposed. The grade of metamorphism increases gradually from greenschist facies (the Hundoli Group) in the east, to granulite facies in the Sand Mata region, near Karera, in the west. The Bhilwara Supergroup consists of shales, slates, quartzites, dolomites, marbles, cherts, graywackes, and hornblende- and mica-bearing schists. A shallow platformal environment of sedimentation is indicated by current bedding, ripple marks, and flute casts.

The existence of a definable Bhilwara Supergroup is controversial. The structure and mineralogy of the rocks are similar to those of small metasedimentary enclaves in the BGC that were regarded by Heron (1953) as pre-Aravalli. The 2,950 m.y. age of the Untalla Granite, which presumably intrudes the equivalents of the Bhilwara Group, indicates that supracrustal sequences older than the Aravalli sequence and the granitic component of the BGC are present in the terrain covered by the BGC. Naha et al. (1984), however, found no intrusive contact for the Untalla Granite. Furthermore, Roy et al. (1981) demonstrated that, at least locally, separation of the Bhilwara Supergroup from the BGC is not tenable on structural grounds.

The Geological Survey of India (1977) defined the Bhilwara Group (Supergroup) as the BGC exposed in the Karera area plus the Hundoli Group, which was earlier mapped by B. C. Gupta (1934) as Aravalli. The original Bhilwara name used by Raja Rao (1976) was retained, but the concept

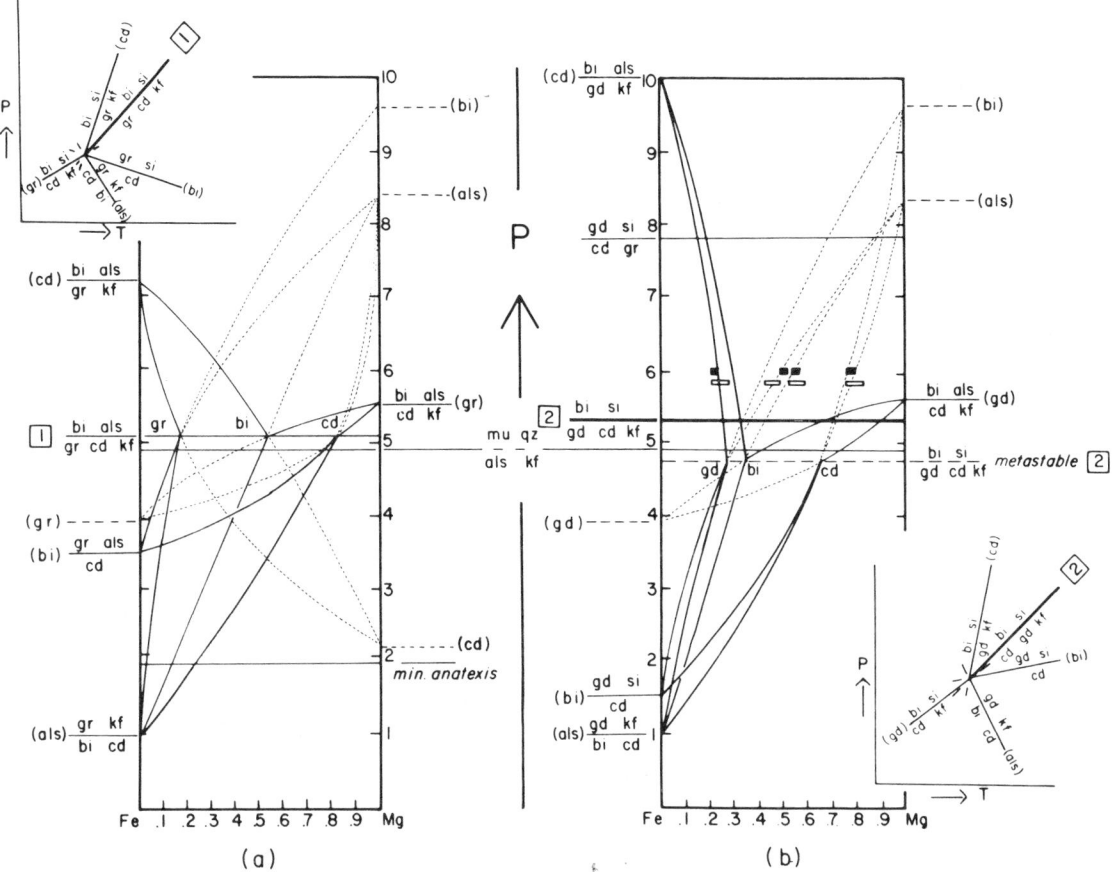

Fig. 9.3. Metamorphic conditions in Banded Gneissic Complex (R. S. Sharma and Macrae, 1981). The diagrams are pressure-composition projections at 700° in compositional systems that allow the mineral assemblages: (a) garnet, cordierite, biotite, sillimanite, quartz, or K-feldspar; and (b) gedrite, cordierite, biotite, sillimanite, quartz, or K-feldspar. Melt is a possible phase in both diagrams. Open rectangles show the determined compositions of minerals at equilibrium, thus specifying that equilibrium pressure is slighty less than 6 kb for the indicated temperature. Solid rectangles are previously determined compositions at 6 kb for other rocks. Reproduced with permission from Contributions to Mineralogy and Petrology, v. 78, p. 57.

was greatly changed. S. N. Gupta et al. (1980) also redefined the term.

ARAVALLI SUPERGROUP

The Aravalli Supergroup overlies the BGC unconformably in most of Rajasthan and Gujarat. Complex and uncertain relationships with the BGC, however, have been discussed previously. The Aravalli suite consists predominantly of gray-wackes and orthoquartzite-carbonate sediments with minor volcanic constituents. Rocks belonging to the suite have undergone polyphase greenschist-facies metamorphism and have been intruded by syn-, late-, and posttectonic granites and ultramafic/mafic bodies. C. A. Sastry et al. (1984) inferred that the Aravalli Supergroup is Early Proterozoic (2,500 to 2,000 m.y.).

The problem of the lower boundary of the Aravalli suite has been discussed in preceding sec-

tions. The upper boundary is also problematic. Local discordance occurs between the Aravalli and overlying Delhi rocks. R. S. Sharma (1983), however, found no distinct "Aravalli orogeny," and stated that the Aravalli rocks were deformed only by a post-Delhi orogeny. Raja Rao (1976) also did not propose a break between the Aravalli and Delhi suites.

Stratigraphy

A synthesis of the stratigraphy of the Aravalli Supergroup is shown in Table 9.2 (Naha and Halyburton, 1974b; Geological Survey of India, 1977). The Geological Survey of India designated the Udaipur Group as the lowest unit, with a basal section of quartzites, metavolcanic rocks, and arkosic grits and conglomerates. These basal rocks are followed upward by phyllites, which are carbonaceous at places and contain bands of dolo-

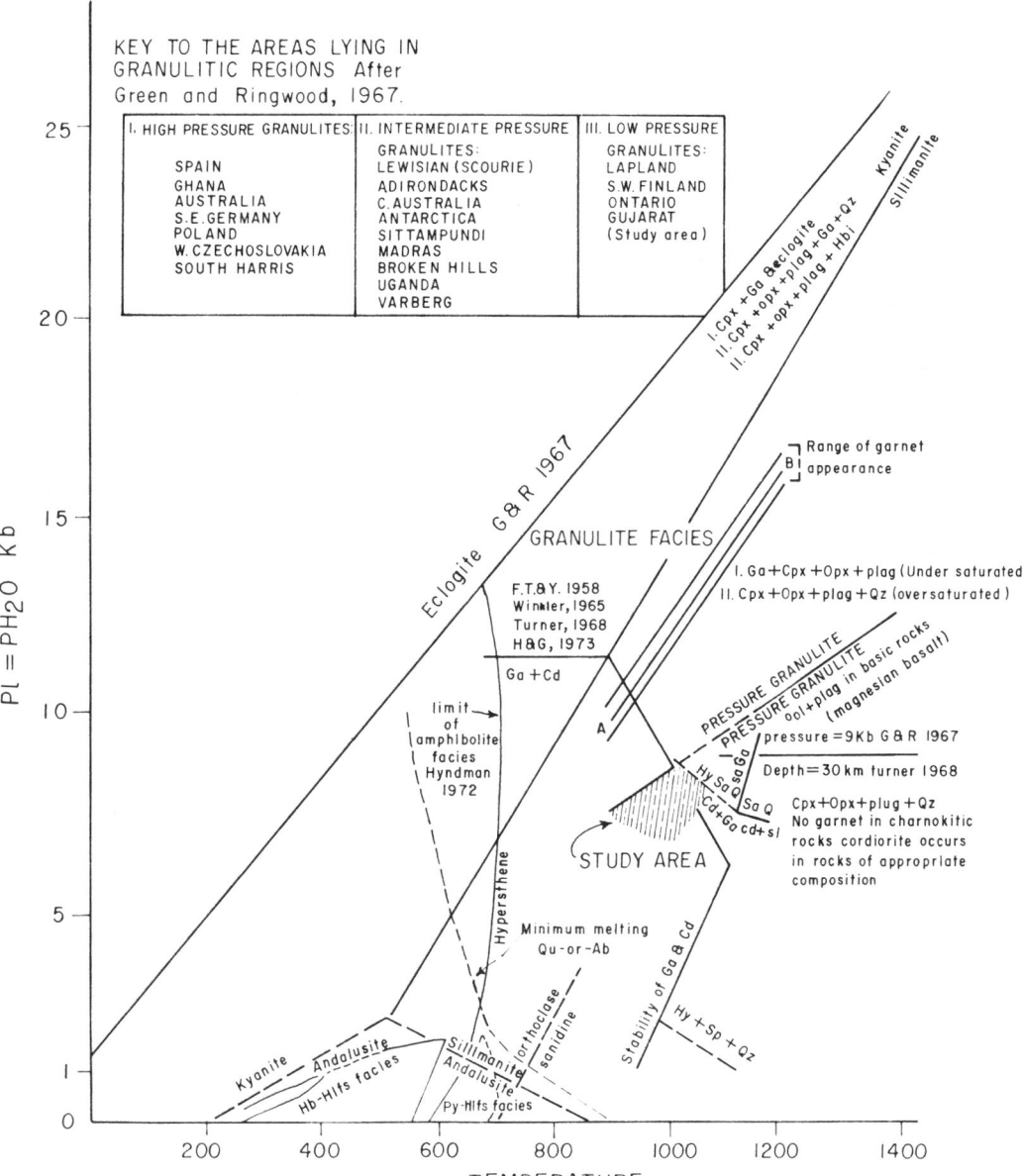

Fig. 9.4. Metamorphic conditions at Abu Roads (Desai et al., 1978). Boundaries of fields and reactions are labeled by references to previous work that established them (references in Desai et al.). The mineral assemblage at Abu Roads is shown in the shaded area and indicates very high temperatures and pressures. Reproduced with permission from the Journal of the Geological Society of India, v. 19, p. 391.

mites and quartzites. Stromatolites occur in the dolomitic beds. A sequence of graywackes and phyllites with bands of quartzites and dolomites overlies the phyllitic unit. The upper part of the Aravalli Supergroup was designated the Jharol Group, a thick sequence of phyllites and quartzites that lies unconformably on the Udaipur Group. The Jharol Group includes rocks formerly designated as the Raialo series (Heron, 1953).

Roy and Paliwal (1981) recognized two "facies sequences" in the Aravalli Supergroup, an eastern suite of shelf sediments and a western suite of deep-sea sediments (Fig. 9.5). The deep-sea assemblage consists of metapelites with some quartzites. The eastern shelf sequence contains the following associations:

Quartzite-association: Orthoquartzites are the basal member of the sequence, although a thin band of amphibolite or chlorite-biotite schist occurs below the quartzite at many places. Thin lay-

TABLE 9.2. *Stratigraphy of Aravalli Supergroup (Geological Survey of India, 1977)*

Jharol Group	Quartzite
	Mica schists and phyllites with quartzite at the base
--- Local structural discordance ---	
Udaipur Group	Phyllites, graywackes, metavolcanics, mica schist, marbles, and dolomites
	Metavolcanics, quartzites, and arkosic conglomerates
	Basal conglomerates, grits, and quartzites
---Unconformity---	
Banded Gneissic Complex/Bhilwara Group	

ers of greenschist also occur between the layers of quartzites. These greenschists and amphibolites (amygdular metabasalts) become dominant members of the association westward, away from the contact with the basement.

Carbonate association: The basal formation grades upward into carbonate, mostly dolomitic, and shows facies changes to orthoquartzite, dolomite, and phyllite. This horizon contains a persistent stromatolitic rock phosphate sequence at Jhamarkotra.

Graywacke-phyllite litharenite association: A thick sequence of graywackes, rhythmically interbedded with phyllites, shows graded bedding, ranges from feldspathic to lithic types, and passes upward into massive lithic arenites or minor graywackes/conglomerates.

Conglomerate-arkose-orthoquartzite association: This sequence begins at its base with a polymictic conglomerate containing pebbles that have a large size range and are generally deformed to ellipsoidal shapes. The matrix consists of schistose arkosic quartzite. The conglomerate grades upward through arkosic quartzites into quartzites with ferruginous cement. The ferruginous quartzites consist of clastic quartz cemented by iron oxide and are not the same as chert-iron oxide rocks (banded iron formations) found in other parts of India.

Quartzite-silty arenite association: The graywacke-phyllite association passes upward through polymictic conglomerate into a quartzite-silty arenite association. The quartzites are massive to current bedded.

Supracrustal rocks of northern Gujarat were, at one time, referred to as the Champaner Series

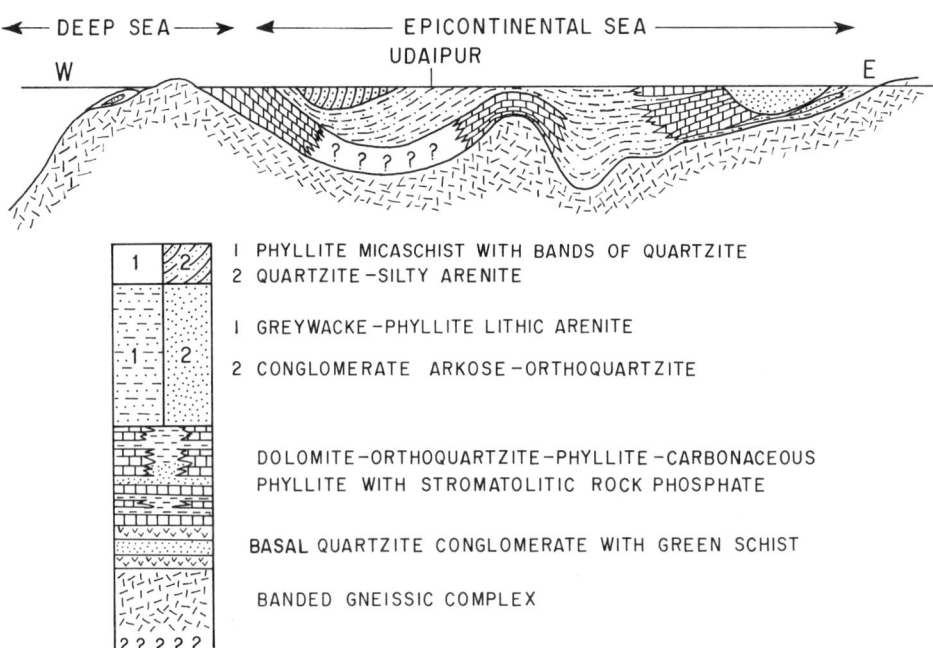

Fig. 9.5. Depositional environments of Aravalli Supergroup (Roy and Paliwal, 1981). Reproduced with permission from Precambrian Research, v. 14, p. 71.

(Fig. 9.2), but this term was later abandoned by Fermor (1936), who stated that the Champaner and Aravalli sequences are coeval. Recently, Gopinath et al. (1977) used the term again, classifying some rocks in Gujarat into a younger Champaner Series and an older Aravalli sequence. The Champaner suite, as now defined, is an interbedded sequence of argillaceous, arenaceous, and impure calcareous rocks that rest unconformably over older metasediments. Other metasediments in Gujarat also have been correlated with the Aravalli Supergroup (Narayana, 1974).

Structure and metamorphism

Heron (1953) regarded the structure of the Aravalli rocks, particularly around Udaipur, to be simple, with upright folding on a NNE to NE trend. More recent studies (Mukhopadhyay and Ghosh, 1980; Mohanty, 1982; Naha and Roy, 1983; Sen Gupta, 1983; and Naha et al., 1984) have shown structural complexities and demonstrated that the Aravalli Supergroup has been affected by as many as four generations of folding. The principal studies are summarized as follows.

In central Rajasthan, between Nathdwara and Amet (Fig. 9.2), all early Precambrian rocks (the BGC, Aravalli Supergroup, and "Raialo Formation") have been affected by three to four stages of superposed deformation (Naha et al., 1966, 1969). First-generation structures are reclined, inclined, and rarely upright isoclinal folds with a westerly trend. These folds are commonly displaced to form rootless hinges. Second-generation coaxial folding of the axial surfaces of the isoclinal folds and their axial planes has caused wide dispersion of the axial surfaces of the first-generation folds and associated lineations and produced interference patterns (Plate 9.1a,b). Complex outcrop patterns in "Hammerhead syncline" and the "Hook syncline" are a result of variable angular relations between the trend of the fold axes and the strike of the axial planes of first-generation folds, spatial variation in the acuteness of the second folding, and angular variation between the attitudes of the first set of folds and that of the axial planes of the second folds.

The Jharol Group (Raialo Formation) was also affected by four generations of folding (Naha and Halyburton, 1977a, 1977b). The structural features are identical to those in the remainder of the Aravalli and BGC suites, confirming opinions that the Raialo Formation does not have an independent stratigraphic status.

In south-central Rajasthan, west of Udaipur, structural complexities are similar to those in central Rajasthan (Roy et al., 1971). The first-generation folds are tight to isoclinal, with drawn-out hinges accompanied by well-developed axial plane schistosity. These folds were initially recumbent with E-W to NW-SE axial trends and

were refolded by second-generation, upright and open to locally isoclinal folds. The second-generation folds have variable plunge and an overall northerly axial trend. The third generation of folding produced tight to isoclinal folds and a set of axial plane crenulation cleavages.

Structural studies in other areas can be summarized briefly. At Jhamarkotra (Fig. 9.2), the outcrop pattern of phosphate-bearing rocks is a result of interference of F2 and F4 folds on the limbs of a large Fl isoclinal fold with coaxially refolded axial planes (Roy et al., 1980). The Pb-Zn-bearing rocks of Zawar, near Udaipur, have also been shaped by four generations of folds, mostly isoclinal anticlines showing variable plunge and trend (Roy and Jain, 1974).

Metamorphic conditions have not been determined quantitatively except in a few areas. Calcareous sediments with the assemblages scapolite-garnet-epidote-plagioclase and calcite-plagioclase-hornblende-pyroxene occur in thin bands in paragneisses of the BGC near Karera (Fig. 9.2). R. S. Sharma (1982) considered them part of the Aravalli Supergroup, but they could be pre-Aravalli Bhilwara sediments. Scapolite occurs in granoblastic, polygonal aggregates. Hornblende forms ragged prisms. The Na-Ca fractionation between scapolite and plagioclase indicates a recrystallization temperature around 700°.

Phosphorites and stromatolites

Rock phosphate/phosphorite and associated stromatolites (Plate 9.1c) are one of the most important rock types of the Aravalli Supergroup (summary of earlier work by D. M. Banerjee et al., 1980). The phosphorite- and stromatolite-bearing horizons are confined to the middle formations of the Udaipur group (the Matoon Formation of D. M. Banerjee, 1971a), and phosphatic minerals are abundant in the brecciated cherty rocks and bluish gray dolomite. The P_2O_5 content varies from 10% to 37%. The phosphorites are associated with stromatolites (*Collenia, Baicalia,* and *Minfaria*) and are considered biogenic. In the Matoon (Udaipur) and Jhamarkotra areas, manganiferous limestones and quartzites, dolomites, slates, and calcareous phyllites commonly occur below, above, and within phosphorite beds. The principal constituents are crypto- and microcrystalline apatite, quartz, dolomite, calcite, dolomieuite, and clays (D. M. Banerjee, 1971a, 1971b; Chauhan, 1979, 1980).

Stromatolitic phosphorite is composed of angular, elongated to wedge-shaped fragments of stromatolites that are partially replaced by apatite, francolite or dahllite (D. M. Banerjee, 1978). The phosphatic stromatolites occur as layered mats or as branching to unbranching columns of variable shape. Stromatolite columns show rare apatite needles, formed during metamorphism,

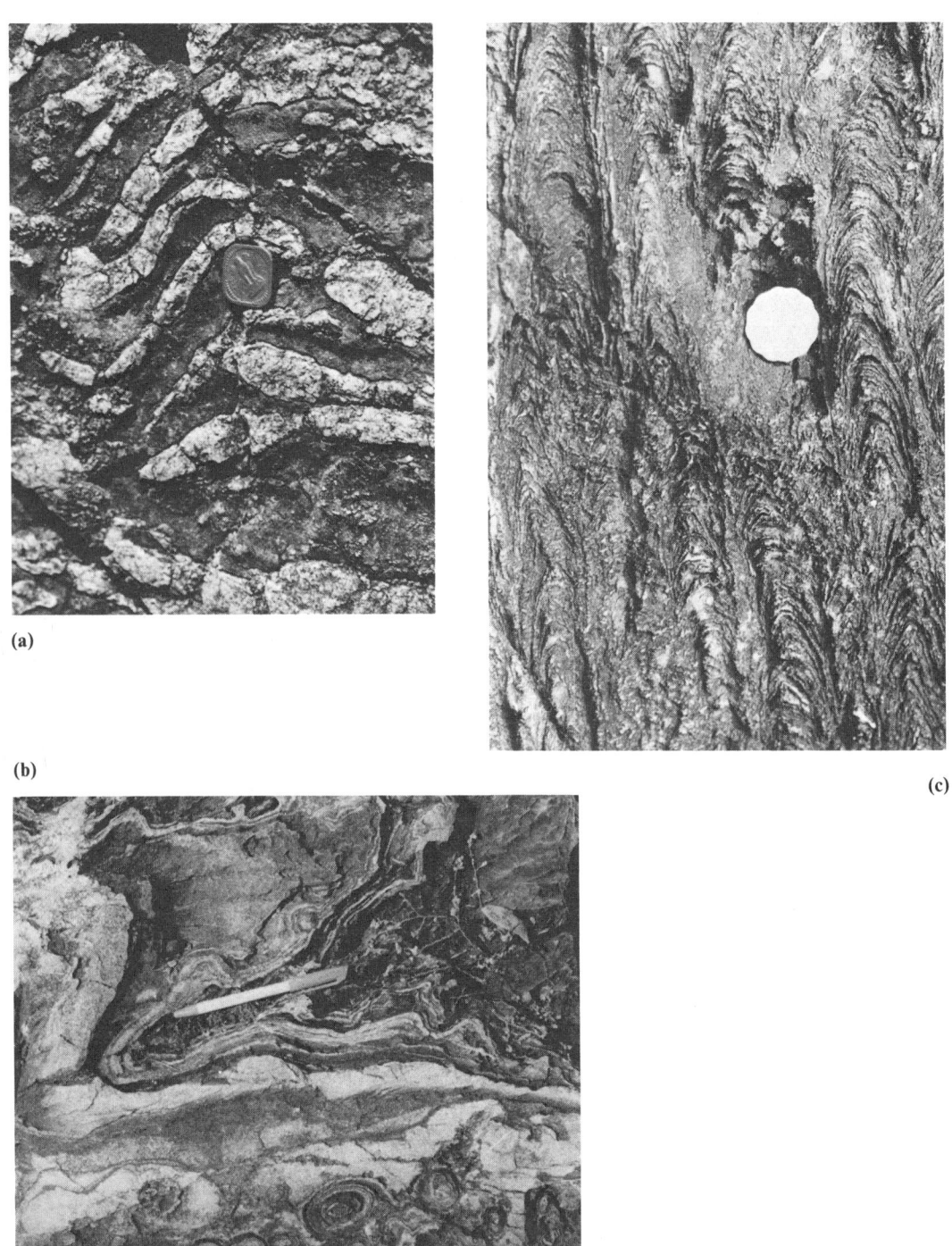

(a)

(b)

(c)

Plate 9.1. *(a)* Hook-shaped folds caused by interference of folds of two systems. Folding is shown by silicate layers in calcareous rocks (courtesy of K. Naha). *(b)* Irregular dome-and-basin structure in Aravalli metasedimentary rocks northeast of Jhamarkotra (courtesy of A. B. Roy). *(c)* Stromatolites in Aravalli phosphorite-carbonate rocks at Jhamarkotra (courtesy of A. B. Roy).

surrounded by amorphous, dark gray or black masses of collophane. Most of the detrital features of the carbonate matrix are well preserved and suggest extensive reworking in warm shallow water and tidal environments (D. M. Banerjee, 1978). The phosphorites were formed by accumulation of excess P_2O_5 in the organic-rich laminae of the stromatolites during their growth (Verma and Barman, 1975). D. M. Banerjee and Klemm (1985) used ratios of *n*-alkanes and pristane/phytane to demonstrate reducing microenvironments within the cyanobacterial-algal stromatolites despite overall oxidation in the waters of the Aravalli basin.

The rhythmic pattern of distribution of Si, Ca, Mg, and P across the stromatolite laminae has been related to pH variations on microenvironmental levels, which in turn appears to have been controlled by diurnal processes of photosynthesis in the intertidal to subtidal photic water. Roy and Paliwal (1981) suggested that the most important tectonosedimentary factor controlling the growth of stromatolites in the Aravalli basin was the development of epicontinental conditions within an eastern shelf. Because of a uniformly low relief of source area, the basin did not receive much terrigenous debris, thus permitting algal growth.

Stromatolites (possibly *Gruneria*) are also found in siliceous dolomite and chert of the Aravalli Supergroup in the ore deposits of Rajpura-Dariba (Deb et al., 1978). The ore minerals are concentrated along the siliceous stromatolitic laminae.

DELHI SUPERGROUP

The rocks of the Delhi Supergroup constitute the main Aravalli mountain ranges over a strike distance of nearly 700 km from Gujarat to Delhi (Fig. 9.2). As a generalization, they form a broad synform. In the northeastern portion, the Delhi rocks rest on older Bhilwara Group/BGC with a marked unconformity. In the southwestern region, there is structural discordance to the Aravalli Supergroup. The Delhi suite is divided by some authors into Raialo, Alwar, and Ajabgarh formations (Table 9.3). Other workers place the Raialo Formation in the Jharol Group of the Aravalli Supergroup.

The main rock types of the Delhi Supergroup are quartzites, conglomerates, arkoses, phyllites, slates, various types of schists, limestones, marbles, calc-gneisses, and mafic flows and amphibolites. They have undergone polyphase deformation and greenschist- to granulite-facies metamorphism, with syntectonic granitic activity.

The northeastern part of the Delhi Supergroup is older than Middle Proterozoic (C. A. Sastry et al., 1984). The Ajmer, Bairat, Dadikar, Harsora, and Seoli Granites, all of which intrude the Delhi

Supergroup in this area, show Rb/Sr isochrons indicating ages older than 1,600 m.y. (Choudhary et al., 1984). Crawford (1970) estimated an age of 1,600 m.y. for the Bairat Granite. The Khetri copper belt, in the Delhi Supergroup, is intruded by the Udaipur and Saladipura Granites, which have an isochron age of 1,480 m.y. (Gopalan et al., 1979a).

Although granites intrusive into the Delhi Supergroup in the northeastern region yield an age of about 1,700 to 1,500 m.y., granites intrusive into the main Aravalli mountain range farther south give a much younger age of 850 ± 50 m.y. (C. A. Sastry et al., 1984). The widely separated ages could be interpreted as two tectono-thermal episodes within apparently similar Delhi rocks or development of the Delhi suite in two independent basins that were widely separated in space and time. Young Delhi rocks may be related to Lower Vindhyan rocks, which are dated as 1,400 to 1,200 m.y. old (see discussion following).

Stratigraphy

The general stratigraphic sequence of the Delhi Supergroup proposed by Sant et al. (1980) is given in Table 9.3. A major controversy concerns the stratigraphic position of the Alwar quartzite. Raja Rao (1976) pointed out that "the so called Alwar quartzite" in parts of Rajasthan cannot be separated from the underlying schists, that no stratigraphic break occurs in Gujarat between Aravalli and Delhi suites, and thus that the Aravalli and Delhi suites represent one geosynclinal sequence without any break. Mukhopadhyay and Dasgupta (1978), however, demonstrated a structural break between the Delhi Supergroup and pre-Delhi rocks near Beawar (Fig. 9.2). R. S. Sharma (1983) proposed that the Aravalli, Raialo, and Delhi suites were involved in a single Delhi orogeny that gave rise to the Aravalli mountains. R. S. Sharma thus proposed only two Precambrian orogenies in Rajasthan, one of which occurred during early Precambrian (pre-Aravalli) time and one during middle Precambrian (Aravalli-Delhi) time.

In the Alwar District the Delhi (including Raialo) rocks have a cumulative thickness of more than 6000 m (Sant et al., 1980). The Delhi sequence starts with a basal conglomerate and shows a facies change from calcareous members in the lower part to arenaceous members in the middle and predominantly argillaceous rocks toward the top. The conglomerate consists mostly of pebbles of various types of quartzites, with minor granites, gneisses, cherts, and mica schists. The pebbles are internally deformed (Gangopadhyay and Lahiri, 1983). Current bedding and graded bedding are both present in the quartzites and arkoses. The source rock for the sediments was mixed orthoquartzites, low- to high-grade

TABLE 9.3. *Stratigraphy of Delhi Supergroup in Alwar District (Sant et al., 1980).*

Upper Proterozoic (?)	Post-Delhi		Acid intrusive: granite, pegmatite and quartz veins Basic intrusives: amphibolite and metadolerite	
Middle Proterozoic	Delhi Supergroup (6000 m)	Ajabgarh Group	Arauli Formation	Quartzite interlayered with staurolite-garnet schist, slate, etc.
			Bharkol formation	Quartzite with interlayered carbon phyllite
			Thana-Ghazi Formation	Carbonaceous phyllite, Marble
			Seriska Formation	Brecciated quartzite and ferruginous quartzite
			Kushalgarh Formation	Impure marble
			Pratabgarh Formation	Massive quartzite with minor schists and marble
		Alwar Group	Kankwarhi Formation	Schists (sericite, biotite, garnet, and andalusite bearing) with quartzites and conglomerates.
			Rajgarh Formation	
				Quartzite, marble, gritty quartzite, conglomerate and pebbly quartzites
		--------- Local unconformity ---------		
		Raialo Group	Tehla Formation (Upper Raialo)	Felspathic quartzite and schists, quartzite
			Dogeta Formation (Lower Raialo)	Marble, dolomitic at places, felspathic quartzite, pebbly and conglomeratic at places
		--------- Unconformity ---------		
Lower Proterozoic to Archaean	Pre-Delhi	(Pre-Aravallis ?)	Granites and gneisses Quartzite with interlayered schist and phyllite Impure marble and associated quartzite	

metamorphic rocks, and acid plutons (Shukla and Sharma, 1970).

Volcanism occurred locally in the Ajabgarh Group (S. P. Singh, 1983). The volcanic rocks consist of tuffs and tuffaceous sediments with amygdular lavas (now amphibolites), pillow breccias, and hyaloclastites.

Structure and metamorphism

The Delhi Supergroup in the north-central Aravalli mountain ranges (Fig. 9.6) has been subjected to two main deformations (R. S. Sharma, 1977). The first deformation affected the bedding planes and produced isoclinal folds with axial plane schistosity parallel to bedding. The fold axes and accompanying lineations plunge 5° to 20° toward the SW and NW. The second deformation affected the axial plane cleavage and the limbs of the first-generation folds. The second-generation folds are tight to open, asymmetrical, generally plunge toward the N to NNW at 30° to 80°, and exhibit a slight scatter due to the variable attitude of hinges and limbs of the first-generation folds.

The Delhi rocks of Ajmer also exhibit large-scale first-deformation structures and superposition of second-generation folds (R. S. Sharma and Upadhyaya, 1975). Roday (1979) found three generations of folds, but in the Alwar region, only the second phase involved the formation boundaries and shaped the map pattern (Gangopadhyay and Sen, 1972, 1975). Sant et al. (1980) stated that F2 formed the major trend of the Aravalli belt.

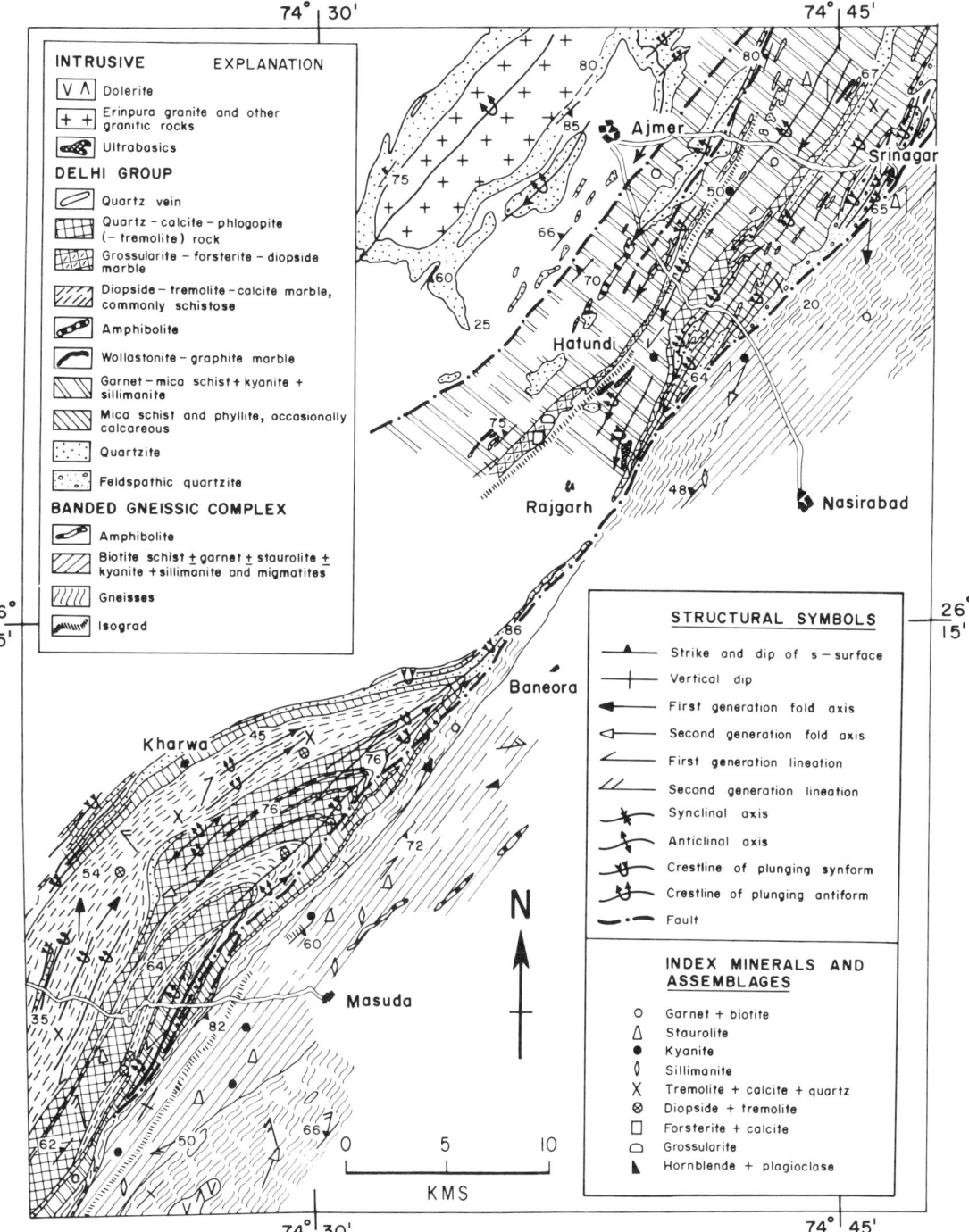

Fig. 9.6. Part of Delhi Supergroup (R. S. Sharma, 1977). Reproduced with permission from Precambrian Research, v. 4.

Two distinct metamorphic facies have been recorded in the Delhi Supergroup: the northeastern part is characterized by an andalusite-staurolite association with sporadic sillimanite near the post-Delhi granites; and the southwestern part shows a medium-pressure, Barrovian metamorphism in which kyanite, sillimanite, and staurolite are widespread (Roy Chowdhary and Das Gupta, 1965; Gangopadhyay and Sen, 1972).

In the Khetri copper belt, the Alwar Group is represented by magnetite quartzites, feldspathic quartzites with or without cummingtonite, and anthophyllite quartzites. The Ajabgarh Group is composed of mica schists and phyllites containing andalusite ± staurolite, chlorite schists (± garnet), and calc-silicate rocks with calcite and actinolite. Other index minerals in these rocks are garnet, biotite, gedrite, cummingtonite, and cordierite. Lal and Shukla (1975a, 1975b, 1976) estimated that metamorphism took place between 500° and 600° at 3 to 5 kb pressure, within the hornblende-hornfels facies of regional metamorphism. In the Saladipura part of the Khetri copper belt, however, metapelitic rocks at a distance from amphibolitic bodies are muscovite and/or biotite rich ± andalusite and staurolite; schists near amphibolites are chloritic ± garnet, cordierite, and orthoamphiboles (S. C. Sarkar et al., 1980).

Ore-bearing rocks of the Ajabgarh Group at Ambaji-Deri (Fig. 9.2) are argillaceous or arenaceous metasediments or amphibolites that contain lensoid bodies of metamorphosed magnesian and calc-magnesian rocks, such as cordierite-anthophyllite-chlorite rocks, quartz-chlorite-tremolite schists and diopside-forsterite marbles. Metamorphism took place in the temperature range of 525° to 625° at 2 to 4 kb pressure (Deb, 1980). The mineral assemblages at Ambaji-Deri were developed during two deformational phases (R. S. Sharma, 1977). In the first deformation, micas were recrystallized and oriented to form schistosity, and garnets with spiral cores were formed. Staurolite, kyanite, and sillimanite were formed in metapelites in the syn- to late-kinematic phase of the first deformation, and garnet and hornblende were formed in the basic rocks. The second deformation took place after the formation of all index minerals (R. S. Sharma, 1977; Sant et al., 1980).

Life

Small disc-shaped fossils from the Alwar quartzite include trilete (pteridophytic), monolete (hemicycadelic), and fungal spores (Dutt and Srivastava, 1975). Some acritarchs and hystrichospheres related to *Leiospheridia, Deunffia downie,* and *Sphaeraporalites* have also been reported. Stromatolites are abundant (Negi and Ravindra, 1980; K. K. Verma and Barman, 1980). Organic matter has been described by Sastri and Krishna Rao (1982).

Sedimentation

A shallow marine sedimentary environment is shown for the Delhi Supergroup by a number of features (Sant et al., 1980). The cross-laminations in the Alwar Group quartzites are mostly a torrential type with asymmetrical ripple marks. The calcareous sediments of the Ajabgarh Group contain well-preserved, penecontemporaneous folds, load casts, and slump structures. The quartzites of the Khetri copper belt contain current bedding, mud crack fillings, load structures, and ripple marks. Although the entire Delhi basin was shallow, the Ajabgarh Group was apparently deposited in relatively deeper water than the older Alwar Group.

Facies changes occur both across and along strike. In the northernmost part of the Alwar district, there is a gradual increase in argillaceous content from east to west. Along strike, the calcareous content of the rocks decreases from south to north and the arenite/argillite ratio increases (S. P. Singh and Ravindra, 1975; Sant et al., 1980).

Lithostratigraphic units of the Delhi Supergroup grade into one another without a major stratigraphic break, although local minor pauses in sedimentation are indicated by discontinuous small lenses of conglomerate at several stratigraphic levels within the Alwar Group. The presence of syngenetic sulfides at various places in the basin (Saladipura, in the Khetri belt, and at Deri and Ambaji) indicate deposition of the older part of the suite in small basins with restricted circulation and reducing conditions (Sant et al., 1980). Particularly at the start of sedimentation, there was restricted deposition in small, localized basins. S. P. Singh (1985) proposed initiation of Delhi sedimentation in small fault troughs that later coalesced into a broad subsiding basin.

Paleocurrent studies of the Delhi Supergroup rocks at Alwar and adjoining areas indicate the presence of two subbasins within the Delhi basin (Sant et al., 1980). These two basins were probably divided by a bottom topographic high (the Ferozepur-Jhirka ridge). A saucer-shaped Delhi basin is likely, as the paleoslope was south to north in the southern and southwestern parts and north to south in the northern and northeastern parts (Sant et al., 1980).

BUNDELKHAND AREA

The Bundelkhand region/massif is a triangular segment of the Indian Shield bounded by the Narmada/Son lineament in the south, the Great Boundary Fault of the Aravalli Range in the west, and the Himalayas in the north (Fig. 9.1). The outcrops in this region are scattered and detached. The area contains four groups of rocks with highly uncertain ages (Prakash et al., 1975; R. P. Sharma, 1982, 1983): the Bundelkhand complex/Group (older than 2,600 m.y.), the Bijawar Group (2,600 to 2,400 m.y.), the Gwalior Group (1,800 m.y.), and the Vindhyan Supergroup (1,400 to 900 m.y.).

Bundelkhand complex/Group

The Bundelkhand complex/Group is composed of volcanosedimentary schists and quartzofeld-

TABLE 9.4. *Stratigraphy of southern Bundelkhand (D. M. Banerjee, 1982)*

Mehroni Group	Madaura Formation	Granite member	Quartz veins, pegmatites and graphic granites
		Gabbro member	Gabbro; pillow lavas
		Ultramafic member	Peridotites, pyroxenites (partially altered to serpentines and talc actinolite schists)
	--- Unconformity ---		
	Berwar Formation		Iron formations, carbonate rocks and gray-green slates, quartzites, and conglomerates
	--- Unconformity ---		
	Rajaula Formation		Biotite-feldspar gneisses, chlorite-biotite-schists, granodiorite, metabasalts

spathic granite gneisses, which are discussed separately as follows.

Schistose rocks in the Bundelkhand complex occur as enclaves of various dimensions in the gneisses and were designated the Mehroni Group (Table 9.4; D. M. Banerjee, 1982) in the Lalitpur area in the southern part of the complex. D. M. Banerjee divided the group into lower (Rajaula) and upper (Berwar) formations. The Rajaula Formation contains amphibolites, chlorite-biotite schists, garnet-sillimanite gneisses, and feldspathic biotite gneisses. Some of the schists are vesicular metavolcanic rocks. The Rajaula Formation is unconformably overlain by the Berwar Formation, which consists of ferruginous quartzites, carbonates, and gray/green slates. These two formations have been intensely invaded, fragmented, and migmatized by a granite complex dated as 2,555 ± 55 m.y. old (Crawford, 1970).

Misra and Sharma (1975) and R. P. Sharma (1982) worked in the northern Bundelkhand area and proposed a different stratigraphic subdivision of the Bundelkhand complex (Table 9.5). The Palar Formation contains extensive aluminous rocks formed during a period of intense Precambrian laterization (R. P. Sharma, 1979a; 1979b) and was tentatively correlated with the Bijawar suite by R. P. Sharma (1982). The older Kuraicha Formation may be the equivalent of the Mehroni Group.

Intrusive rocks are abundant in the Bundelkhand area. Granites are the main rock type of the complex. Principal ones are the Garhmau Granite, which is gray, massive to gneissic, and contains porphyroblasts of feldspars and quartz plus xenoliths of metasediments and recrystallized resistates; and the Matatila granite suite, consisting of pink, massive granite, monzonite, and adamellite (R. P. Sharma, 1983). In addition to granites, other intrusive rocks in the Bundelkhand complex include young, mafic to silicic dikes and the Madaura basic/ultrabasic rocks, which consist of metamorphosed peridotites, pyroxenites, and gabbros (R. P. Sharma, 1982).

The Bundelkhand complex appears to have been affected by five phases of folding (R. P. Sharma, 1982). The Banded Gneissic Complex, west of the Great Boundary Fault, also shows two major and two minor phases of folding (Naha and Halyburton, 1974a). Therefore, the entire Aravalli craton may have been affected by a large number of similar deformational events.

Bijawar Group

The Bijawar Group forms isolated outcrops unconformably overlying the Mehroni Group and the Bundelkhand gneisses/granites along the southern edge of the Bundelkhand massif, north of Vindhyan outcrops (Fig. 9.1; Prakash et al.,

TABLE 9.5. *Stratigraphy of northern Bundelkhand (R. P. Sharma, 1983)*

Madaura ultrabasics	Pyroxenite, gabbroic serpentinite, metabasite
Mahoba Dolerite	Dolerite dikes, keratophyres, carbonatites, lamprophyres
Matatila Granite	Pink, coarse- to fine-grained, massive granite
Garhmau Granite	Gray, coarse- to fine-grained, porphyroblastic gneisses
Paron volcanics	Porphyroblastic, sheetlike, granitic rocks
Palar Formation	Quartzite, ferruginous quartzite, phyllite, black shale, and limestone; pyrophyllite-diaspore deposits at the base
--- Unconformity ---	
Kuraicha Formation	Migmatites, gneisses, augen gneisses, chlorite-biotite schists, quartzites, and metaarkoses; some garnet-bearing rocks

1975; D. M. Banerjee, 1982). At the southern margin of the Vindhyan basin, in the Narmada-Son valley, rocks correlated with the Bijawar suite occur in scattered outcrops over a stretch of 800 km in a north-south direction (Iqbaluddin and Moghni, 1979). Locally these rocks are found on both sides of the Narmada lineament (Roday and Bhatt, 1980). None of the suites in the Narmada valley has been dated, and the correlation with the type Bijawar outcrop is unclear.

The Bijawar Group of the type area consists of a terrigenous sequence of basal carbonates and shales with greenschists or pillow basalts; chloritic shales, ferruginous quartzites, and banded iron formations; and an upper sequence of pelitic rocks. Total thickness is about 200 m. The group is intruded by gabbros and granites older than the overlying Vindhyan sediments. Pillow lavas near Bijawar town are dated at 2,600 to 2,400 m.y. old (Crawford and Compston, 1970). The sequence is mostly flat lying but is locally deformed (Mathur and Mani, 1975).

Presumed Bijawar equivalents in one area of the Narmada valley (Roday and Bhatt, 1980) consist of sandstones, shales, impure limestones, and cherts. The lava flows found at Bijawar are absent. Three generations of deformation have been observed, but only the first two are well developed. The first deformation consisted of shearing of partly consolidated rocks. The second deformation produced asymmetrical or overturned folds that plunge roughly ENE at 22°.

Stromatolites occur in postulated Bijawar correlatives at one location in the Narmada valley (Balasundaram and Mahadevan, 1972; Krishna Murty, 1972). In this area a lower dolomite horizon containing *Collenia* is overlain by an upper siliceous horizon characterized by progressive development of cylindroidal conophytons. Cylindrical columns of conophytons (*Conophyton jugensis*), about 24 cm long and 8 to 12 cm in diameter, with nearly perfect circular cross sections, occur in the central parts of bioherms. Along the outer fringes of the bioherms these cylinders are replaced by stunted conophytons a few decimeters in both height and base diameter.

Choubey (1971) described a "Transitional Series" of quartzites, phyllites, dolomitic limestones, and banded iron formations in thrust contact with mildly deformed Vindhyan sediments along the Narmada valley near Jabalpur (Fig. 6.1). This series is clearly older than the Vindhyan suite, as shown by more extensive metamorphism and deformation, and has been thrust upward along a south-dipping fault. Because of its intermediate age between that of the Vindhyan rocks and that of the crystalline rocks of the Bundelkhand massif, the Transitional Series has been regarded as Bijawar equivalent, but no actual age data are available.

Gwalior Group

The Gwalior Group (Fig. 9.1) is exposed around the city of Gwalior and is chiefly composed of sedimentary and volcanic rocks unconformably overlying the Bundelkhand complex (Bajpai, 1935; Crawford and Compston, 1970). The basic lavas of this group have been dated at 1,830 ± 200 m.y. with an initial $^{87}Sr/^{86}Sr$ of 0.7047 ± 0.007 (Crawford and Compston, 1970). If this age is correct, the Gwalior Group is younger than both the Aravalli and Bijawar Groups and more similar to the Delhi Group. Gwalior rocks have an east-west strike and northerly dip of 5°.

The main constituents of the Gwalior Group are quartzites, limestones, banded ferruginous rocks, jasper, and lava flows. The Group is divided into two formations. The lower, Pur Formation consists of 50 to 200 ft of thin-bedded sandstones and conglomerates. The upper, Morar Formation consists of thin-bedded, siliceous, and ferruginous shales interbedded with ironstone and jasper. Two horizons of thin limestones and two principal horizons of basic lavas are present. The upper flows are about 500 ft thick. The volcanic rocks are quartz-normative tholeiites (Bajpai, 1935). The Gwalior Group is overlain by the upper Vindhyan Supergroup.

GRANITES OF THE ARAVALLI-DELHI BELT

The Aravalli-Delhi belt contains a large number of granites that were emplaced from 3,500 to 750 m.y. ago (Fig. 9.7). Two orogenies, one designated as Delhi and the other as pre-Delhi (R. S. Sharma, 1977), are shown by field data, but according to C. A. Sastry et al. (1984), Rb/Sr dates have shown five events of acid magmatism: 3,000 to 2,900 m.y., 2,600 to 2,500 m.y., 2,000 to 1,900 m.y., 1700 to 1500 m.y., and 850 to 750 m.y. The granites occur within, and on both sides of, the Aravalli mountain belt. Table 9.6 summarizes information on the granites in the region.

Terminology and classification of the plutonic suites has changed repeatedly through time. Almost all of the granitic rocks to the west of the Aravalli Mountains (Fig. 9.7) were included in the Erinpura Granite by Heron (1953), but later studies have shown that these granites represent various events. Most of the originally named Erinpura Granite is now referred to as the Godhra Granite. The Mt. Abu pluton, which was correlated with the Erinpura Granite, is now grouped with the Malani igneous suite on the basis of a Rb/Sr date of 735 m.y. by Crawford (1975). Other granites included in the Malani event are Jalor, Siwana, Idar and Mirpur (C. A. Sastry et al., 1984). Several major granite suites are discussed as follows.

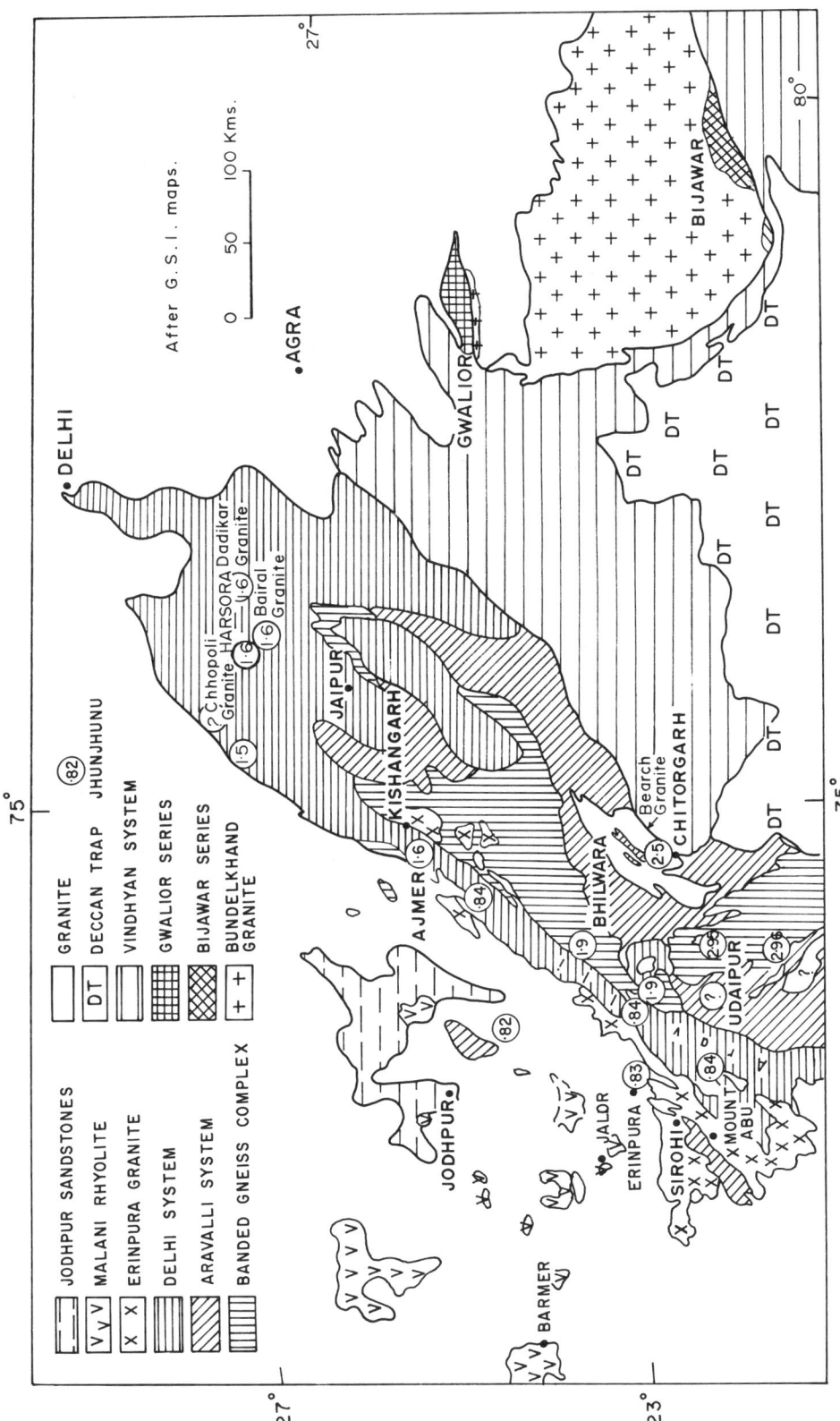

Fig. 9.7. Northern Aravalli-Delhi belt and Bundelkhand (Choudhary et al., 1984). Numbers in circles show radiometrically determined ages of granites. Reproduced with permission from Tectonophysics, v. 105, p. 133.

JODHPUR SANDSTONES

MALANI RHYOLITE

ERINPURA GRANITE

DELHI SYSTEM

ARAVALLI SYSTEM

BANDED GNEISS COMPLEX

GRANITE

DT DECCAN TRAP

VINDHYAN SYSTEM

GWALIOR SERIES

BIJAWAR SERIES

BUNDELKHAND GRANITE

After G.S.I. maps.

0   50   100 Kms.

DELHI

AGRA

JHUNJHUNU

GWALIOR

BIJAWAR

JAIPUR

? Chhopoli Granite HARSORA Dadikar
1·6 Granite

1·6 Bairal Granite

KISHANGARH

AJMER

CHITORGARH

Beach Granite

BHILWARA

UDAIPUR

JODHPUR

JALOR

ERINPURA

SIROHI

MOUNT ABU

BARMER

183

TABLE 9.6. *Ages of granites in Aravalli-Delhi belt and farther west (numerical values from Choudhary et al., 1984)*

Name of Granite	Age in b.y.
Granites associated with Malani suite	
Fhirtal	—
Gabbar	—
Jalor	—
Jhunjunu	0.82
Pali	0.82
Siwana	—
Tusham	0.94
Granites intrusive into Delhi Supergroup	
Ajitgarh	1.6
Ajmer	1.6
Ambaji	0.84
Anasagar	1.6
Bagan River	—
Bairat	1.6
Chapoli	1.6
Dadikar	1.6
Erinpura	0.83
Godhra	0.95
Harsora	1.6
Idar	0.74
Mount Abu	0.74
Sadri-Ranakpur	0.84
Sai	0.84
Saladipura	1.5
Sendra	0.84
Seoli	1.6
Granites intrusive into Aravalli Supergroup	
Ahar River	—
Amet	1.9
Dharwal	1.9
Granites associated with Banded Gneissic Complex	
Berach	2.6
Sarara	—
Untala-Gingla	2.95

The Berach Granite is dated as 2,600 ± 150 m.y. old by Rb/Sr methods (Choudhary et al., 1984) and 2,610 m.y. by Pb/U methods (Sivaraman and Odom, 1982). This granite also exhibits the effects of another thermal event around 710 m.y. ago. The pluton is a complex and heterogeneous body having granitic and gneissic components with compositional variations from tonalites to granites. It is regarded as part of the Banded Gneissic Complex by some workers.

The presently defined Erinpura granitic event, about 800 m.y. old, is widespread to the west of the Aravalli ranges. C. A. Sastry et al. (1984) found no Sr isotopic differences between the type Erinpura Granite and granites within the Aravalli mountain ranges that have undergone Delhi deformation. Similarly, the Erinpura granites and granites deformed in the Delhi orogeny both yield secondary mineral ages of about 550 m.y. This

similarity has not been explained. The Erinpura Granite was emplaced as sheet complexes and massive stocks (Heron, 1953), and enclaves of calc-schists and calc-gneisses occur within it.

The Bairat Granite and its equivalents (Dadikar, Harsora, Chapoli, Seoli, and Ajitgarh) are dated at 1,700 to 1,500 m.y. old (Choudhary et al., 1984). Crawford (1970) previously estimated a model age of 1,650 m.y. A mineral isochron from Bairat and equivalent granites shows reequilibration of Sr isotopes at 800 ± 50 m.y., presumably due to a secondary thermal event. The Bairat Granite was emplaced in the metasediments of the Delhi Group. Two varieties of Bairat Granite are present: foliated and lineated granite with mafic constituents, and massive pink granite with a negligible amount of mafic constituents. Both are alkali granites with abundant K-feldspars. The deformational fabric of the granites is parallel to, and similar to, that of surrounding Delhi sediments, indicating their development in a common deformational event (Goswami and Gangopadhyay, 1971).

The Rb/Sr age of the Godhra Granite is 955 ± 2 m.y. (Gopalan et al., 1979b). This granite is considered to be the major igneous intrusion into metasediments of the Aravalli Supergroup of Gujarat (or Champaner sequence) and is correlated with the post-orogenic Erinpura Granite.

The Udaipur, Saladipura, and Seoli Granites of the Khetri copper belt are dated at 1,480 ± 40 m.y. (Gopalan et al., 1979a). The Seoli Granite is a massive body trending NNE-SSW along an anticlinal core with sharp contacts with the Delhi sediments. No secondary structure is seen in the Seoli Granite, but the constituent minerals bear evidence of weak deformation. The Seoli Granite is rich in K_2O and Al_2O_3, and on a normative Ab-An-Qtz-H_2O diagram, it falls in the quartz field and outside the low-temperature valley of magmatic crystallization.

Specific studies have been made of a variety of other granites in the Aravalli area. Bodies studied include Shahgarh (Rajarajan, 1971), Jhirgadandi (A. K. Srivastava, 1977), Ojagar (K. K. Sharma, 1980), and Dadikar (Gangopadhyay and Das, 1974).

MALANI IGNEOUS SUITE

The Malani igneous suite covers a wide area to the west of the Aravalli Mountains (Fig. 9.1). It has been dated at 740 ± 10 m.y. old (Crawford, 1970). The Jalor, Siwana, Idar, and several other granites (Table 9.6) are considered equivalents of the Malani volcanic rocks and hence included in the Malani igneous suite (Mukherjee and Roy, 1981). Other reported occurrences of possibly equivalent rhyolites and associated granite porphyries and tuffs are from Jhunjhunu, in western Rajasthan, and farther north in the Tusham and related complexes in Haryana and Punjab (C. A. Sastry et al., 1984; Fig. 9.7). The Tusham ring complex has been studied by Kochhar (1983) and dated as 940 ± 20 m.y. old. Granite porphyries from Jhunjhunu yield a Rb/Sr isochron age of 823 ± 18 m.y. (Crawford, 1975), and Crawford correlated the Mt. Abu Granite, of 735 m.y. age, with the Malani suite. The Malani rhyolites overlie the Aravalli Supergroup and underlie, with erosional disconformity, the Jodhpur Sandstone member of the Marwar Group (Awasthi and Prakash, 1981).

The Malani rhyolites have been the subject of a number of studies (Pareek, 1981a; Kochhar and Pareek, 1982; and Kochhar, 1984b). They contain euhedral phenocrysts of quartz, orthoclase, and oligoclase in a cryptocrystalline to microgranitic groundmass that is mainly composed of K-feldspar and quartz. Euhedral phenocrysts of hornblende are present in some rocks. The extensive silicic magmatism in this part of India in the Late Proterozoic has been attributed by Kochhar (1984b) to the activity of a hot spot, leading to cratonization of the shield, but the significance of a crust-forming event at this time is controversial (Kochhar, 1984a; K. K. Sharma et al., 1984). The Malani suite also contains olivine basalts.

ALKALINE ROCKS, KIMBERLITES, AND CARBONATITES

Syenites

Nepheline syenites northeast of Kishangarh City (Fig. 9.1) consist of one large intrusion and seven small sill-like intrusions in pre-Aravalli quartzite. Crawford (1970) dated them as 1,490 ± 150 m.y. old. Alkaline rocks were also emplaced within the Aravalli Supergroup (Niyogi, 1966). Chaudhuri and Mukherjee (1978) proposed that foliation in the concordant nepheline syenites shows identical development of superposed folds as in the Aravalli rocks. This structural concordance is inconsistent with an age as young as 1,500 m.y.

The alkaline rocks at Kishangarh are generally associated with the Mandaoria Gabbro (Niyogi, 1966). The syenites contain equal proportions of nepheline and plagioclase (andesine) with lesser amounts of magnesian amphibole (magnesioriebeckite), biotite, and sodic pyroxene. Niyogi concluded that the syenites originated by syn- to late-tectonic alkali metasomatism of gabbro.

Syenites and similar alkaline rocks have been reported elsewhere in the Aravalli craton. A major occurrence is at Sirohi (Fig. 9.2; M. K. Chakraborty and Bose, 1978; M. K. Chakraborty, 1979).

Kimberlites

Kimberlitic pipes occur in two main areas in the Aravalli craton (Fig. 9.1). The Majhgawan, Hin-

ota, and Agnor pipes are located in the Panna region of Madhya Pradesh, and a number of pipes occur in the Jungel Valley of Uttar Pradesh (Chattopadhyay and Venkataraman, 1977).

The diamond-bearing kimberlitic pipe of Majhgawan has been described by Venkataraman (1960), Mathur (1962), and Mathur and Singh (1971). It was emplaced in the Vindhyan Supergroup. The Rb/Sr age of the pipe was estimated as 1,140 ± 12 m.y. by Crawford and Compston (1970) and 1,630 m.y. by Paul (1979), although scatter on the isochron could be the result of alteration of the pipe material. Paul et al. (1975b) found K-Ar ages of the pipes from the Panna area between 974 ± 30 m.y. and 1,120 ± 45 m.y.

The Majhgawan pipes contain three types of rocks: yellow-green pipe rock, weathered tuffaceous material, and dark-colored volcanic rock (Mathur and Singh, 1971). The yellow-green variety is brecciated and consists of rock fragments and groundmass containing serpentine, phlogopite, chlorite, carbonate minerals, and oxides. Serpentine is pseudomorphous after olivine (and possibly pyroxene) phenocrysts. The weathered tuffaceous material consists of fragments of serpentine and shales in a fine-grained groundmass composed of serpentine, carbonate, hematite, and leucoxene. The dark-colored rock occurs at depth and is mineralogically similar to the yellow-green pipe rock. Phenocrysts include garnet and possibly pseudomorphs of primary melilites (Venkataraman, 1960). The dark-colored rock is primary and the other two varieties are its weathered product. The Hinota pipe is situated about 3 km northwest of Majhgawan. It is a basaltic kimberlite and consists of serpentine, chlorite, smectite, phlogopite, and magnesian vermiculite (Kresten and Paul, 1976). The Agnor pipe consists of several bodies within the Bundelkhand Granite. The pipe has a variable composition, with peridotite at lower levels passing upward through olivine pyroxenite to pyroxene-rich gabbro at higher levels. This rock is not a kimberlite, but diamonds occur in it (Mathur, 1981).

Paul et al. (1975a) found that the REE abundances in pipes of the Panna region are higher than those in similar rocks in the Eastern Dharwar craton (Ch. 3). The initial $^{87}Sr/^{86}Sr$ ratio in the Panna rocks shows a variation from 0.7027 to 0.7102 (Paul, 1979). The whole-rock uranium content ranges from 1.87 to 3.93 ppm., with an average Th/U ratio of 8.80 for Majhgawan to 6.88 for Hinota kimberlites (Paul et al., 1977).

Five elliptical, plug-shaped kimberlitic bodies occur in Jungel Valley (Chattopadhyay and Venkataraman, 1977). Balasubrahmanyam et al. (1978) reported a K/Ar age of 919 m.y. The pipe rocks in the weathered zone are mainly serpentinous clay, tremolite, actinolite and calcite with pseudomorphs of olivine and pyroxene and xenoliths of magnetite-rich rock, limestone, and basalt. They resemble kimberlites and have yielded a few diamonds.

Newania Carbonatite

At Newania, near Udaipur, a deeply eroded carbonatite complex occurs as an intrusive into the Untala Granite (Deans and Powell, 1968; Phadke and Jhingran, 1968). It is composed predominantly of ferruginous dolomitic carbonate and is fringed by fenites. Apatite deposits occur in the center of the complex. The principal accessory is an alkali amphibole resembling crossite. K-Ar dating of this amphibole has yielded an age of 959 ± 24 m.y. The initial $^{87}Sr/^{86}Sr$ ratio is about 0.705 (Deans and Powell, 1968).

VINDHYAN SUPERGROUP

The Vindhyan suite was once regarded as having been deposited in two separate basins on either side of the Aravalli mountain range (Fig. 9.1): the main Vindhyan basin in parts of eastern Rajasthan, Haryana, Madhya Pradesh, Uttar Pradesh, and Bihar; and the trans-Aravalli Vindhyan basin in western Rajasthan (Heron, 1953; Krishnan and Swaminath, 1959). Later work (S. N. Gupta et al., 1980; C. A. Sastry et al., 1984), however, has shown that the so-called "trans-Aravalli Vindhyans" are younger than the sediments of the main basin, and they are now designated as a separate Marwar Supergroup. The main Vindhyan basin occupies a large area surrounding the Bundelkhand massif (Fig. 9.8).

Vindhyan sediments occur primarily north of the Narmada-Son lineament. Surprisingly, seismic sounding profiles across the lineament do not show its subsurface presence (Kaila et al., 1985); instead, a deep-seated fault is shown about 50 km south of the lineament. This fault, discussed in Chapter 10, apparently bounds the Vindhyan basin on the south in the same fashion that the Great Boundary Fault bounds the basin on the west. Vindhyan sediments cover more than 104,000 sq km of central India, and another 78,000 sq km of Vindhyan rocks are concealed by the Deccan basalts where they cover the Narmada valley and the southern extension of the Great Boundary Fault (Misra, 1969). The thickness of Vindhyan deposits below the Deccan plateau is estimated to be about 600 m by DSS data (Kaila, 1984). The area of the Vindhyan deposits hidden under the Gangetic plain is possibly much larger than the exposed area (Fuloria, 1969; Misra, 1969; I. Banerjee, 1982; B. N. Srivastava et al., 1983), and the suite extends into the Himalayas (Ch. 11).

Palynological assemblages obtained from wells in the Ganges basin are comparable to those from the upper Vindhyan sequence of the Son Valley (Salujha, 1982). Seismic and gravity data also indicate that the Vindhyan basin extends to the

north under the Gangetic plain (Fig. 9.9), possibly divided by the Faizabad ridge into western and eastern basins (Fig. 9.1; Narain and Kaila, 1982). This separation is also shown by deep well data (V. V. Sastry et al., 1971). Paleocurrent data indicate that the Vindhyan basin submerged the Bundelkhand massif.

The first isotopic dating of the Vindhyan rocks was by Tugarinov et al. (1965), who proposed a 1,400 m.y. age for the lower Semri Group and 910 m.y. for the Kaimur Group. The Majhgawan kimberlite, which intrudes Kaimur sandstones, gives a Rb/Sr isochron age of 1,140 m.y. (Crawford and Compston, 1970). Based on these data, Misra (1969) assigned the Vindhyan suite an age of 1,400 to 900 m.y., but Crawford and Compston (1970) put a younger age limit at 550 m.y.

Stratigraphy

The Vindhyan Supergroup shows differences in the lithology of different groups and formations at different places. Early stratigraphic work proposed a division into Lower (Semri) and Upper (Kaimur, Rewa, and Bhander) Groups separated by a major unconformity (summaries by Chanda and Bhattacharyya, 1982; and Prakash and Dalela, 1982). Many workers, however, have not found an unconformity and do not recognize Lower and Upper suites (Auden, 1933; Ahmad, 1971; Mathur, 1981). Stratigraphic sections are shown in Table 9.7 for eastern Rajasthan and in Table 9.8 for the type Vindhyan area of the Son valley, where the Bhander Group is not developed.

Argument about separation of the upper and lower parts of the Vindhyan sequence is based on the nature of the lower contact of the Kaimur Group. The basal Kaimur unit is a conglomerate and is unconformable throughout a large part of its outcrop, either on Bijawar or related metamorphic rocks or on Bundelkhand granites and migmatites. Prakash and Dalela (1982) proposed a vast erosional unconformity between the Rohtas limestones of the Semri Group and the Kaimur sandstones, with the Kaimur rocks overlying folded, deformed, and eroded Rohtas limestones. Chanda and Bhattacharyya (1982) stated that the contact between lower and upper Vindhyan sequences, although locally conformable, is mostly discordant and marks a major break in sedimentation. According to them, the lower Vindhyan rocks display greater tectonic and magmatic activity compared to the upper Vindhyan sequences. A contrary point of view, that no unconformity exists at the base of the Kaimur Group, has been stated by other workers (referenced above).

The aggregate thickness of the Semri and Kaimur Groups is estimated as between 4,993 and 3,477 m at different places. Thicknesses of the

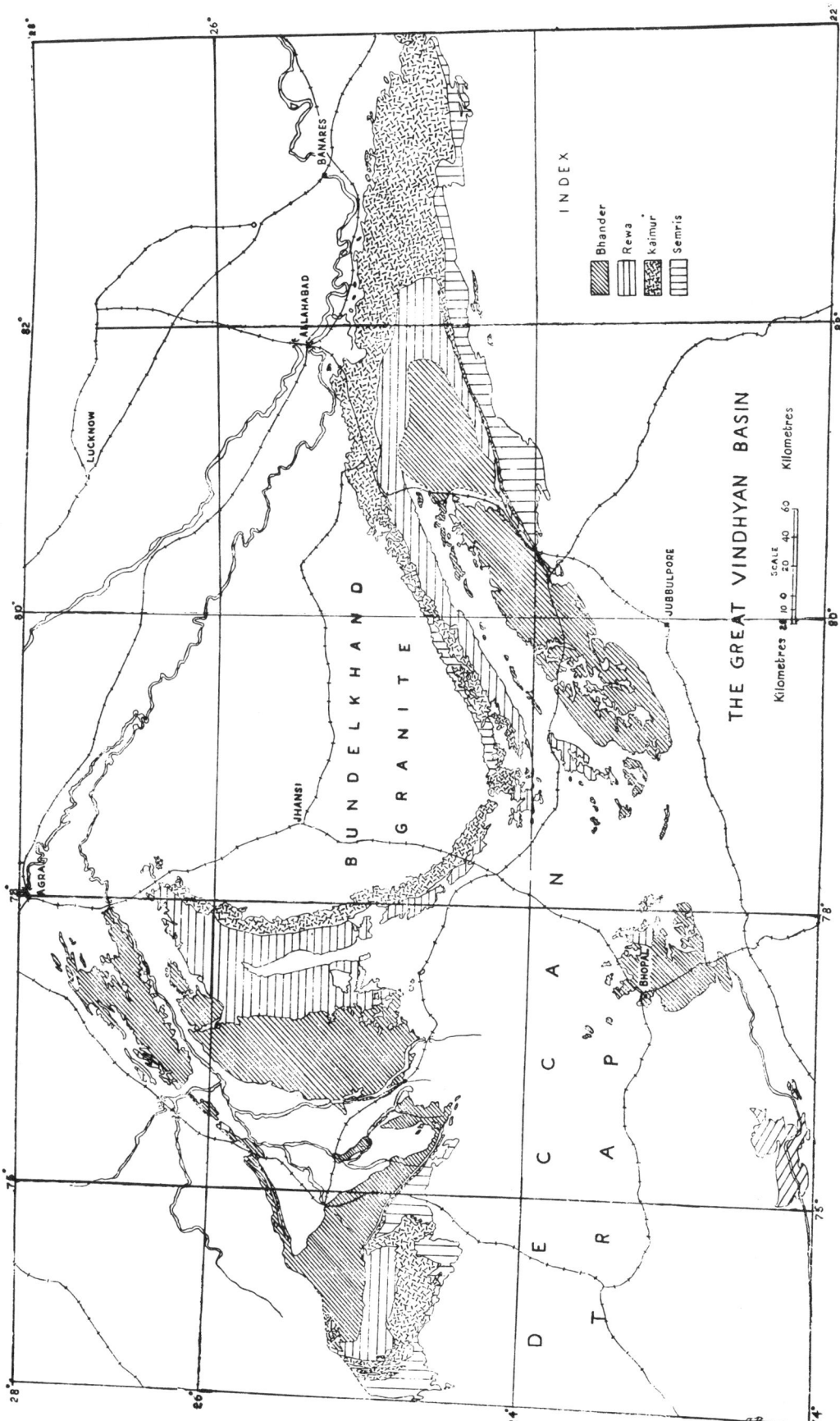

Fig. 9.8. Map of Vindhyan Supergroup (Krishnan and Swaminath, 1959). Reproduced with permission from the Journal of the Geological Society of India, v. 1.

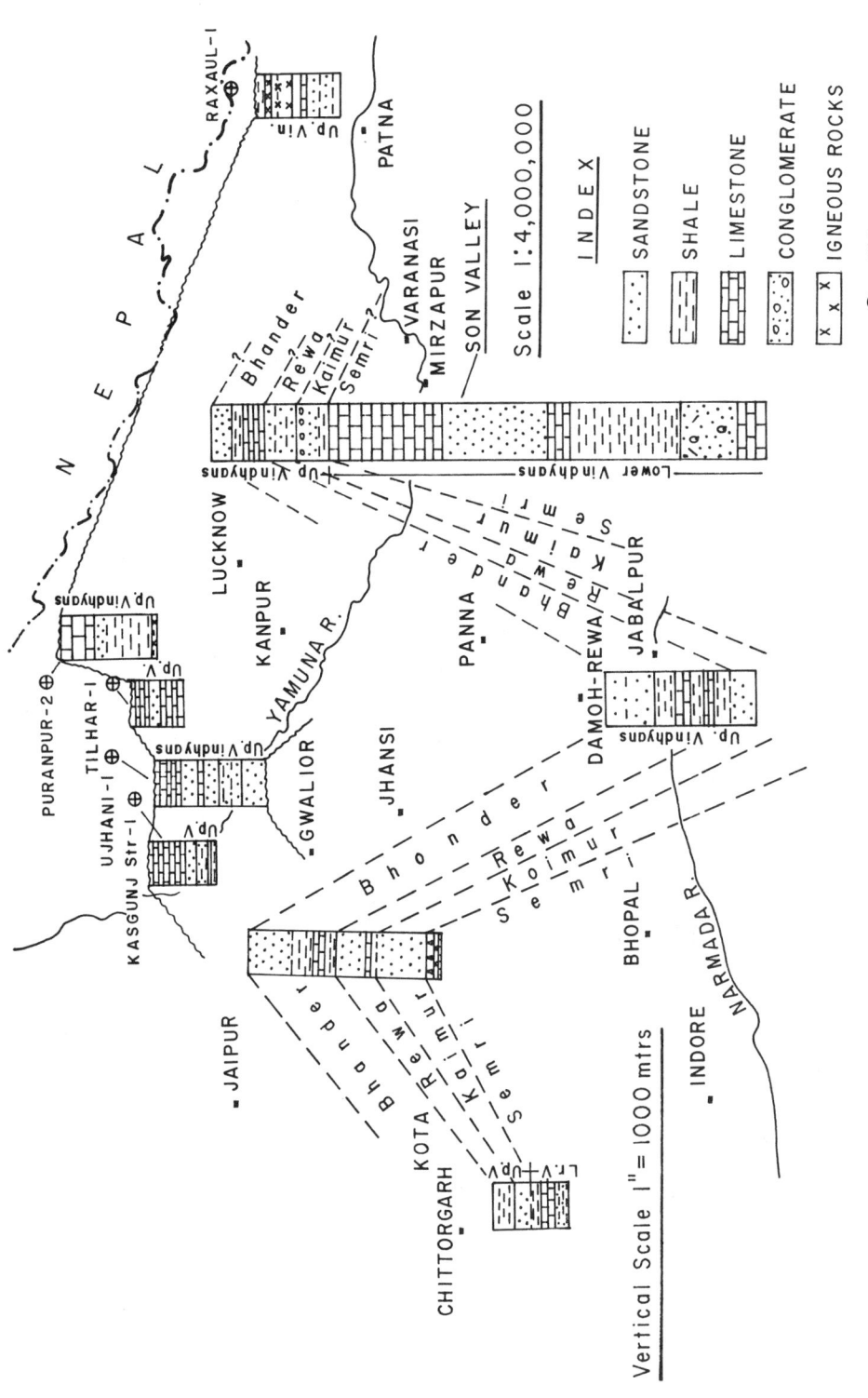

Fig. 9.9. Fence diagram showing relationships between various horizons in Vindhyan Supergroup, mostly as determined by drilling in the Ganges plain (B. N. Srivastava et al., 1983) (courtesy N. K. Verma).

TABLE 9.7. *Stratigraphy of Vindhyan Supergroup in Rajasthan (western part of basin) (Geological Survey of India, 1977)*

Bhander Group	Upper Bhander sandstones, shales and limestones
	Sirbu shales and limestones (stromatolitic)
	Lower Bhander sandstones and shales
	Samria shales and limestones (stromatolitic)
	Lower Bhander limestones
	Ganugarh shales
Rewa Group	Upper Rewa sandstones
	Jhiri shales
	Lower Rewa sandstones
	Panna shales
Kaimur Group	Kaimur sandstones, conglomerates, and shales
Semri Group	Tirohan limestones, breccia, and hematite shales
	Suket shales
	Jhaira Patna sandstone
	Nimbahara limestone
	Nimbahara shales
	Jiran sandstones
	Khori Malan conglomerate
	Binota shales and sandstones
	Sawa conglomerate, sandstones, shales, and porcellanite
	Bhagwanpura limestones
	Khardeola sandstones and shales
	Khairmalia flows (andesites)

Rewa Group are 296 to 175 m. The thickness of the Bhander Group varies from 70 to 1,000 m (Misra, 1969; Srinivasa Rao and Neelakantam, 1978).

The Vindhyan Supergroup is largely undisturbed, flat lying, and affected only by faulting. In the western and southern margins against the Aravalli and Satpura mountain ranges, however, they are highly sheared, brecciated, and severely folded (Ch. 10). A. K. Sharma (1980) found that the Semri and Kaimur Groups in the Son Valley underwent two phases of deformation. The early folds are recumbent, inclined, and overturned, whereas the later folds are congruous and incongruous shears. The first deformation resulted from E-W compression, and second-phase structures are the result of N-S compression. The Semri Group near the Great Boundary Fault was also locally deformed into open, discontinuous, and generally asymmetrical folds (Prasad, 1975). The western limbs of these folds are steeper than the eastern. The strike of the fold axes is parallel to the strike of the beds and plunges at low to moderate angles toward the north and south.

Lithology and petrology (Lower Vindhyan)

Srinivasa Rao and Neelakantam (1978) provided information on the Lower Vindhyan (Semri Group) in the type area of the Son Valley (Table 9.8). The basal part of the lowest unit (Deolond quartzite) is a conglomerate or pebbly quartzite, containing pebbles of vein quartz, quartzite, amphibolite, and jasper. It grades into quartz arenite laterally and vertically. The conglomerate separates Vindhyan from preVindhyan rocks continuously along the southern margin in the Son Valley, but it is scattered on the northern side of the Vindhyan outcrop along the Bundelkhand massif (Fig. 9.8). It has been regarded as glacial (Ahmad, 1971), a shoreline accumulation (Misra, 1969; Ghosh, 1973), and a deposit of coastal rivers of high gradient (I. B. Singh, 1973).

Higher in the Son Valley section, the Khajrahat Limestone is dark gray or yellow and locally argillaceous. The Deonar Porcellanite is a sequence of rhythmically interbedded shale and porcellanite, with sparse coarse-grained arenites that contain authigenic glauconite. Tuffs and volcanic ag-

TABLE 9.8. *Stratigraphy of Vindhyan Supergroup in Son Valley (Srinivasa Rao and Neelakantam, 1978)*

Names of Units	Thicknesses (in m)
Rewa Group	
Govindgarh Sandstone	120–125
Rewa Shale	25–125
Kaimur Group	50–350
Semri Group	
Rohtas Subgroup	
Kaimur Sandstone	50–350
Baghwar Shale	20–125
Rohtas Limestone	200–280
Kheinjua Subgroup	
Rampur Shale	180–800
Chorhat Sandstone	385–525
Koldaha Shale	70–120
Porcellanite Subgroup	
Deonar Porcellanite	870–1,525
Basal Subgroup	
Khajrahat Limestone	135–145
Basal Shale	280–460
Deolond Quartzite	45–200

glomerates are commonly associated with the porcellanite. The Koldaha Shale is a laminated olive-green shale with beds of fine-grained sandstone, limestone, and porcellanite. The Chorhat Sandstone is a sequence of alternating shales and sandstones that also contains rhythmically interbedded calc-arenites, argillaceous limestones, calcareous shales, and calcareous sandstones. The Rampur Shales contain calcareous shale, feldspathic arenite (partly glauconitic), and limestone. The Rohtas Limestone includes micritic, stromatolitic limestones, calcareous shales, porcellanic shales, and sandstones.

The Semri Group of eastern Rajasthan (Prasad, 1975) is slightly different from that of the type area of the Son Valley. In eastern Rajasthan, Vindhyan sedimentation commenced with volcanic activity that is preserved as the Khairmalia andesite (Prasad, 1975) and pyroclastic rocks resting unconformably on granite that some workers correlated with the Berach Granite, west of the Great Boundary Fault. The thickness of the flows and pyroclastic rocks varies between 40 and 100 m. Other units in eastern Rajasthan (Prasad, 1975) include the subarkosic Khardeola Sandstone, which is locally conglomeratic, and the polymictic Bhagwanpura Conglomerate. The Bhagwanpura Conglomerate is overlain by dolomitic limestone, which is locally stromatolitic; the Sawa Sandstone, which contains minor grains of oolitic chamosite; and the Pari and Binota Shales. The other formations of the Semri Group of eastern Rajas-

than also show significant petrological differences from those of the Son Valley.

Lithology and petrology (Upper Vindhyan)

The basal part of the Kaimur Quartzite in the Son Valley section (Srinivasa Rao and Neelakantam, 1978) is a medium- to coarse-grained lithic arenite that grades upward into a mature quartz arenite. The Kaimur/Rewa contact is marked by a conglomerate consisting of pebbles of the underlying Kaimur Sandstone in a sandy matrix. The Rewa shales are interbedded with fine-grained sandstones and pass upward into the Govindgarh Sandstone, whose basal part is a shale-pebble conglomerate.

The Bhander Group is extensively developed in southeastern Rajasthan and adjoining Madhya Pradesh, west of the Son Valley section (Akhtar, 1976, 1978). Here this group is represented by a 335-m-thick sequence comprising the Ganugarh Shale, Bhander Limestone, Sirbu Shale, and Bhander Sandstone. The Ganugarh Shale, the basal formation, is 20 m thick and shows a conformable relationship with the underlying Rewa Sandstone and the overlying Bhander Limestone (Akhtar and Srivastava, 1976). The Bhander Limestone contains micritic limestone, crystalline dolostone, siltstone, and shale. The rocks show desiccation structures, algal laminations, palisade structures, sporadic ripple marks, ripple laminations, and micro-cross-laminations (Chanda and Bhattacharyya, 1974). The major petrographic constituents (Akhtar, 1976) are micrite, intraclasts, sparry calcite cement, pseudospar, and replacement dolomite. It is stromatolitic at places, and deformed ooids occur in the upper part (Chanda et al., 1977; B. Sarkar, 1983). Halite casts occur in abundance in the horizon adjacent to the carbonates (Mathur, 1965).

Three diamond-bearing conglomerates (lower Rewa, basal Jhiri, and upper Rewa) occur in the Vindhyan basin around kimberlitic pipes near Panna (Fig. 9.1; Mathur, 1962, 1981). They confirm existence of the pipes before Upper Vindhyan sedimentation. The lower Rewa conglomerate is found at various levels in the lower Rewa Sandstone. The basal Jhiri conglomerate is composed of well-rounded pebbles (mostly jasper and quartz) in a siliceous or ferruginous matrix. The upper Rewa Conglomerate is a boulder bed with well-rounded pebbles (4″ to 12″) of sandstone embedded in a fine conglomerate that consists of pebbles of jasper, chert, vein quartz, quartzite, and sandstone. All conglomerates are intraformational and do not show any unconformity between the Kaimur, Rewa and Bhander Groups.

Paleogeography, sedimentation, paleocurrents

The Vindhyan suite has a sheetlike geometry and is conformable with the present margins of the

TABLE 9.9. *Depositional environments of Vindhyan sediments (from I. B. Singh, 1973)*

Unit	Environment
Kaimur Series	
Dhandhraul Quartzite	Shoals
Scarp Sandstone	Shoals
Bijaigarh Shale	Lagoon and lagoonal beach
Upper Quartzite	Subtidal
Lower Quartzite	Subtidal to intertidal
Semri Series	
Rohtas Limestone	Subtidal to intertidal
Glauconitic Sandstone	Intertidal
Fawn Limestone	Intertidal to supratidal
Olive Shale	Lagoon
Porcellanite	Lagoon
Upper Kajrahat Limestone	Supratidal (wave affected)
Middle Kajrahat Limestone	Protected intertidal
Lower Kajrahat Limestone	Subtidal
Shale	Lagoon
Sandstone	Subtidal to intertidal
Stromatolitic Limestone	Subtidal to intertidal
Basal Conglomerate	High-gradient rivers

basin, suggesting that the original basin configuration has been preserved more or less intact. The dominant lithology changes vertically, however. The Semri series is predominantly calcareous, the Kaimur and Rewa Groups are mostly arenaceous, and the Bhander Group is a mixed clastic/carbonate sequence. The beginning of Kaimur sedimentation represented a change from predominantly carbonate to clastic sedimentation (Chanda and Bhattacharyya, 1982). The depositional environments of Vindhyan sediments were reviewed by Sahni (1975), and a summary is given in Table 9.9.

I. Banerjee (1982) described Vindhyan sediments as mostly tidal deposits. Tidal current patterns should yield bimodal and bipolar paleocurrent indicators, but analysis of paleocurrent data over the Vindhyan basin shows dominance of a unimodal northwesterly pattern (Misra, 1969). The unimodal paleocurrent data, however, may be apparent, and subdivision into different environments may show a tidal bimodality (P. Chakraborty, 1976).

Life during Vindhyan times

Stromatolites in the Vindhyan Supergroup have been extensively studied. Well-preserved columnar stromatolites occur as bioherms and bios-

tromes in both the Semri Group (Basal Formation, Kheinjua Formation, and Rohtas Formation) and the Bhander Limestone. A microcrystalline limestone in the basal conglomerate has yielded *Kussiella kussiensis* (Kumar, 1977). The uppermost part of the Khajrahat Limestone (Basal Formation) is characterized by *Kussiella kussiensis, Kussiella dalaeusis, Colonella rajrahatensis, Conophyton vindhyaensis,* and *Collenia symmetrica* (Kumar, 1977). The Fawn Limestone of the Kheinjua Formation has yielded stromatolites that include *Colonella columaris, Conophyton garganicus,* and *Collenia clappii* (Kumar, 1977).

Although the development of stromatolites in the Rohtas Limestone is poor, oncolites and *Collenia cliappii* have been recorded (Kumar, 1977). The Semri Group also has yielded *Collenia symmetrica, Kussiella kussiensis, Colonella lodhwarensis, Colonella columnaris, Baicalia baicalica, Collenia frequence, Cryptozoon occidentale,* and *Conophyton* (Raja Rao and Mahajan, 1965).

The Bhander Limestone in Madhya Pradesh contains *Baicalia baicalica, Tungussia colonella, Boxonia, Collenia symmetrica, Stratifera,* and oncolites (Kumar, 1977; K. S. Rao et al, 1977). The Sibru shales of the Bhander Group contain an arenaceous/argillaceous horizon showing evaporitic conditions of deposition and the stromatolite *Maiharia maiharensis* (Kumar, 1977). Stromatolites also have been recorded in the calcareous members in the Sirbu and Samria shales (Bhander Group), Jhiri shales (Rewa Group), Bijaigarh shales (Kaimur Group), and Arangi Formation (Semri Group).

In addition to stromatolites, a large number of palynological fossils and other organic remains have been found in Vindhyan sediments (Salujha, 1982; Mathur, 1982). Jacob et al. (1953) reported woody elements and spores from the Suket shales belonging to primitive pteridosperms, pteridophytes, and doubtful primitive gymnosperms. Sahni and Srivastava (1954) studied the genus *Fermoria* (now placed under *Chuaria circularis*). Salujha et al. (1971) studied fossil palynomorphs, including acritarchs, fungal spores, and some dark brown to black bodies of indeterminate nature. The Semri Group in the Son Valley contains a variety of acritarchs, trilete spores, and unindentifiable brown-black bodies (Salujha et al., 1971). A coal seam 1.6 m thick occurs in the Arangi Formation of the Semri Group (Mathur, 1982), indicating that vegetation was luxuriant enough to produce coal at this time.

MARWAR SUPERGROUP

Rocks of the Marwar Supergroup are exposed west of the main Aravalli Mountain range (Fig. 9.1) and were earlier described as "Trans-Aravalli Vindhyans" by Heron (1953). The supergroup rests unconformably on the Erinpura Granite,

Malani rhyolites, and Delhi Supergroup. The
Marwar suite is about 1000 m thick and divided
into three groups (Jodhpur, Bilara, and Nagaur).
It consists mainly of sandstones, limestones, and
siltstones interbedded with shales, gypsum, and
anhydrite.

Awasthi and Prakash (1981) divided the Jodhpur Group into five lithostratigraphic formations
with a total thickness of 185 m. Sandstones of
these formations are mostly mature quartz arenites with minor subfeldspathic, feldspathic, and
subarkosic arenites. Awasthi and Prakash recognized six lithofacies: braided alluvial facies, low-
energy beach facies during depositional regression, storm-influenced beach facies during depositional regression, low-energy beach facies during
depositional transgression, lagoonal facies, and
shelf facies. Paleocurrent measurements show
that streams (mostly braided) flowed west-north-
west into an intracratonic basin. The rocks are unfossiliferous and undeformed. Pareek (1983) used
a different stratigraphy and divided the Jodhpur
group into the Sonia Sandstone and Girbhakar
Sandstone. The Pokaran Boulder Bed forms the
basal part of the Sonia Sandstone and marks the
outer margin of the basin (Pareek, 1981a, 1981b,
1983).

The Bilara Group is dolomitic and cherty, and
stromatolites are present at several horizons (Pareek, 1981a, 1981b, 1983). Poorly preserved la-
mellibranch and brachiopod shells are also present. The thickness of this group varies from 100
to 300 m. The Pandlo Dolomite is overlain by the
Khichan Conglomerate of the Nagaur Group,
consisting of pebbles and fragments of chert, dolomite, feldspar, quartz, and sandstone embedded
in a pink to brown, sandy to calcareous matrix.
This conglomerate is followed upward by a thick
sequence of Nagaur Sandstone and a thick eva-
porite sequence, containing gypsum, anhydrite,
halite, polyhalite and other potassium salts. Halite exceeds 100 m in thickness. Ripple marks,
mud cracks, and current bedding are present. The
overlying Tuaklian Sandstone is composed of
claystone, siltstone, clay, and gritty to pebbly
sandstone.

MINERAL RESOURCES

The Aravalli craton is extremely rich in minerals
except for the virtual absence of banded iron formation. Deposits of nonmetallic minerals include
apatite, asbestos, barite, beryl, phosphorite, fluorite, gemstones, gypsum, and kyanite. Diamonds
are mined at the Panna diamond field, both in
kimberlite pipes and Vindhyan conglomerate.
The Khetri Copper belt is one of the major cop-
per-producing areas of India, and the Pb-Zn
mines of Zawar are situated in this region. Due to
space limitations, only a few important deposits
are included here.

Uranium

Uranium mineralization, consisting mostly of
uraninite and less commonly of braunerite, occurs in pink aplitic emplacements along shear
zones in a 16-km-long belt in the Sikar District,
near Khetri (Narayan Das et al., 1980). The shears
are parallel to the main tectonic trends. At some
places the uranium-bearing aplites are folded
along with the host quartzites and other metasediments of the Delhi Supergroup.

Khetri Copper Belt

The Khetri copper belt extends over a strike
length of about 80 km near Khetri (Fig. 9.2). The
important deposits are at Madhan-Kudhan, Kolihan, Akwali, Satkui, and Dhanota (S. P. Das
Gupta, 1964; Roy Chowdhary and Das Gupta,
1965; N. K. Rao and Rao, 1968; and Chandra
Chowdhury et al., 1977).

The Madhan-Kudhan, Kolihan and Akwali deposits are near the northern end of the belt. Copper mineralization occurs in garnet-, chlorite-,
biotite-, and andalusite-schists and quartzites of
the Alwar and Ajabgarh Groups of the Delhi Supergroup. Minerals formed hydrothermally along
fissures and shears, and the lodes and individual
minerals are closely sheared and brecciated. To a
minor extent, mineralization has been controlled
by folds and intersections of shear zones and
cross-faults. Mineralization is in the form of im-
persistent stringers, disseminations and patches
of chalcopyrite, pyrrhotite, pyrite, cubanite,
sphalerite, valeriite, and galena. Regional metamorphism and metasomatism occurred before
the ore mineralization, and temperatures of formation of ores were estimated as 438° to 470°.

The major deposits in the southern part of the
Khetri belt are at Satkui and Dhanota. Mineralization is mostly restricted to Alwar quartzites and
phyllites. Most of the workable ore is along the
breccia zones of faults.

Lead-zinc deposits

Lead-zinc deposits of the Aravalli region occur at
Zawar and Dariba-Rajpura, near Udaipur, and at
Deri-Ambaji, farther south (Fig. 9.2).

The Zawar Pb-Zn belt is situated 40 km to the
south of Udaipur and extends for a distance of
about 20 km (A. K. Mukherjee and Sen, 1980; P.
K. Basu, 1981, 1982). Dolomites and dolomitic
rocks are the most favorable host rocks, and mineralization appears to be controlled by fractures
and fissures. The ore minerals at Zawar include
pyrrhotite, pyrite, chalcopyrite, sphalerite, galena,
small quantities of argentite and native silver, and
minor argyrodite. A. K. Mukherjee and Sen suggested a two-stage model of mineralization, with

some of the ore formed hydrothermally in shear zones following original syngenetic deposition of the sulfides.

The Dariba-Rajpura Zn-Pb-Cu belt extends over a strike length of 20 km (Deb and Bhattacharyya, 1980; Pandya et al., 1980). The host rocks are kyanite-graphite-quartz-mica schists, metacherts interbedded with quartzites, and calc-silicates and marbles of the Aravalli Supergroup. The deposit also contains silver, cadmium, arsenic, antimony, mercury, and indium. Minerals include sphalerite, pyrite, galena, chalcopyrite, pyrrhotite, and less common varieties, including the rare mineral owyheeite. The carbonate rocks are stromatolitic (Deb et al., 1978). The structures in the ores suggest that these deposits are synsedimentary (Nair and Agarwal, 1976; Chauhan, 1977; Deb et al., 1978).

Massive, stratiform Zn-Pb-Cu copper deposits occur at Deri and Ambaji, located 8 km apart in Rajasthan and Gujarat (Deb 1979, 1980; Fig. 9.2). They are considered to be controlled by shear zones and strike faults (Geological Survey of India, 1977). The sulfide ores are in the Ajabgarh Group in lensoid bodies of metamorphosed magnesian and calc-magnesian rocks. The country rocks and the ores were subjected to an early greenschist regional metamorphism and later to hornblende-hornfels thermal metamorphism related to the intrusion of granite and alkali syenite plutons. Conditions of the last metamorphism have been estimated at 525° to 625° and 2 to 4 kb pressure. Deb (1980) inferred that the host rocks of the ores were formed by isochemical metamorphism of sedimentary calc-magnesian and magnesian beds. The metals were derived from subaqueous basic volcanic rocks (now metamorphosed to amphibolites) and were concentrated syngenetically or diagenetically with the magnesian sediments. Two phases of deformation produced identical fold patterns in both ores and host rocks.

A stratiform sulfide deposit occurs in the Ajabgarh Group at Saladipura, southeast of Khetri (Fig. 9.1). Pyrite and pyrrhotite are the dominant sulfides, followed by sphalerite (Das Gupta, 1968). Chalcopyrite, galena, and arsenopyrite are minor. The pelitic host rocks have evolved through a low- to intermediate-pressure facies series of metamorphism to conditions for the coexistence of all three aluminosilicate polymorphs (about 600° and 5.5 kb; S. C. Sarkar et al., 1980). Associated amphibolites represent volcanic rocks. The ore is very thinly laminated, exhibits parallelism with the primary laminations in the metapelites, and shows slump structures, buckle folds, pyritic axial plane schistosity, and an absence of replacement features. These features show that the deposit is sedimentary/diagenetic, with formation of sulfide occurring earlier than the first phase of deformation (Ray, 1974). S. C. Sarkar et al. (1980) suggested that sedimentation took place in a near-shore anoxic environment.

Phosphorite deposits

Phosphate deposits occur in the Aravalli craton in both the Aravalli Supergroup and the Bijawar Group (Pant, 1980). The phosphorites of the Aravalli hill ranges form discontinuous outcrops in dolomitic limestones and silicified dolomites (see section on Aravalli Supergroup). D. M. Banerjee et al. (1980) summarized deposits from the Udaipur District (Matoon, Kanpur, Jhamarkotra) and the Jhabua District (Khatma, Kelkna, and Amlamal). Phosphorites at Matoon are stromatolitic, fragmented rocks composed of microphosporite and embedded in a carbonate/chert matrix. Phosphates occur as nodules, intraclasts and irregular fragments. The phosphorite of Kanpur occurs in stromatolitic columns and is composed of microphosphorite grains embedded in a dolomite groundmass. The Jhabua phosphorite also is a microphosphorite. The F/P_2O_5 ratio is higher in Jhabua than in Udaipur deposits. D. M. Banerjee et al. (1980) distinguished three important classes of phosphorites in the Aravalli Supergroup: stromatolite-carbonate biostromal phosphorite, massively bedded phosphorite, and fragmental brecciated phosphorite. Each class is characterized by a distinct range of P_2O_5, CaO, MgO, Fe_2O_3, Al_2O_3 and SiO_2.

The reddish brown phosphorites of the Bijawar Group occur near the base of the Gangau Formation (D. M. Banerjee, 1982). They contain carbonate-fluorapatite, are not associated with stromatolites, and probably originated along restricted shoals in littoral basins of the intracratonic sea that fringed the margins of the Bundelkhand complex.

REFERENCES

Ahmad, F. (1971). Geology of the Vindhyan system in the eastern part of the Son Valley in Mirzapur District, U. P. Geol. Surv. India Records, v. 96, Part 2, 1–41.

Akhtar, K. (1976). Facies analysis and depositional environments of the Bhander limestone (Precambrian), southeastern Rajasthan and adjoining Madhya Pradesh, India. Sediment. Geol., v. 16, 299–318.

―――― (1978). Paleogeography and sediment dispersal patterns of the Proterozoic Bhander Group, western India. Paleogeog., Paleoclimat., Paleoecol., v. 24, 327–357.

――――, and Srivastava, V. K. (1976). Ganugarh shale of southeastern Rajasthan, India: Precambrian regressive sequence of lagoon-tidal origin. Jour. Sed. Petrol., v. 45, 14–21.

Auden, J. B. (1933). Vindhyan sedimentation in the Son valley, Mirzapur District. Geol. Surv. India Mem. 62, Part 2, 141–250.

Awasthi, A. K., and Prakash, B. (1981). Depositional environments of unfossiliferous sediments of the Jodhpur Group, western India. Sediment. Geol., v. 30, 15–42.

Bajpai, M. P. (1935). The Gwalior trap from Gwalior, India. J. Geol., v. 43, 61–75.

Balasubrahmanyam, M. N., Murthy, M. K., Paul, D. K., and Sarkar, A. (1978). K-Ar ages of Indian kimberlites. Geol. Soc. India J., v. 19, 584–585.

Balasundaram, M. S., and Mahadevan, T. M. (1972). Stromatolites from the Bijawars of Joga, Hoshangabad Dist., Madhya Pradesh. Geol. Surv. India Records, v. 99, 127–132.

Banerjee, D. M. (1971a). Aravallian stromatolites from Udaipur, Rajasthan. Geol. Soc. India J., v. 12, 349–355.

—— (1971b). Precambrian stromatolitic phosphorites of Udaipur, Rajasthan, India. Geol. Soc. Amer. Bull., v. 82, 2319–2329.

—— (1978). Chemical rhythmicity in the Precambrian laminated phosphatic stromatolites and its bearing on the origin of algal phosphorite. Ind. J. Earth Sci., v. 5, 102–110.

—— (1982). Lithotectonic phosphate mineralization and regional correlation of Bijawar Group of rocks in central India. In Geology of Vindhyanchal (ed. K. S. Valdiya, S. B. Bhatia, and V. K. Gaur), Hindustan Publ. Co., New Delhi, 47–54.

——, and Klemm, U. (1985). Organo-chemical studies of Proterozoic stromatolitic phosphorite and the intercolumnar dolomite from the Aravalli Group, India. Geol. Soc. India J., v. 26, 245–254.

——, Basu, P. C., and Srivastava, N. (1980). Petrology, mineralogy, geochemistry, and origin of the Precambrian Aravallian phosphorite deposits of Udaipur and Jhabua, India. Econ. Geol., v. 75, 1181–1199.

Banerjee, I. (1982). The Vindhyan tidal sea. In Geology of Vindhyanchal (ed. K. S. Valdiya, S. B. Bhatia, and V. K. Gaur), Hindustan Publ. Co., New Delhi, 80–87.

Basu, K. K., Arora, Y. K., and Naha, K. (1976). Early Precambrian stratigraphy of central and southern Rajasthan, India. Precamb. Res., v. 3, 197–205.

Basu, P. K. (1981). Sulphur isotope studies of Zawar lead-zinc deposits, Rajasthan. Ind. Minerals, v. 35, 19–23.

—— (1982). Zawar lead-zinc deposit, India: A prototype syn-sedimentary Precambrian sulphide deposit. Geol., Min. Metall. Soc. India Quart. J., v. 54, nos. 1/2, 78–93.

Bharktya, D. K., and Gupta, R. P. (1983). Lineament structures in the Precambrian of Rajasthan—As deciphered from LANDSAT images. Recent Res. Geol., v. 10, 186–197.

Chakraborty, M. K. (1979). Alkali syenites of Mundwara suite, Sirohi District, Rajasthan. Ind. Natl. Sci. Acad. Proc., Part A, v. 45, 284–292.

——, and Bose, M. K. (1978). Theralite-melteigite-carbonatite association in Mer ring of Mundwara suite, Sirohi District, Rajasthan. Geol. Soc. India J., v. 19, 454–463.

Chakraborty, P. (1976). Study of paleocurrents in the Vindhyan rocks of Kela Devi, Karaula State, Rajasthan. Geol., Min. Metall. Soc India Quart. J., v. 48, 89–102.

Chanda, S. K., and Bhattacharyya, A. (1974). Ripple-drift cross-lamination in tidal deposits; Examples from the Precambrian Bhander Formation of Maihar, Satna District, Madhya Pradesh, India. Geol. Soc. Amer. Bull., v. 85, 1117–1122.

—— (1982). Vindhyan sedimentation and paleogeography. In Geology of Vindhyanchal (ed. K. S. Valdiya, S. B. Bhatia, and V. K. Gaur), Hindustan Publ. Co., New Delhi, 88–101.

——, ——, and Sarkar, S. (1977). Deformation of ooids by compaction in the Precambrian Bhander Limestone, India: Implications for lithification. Geol. Soc. Amer. Bull., v. 88, 1577–1585.

Chandra Chowdhury, Y. M. K., Banerjee, A. K., and Chande, V. D., (1977). Exploration for copper ore in Kolihan, Khetri copper belt, Jhun-Jhuna District, Rajasthan. Geol. Surv. India Misc. Publ. 27, 321–346.

Chattopadhyay, P. B., and Venkataraman, K. (1977). Petrography and petrochemistry of the kimberlite and associated volcanic rocks of the Jungel Valley district, Mirzapur, U. P., India. Geol. Soc. India J., v. 18, 653–661.

Chaudhuri, A. K., and Mukherjee, D. (1978). Superposed folding in nepheline syenites and associated metamorphism around Kishangarh, Rajasthan, India. Geol. Soc. India J., v. 19, 46–52.

Chauhan, D. S. (1977). The Dariba main lode of Rajpura-Dariba zinc-lead-copper belt, Udaipur District, Rajasthan. Geol. Soc. India J., v. 18, 611–616.

—— (1979). Phosphate bearing stromatolites of the Precambrian Aravalli phosphorite deposits of the Udaipur region—Their environmental significance and genesis of phosphorite. Precamb. Res., v. 8, 95–126.

—— (1980). Stromatolites of Nimach Mata and Baragaon phosphorite deposits and genesis of rock phosphate. Geol. Surv. India Misc. Publ. 44, 314–329.

—— (1981). Amphibolites from the granitic terrane northwest of Udaipur, Rajasthan. Ind. J. Earth Sci., v. 8, 35–43.

Choubey, V. D. (1971). Superimposed folding in the transitional rocks (Precambrian) and its influence on the structure of southeastern margin of the Vindhyan basin, Jabalpur District, Madhya Pradesh. Geol. Soc. India J., v. 12, 142–151.

Choudhary, A. K., Gopalan, K., and Sastry, C. A. (1984). Present status of the geochronology of the Precambrian rocks of Rajasthan. Tectonophysics, v. 105, 131–140.

Crawford, A. R. (1970). The Precambrian geochronology of Rajasthan and Bundelkhand, northern India. Can. J. Earth Sci., v. 7, 91–110.

—— (1975) Rb-Sr age determinations for the Mount Abu Granite and related rocks of Gujarat. Geol. Soc. India J., v. 16, 20–28.

——, and Compston, W. (1970). The age of the Vindhyan system of peninsular India. Geol. Soc. London J., v. 125, 351–371.

Crookshank, H. (1948). Minerals of the Rajputana pegmatites. Mining, Geol., Metall. Inst. India Trans., v. 42, 105–189.

Das Gupta, S. P. (1964). Genesis of sulfide mineralisation in the Khetri copper belt, Rajasthan, India. 22nd Internat. Geol. Cong. Rept., Delhi, Sect. 5, 239–257.

—— (1968). The structural history of the Khetri copper belt, Jhunjhunu and Sikar Districts, Rajasthan. Geol. Surv. India Mem. 98, 170 p.

Deans, T., and Powell, J. L. (1968). Trace elements and strontium isotopes in carbonatites, fluorites and limestones from India and Pakistan. Nature, v. 218, 750–752.

Deb, M. (1979). Polymetamorphism of ores in Precambrian stratiform massive sulfide deposits at Ambaji-Deri, western India. Mineralium Deposita, v. 14, 21–31.

—— (1980). Genesis and metamorphism of two stratiform massive sulfide deposits at Ambaji and Deri in the Precambrian of western India. Econ. Geol., v. 75, 572–591.

——, and Bhattacharyya, A. K. (1980). Geological setting and estimates of conditions of metamorphism of the Rajpura-Dariba polymetallic ore deposit, Rajasthan, India. In Proc. Fifth Quadrennial IAGOD Symposium on the Genesis of Ore Deposits (ed. J. D. Ridge), E. Schweizerbart'sche Verlagsbuchhandlung, Stuttgart, no. 5, 679–697.

——, Banerjee, D. M., and Bhattacharyya, A. K. (1978). Precambrian stromatolites and other structures in the Rajpura-Dariba polymetallic ore deposit, Rajasthan, India. Mineralium Deposita, v. 13, 1–9.

Desai, S. J., Patel, M. P., and Merh, S. S. (1978). Polymetamorphism of Balaram-Abu Road area, North Gujarat and Southwest Rajasthan. Geol. Soc. India J., v. 19, 383–394.

Dutt, G. N., and Srivastava, R. N. (1975). Fossil flora in the Alwar quartzite, Ferozpur-Jhirka, Gurgaon District, Haryana. Geol. Surv. India Misc. Publ. 23, Part 1, 149–157.

Fermor, L. L. (1936). An attempt at correlation of the ancient schistose rocks of peninsular India. Geol. Surv. India Mem. 70, 1–217.

Fuloria, R. C. (1969). Geological framework of the Ganga basin. In Selected Lectures on Petroleum Exploration (ed. S. N. Bhattacharya and V. V. Sastry), Inst. Petroleum Exploration, Oil Natural Gas Comm. India, Dehra Dun, 170–186.

Gangopadhyay, P. K., and Das, D. (1974). Dadikar Granite: A study on a Precambrian intrusive body in relation to structural environment in north-eastern Rajasthan. Geol. Soc. India J., v. 15, 189–199.

——, and Lahiri, A. (1983). Barr conglomerate: Its recognition and significance in the stratigraphy of the Delhi Supergroup in central Rajasthan. Geol. Soc. India J., v. 24, 562–570.

——, and Sen, R. (1972). Trend of regional metamorphism; An example from "Delhi System" of rocks occurring around Bairawas, north-eastern Rajasthan, India. Geol. Rundschau, v. 61, 270–281.

——, and —— (1975). Structural framework in Delhi Group of rocks with special reference to interference patterns; A study around Kushalgarh, north-eastern Rajasthan. Geol. Soc. India J., v. 16, 317–325.

Geological Survey of India (1977). Geology and Mineral Resources of the States of India: Part XII, Rajasthan. Geol. Surv. India Misc. Publ. 30, 75 p.

Ghosh, S. K. (1973). Comparative study of the lower Vindhyan limestones around Churhat, Sidhi District, Madhya Pradesh. Geol., Min. Metall. Soc. India Quart. J., v. 44, 41–53.

Goel, O. P., and Chaudhuri, M. W. (1979). Compositional restraints on the sillimanite paragenesis in metapelites from Kuanthal, District Udaipur, India. Lithos, v. 12, 153–158.

Gopalan, K., Trivedi, J. R., Balasubrahmanyam, M. N., Ray, S. K., and Sastry, C. A. (1979a). Rb-Sr chronology of the Khetri copper belt, Rajasthan. Geol. Soc. India J., v. 20, 450–456.

——, ——, and Merh, S. S. (1979b). Rb-Sr age of Godhra and related granites, Gujarat, India. Ind. Acad. Sci. Proc., Sect. A, v. 88, Part II, No. 1, 7–17.

Gopinath, K., Prasad Rao, A. D., and Murty, Y. G. K. (1977). Precambrians of Baroda and Panchmahals, Gujarat; Elucidation of stratigraphy and structure. Geol. Surv. India Records, v. 108, Part II, 60–68.

Goswami, P. C., and Gangopadhyay, P. K. (1971). Petrology and structure of Bairat Granite in the Districts of Alwar and Jaipur, Rajasthan. Geol. Min. Metall. Soc. India Quart. J., v. 43, 141–147.

Gupta, B. C. (1934). The geology of central Mewar. Geol. Surv. India Mem. 65, 107–169.

Gupta, M. L., Verma, R. K., Rao,, R. U. M., Hamza, J. M., and Venkateshwar Rao, G. (1967). Terrestrial heat flow in Khetri copper belt, Rajasthan, India. J. Geophys. Res., v. 72, 4215–4220.

Gupta, R. P, and Bharktya, D. K. (1982). Post-Precambrian tectonism in the Delhi-Aravalli belt, Precambrian Indian shield—Evidences from LANDSAT images. Tectonophysics, v. 85, 79–120.

Gupta, S. N., Arora, Y. K., Mathur, R. K., Iqbaluddin, P. B., Sahni, T. N., and Sharma, S. B. (1980). Lithostratigraphic map of Aravalli region, southern Rajasthan and northeastern Gujarat. Geol. Surv. India, Calcutta, scale 1:1,000,000.

Gyani, K. C. (1971). Granulitic rocks from Banded Gneissic Complex of Bandanwara region, Ajmer District, Rajasthan. In Second Symposium on Upper Mantle Project, Nat. Geophys. Res. Inst., Hyderabad, 339–348.

Heron, A. M. (1953). The geology of central Rajputana. Geol. Surv. India Mem. 79, 385 p.

Iqbaludddin and Moghni, A. (1979). Stratigraphy of the Bijawar Group in Son Valley, Mirzapur District, U. P., and Sidhi District, M. P. Geol. Surv. India Misc. Publ. 3, 81–93.

Jacob, K., Jacob, C., and Srivastava, R. N. (1953). Evidence for the existence of vascular plants in the Cambrian. Curr. Sci., v. 22, 34–36.

Kaila, K. L., Reddy, P. R., Dixit, M. M., and Koteswara Rao, P. (1985). Crustal structure across the Narmada-Son lineament, central India from deep seismic soundings. Geol. Soc. India J., v. 26, 465–480.

Kochhar, N. (1983). Tusham ring complex, Bhiwani, India. Ind. Natl. Sci. Acad. (Delhi) Proc., Part A, v. 49, 459–490.

——— (1984a). Paleo-uplift and cooling rates from various orgenic belts of India, as revealed by radiometric ages—Discussion (2). Tectonophysics, v. 107, 165–167.

——— (1984b). Malani igneous suite: Hot spot magmatism and cratonization of the northern part of the Indian shield. Geol. Soc. India J., v. 25, 155–161.

———, and Pareek, H. S. (1982). Petrochemistry and petrogenesis of the Malani igneous suite, India (discussion and reply). Geol. Soc. Amer. Bull., v. 93, 926–928.

Kresten, P., and Paul, D. K. (1976). Mineralogy of Indian kimberlites—A thermal and X-ray study. Can. Mineral., v. 14, 487–490.

Krishna Murty, M. (1972). Stromatolites from the Bijawars of the Joga area, Hoshangabad District, Madhya Pradesh. Geol. Soc. India J., v. 13, 181–185.

Krishnan, M. S., and Swaminath, J. (1959). The great Vindhyan basin of northern India. Geol. Soc. India J., v. 1, 10–30.

Kumar, S. (1977). Stromatolites and environment of deposition of the Vindhyan Supergroup of central India. Paleontol. Soc. India J., v. 21/22, 33–43.

Lal, R. K., and Shukla, R. S. (1970). Paragenesis of staurolite in pelitic schists of Kishangarh, District Ajmer, India. Mineral. Mag., v. 37, 561–567.

———, and ——— (1975a). Low pressure regional metamorphism in the northern portion of the Khetri copper belt of Rajasthan, India. Neues Jahrb. Mineral. Abh., v. 124, 294–325.

———, and ——— (1975b). Genesis of cordierite-gedrite-cummingtonite rocks of the northern portion of the Khetri copper belt, Rajasthan, India. Lithos, v. 8, 175–186.

———, and ——— (1976). Regional metamorphism of Madan-Kudhan, Khetri, Rajasthan, and facies series in Rajasthan metamorphic belt. Recent Res. Geol., v. 4, 46–76.

Mathur, S. M. (1962). Geology of the Panna diamond deposits. Geol. Surv. India Records, v. 87, 787–818.

——— (1965). Halite casts and other sedimentary structures in Sirbu Shale (Bhander Group, Vindhyan System). In D. N. Wadia Commemorative Volume, Geol., Min. Metall. Soc. India, Calcutta, 572–582.

——— (1981). A revision of stratigraphy of the Vindhyan Supergroup in the Son Valley, Mirzapur District, Uttar Pradesh. Geol. Surv. India Misc. Publ. 50, 7–21.

——— (1982). Organic materials in the Precambrian Vindhyan Supergroup. In Geology of Vindhyanchal (ed. K. S. Valdiya, S. B. Bhatia, and V. K. Gaur), Hindustan Publ. Co., New Delhi, 126–131.

———, and Mani, G. (1975). Geology of the Bijawar Group in the type area, M. P. Proc. Symposium on Purana Formations of Peninsular India, Sagar Univ., Sagar, 313–320.

———, and Singh, H. N. (1971). Petrology of the Majhgawan pipe rocks. Geol. Surv. India Misc. Publ., v. 19, 78–85.

McDougall, J. D., Gopalan, K., Lugmair, G. W., and Roy, A. B. (1983). An ancient depleted mantle source for Archaean crust in Rajasthan, India (Abstract). In Workshop on a Cross Section of Archaean Crust, Lunar and Planetary Inst., Houston, Technical Report 83–03, 55–56.

Misra, R. C. (1969). The Vindhyan system. 56th Ind. Sci. Cong. Proc., Part 2, 111–142.

———, and Sharma, R. P. (1975). New data on the geology of the Bundelkhand complex of central India. Recent Res. Geol., v. 2, 311–346.

Mohanty, S. (1982). Structural studies around Kathar, Udaipur District, Rajasthan. Geol. Soc. India J., v. 23, 209–218.

Mukherjee, A. K., and Roy, A. (1981). Cooling conditions of the high level Precambrian granite of Siwana: Evidence of experimental melting behaviour and the sodic amphibole-pyroxene reaction relation. Ind. J. Earth Sci., v. 8, 99–108.

———, and Sen, R. N. (1980). Geochemical implications in the genesis of the Zawar lead-zinc deposit, Rajasthan, India. In Proc. 5th Quadrennial IAGOD symposium on the genesis of ore deposits (ed. J. D. Ridge), E. Schweizerbart'sche Verlagsbuchhandlung, Stuttgart, v. 1, 709–718.

Mukhopadhyay, D., and Dasgupta, S. (1978). Delhi-pre-Delhi relations near Badnor, central Rajasthan. Ind. J. Earth Sci., v. 5, 183–190.

———, and Ghosh, K. P. (1980). Deformation of early lineation in the Aravalli rocks near Fatehpur, Udaipur District, southern Rajasthan, India. Ind. J. Earth Sci., v. 7, 64–75.

Naha, K. (1983). Structural-stratigraphic relations of pre-Delhi rocks of southcentral Rajasthan: A summary. Recent Res. Geol., v. 10, 40–52.

———, and Halyburton, R. V. (1974a). Late stress systems deduced from conjugate folds and kink bands in the "Main Raialo Syncline," Udaipur District, Rajasthan, India. Geol. Soc. Amer. Bull., v. 85, 251–256.

———, and ——— (1974b). Early Precambrian stratigraphy of central and southern Rajasthan, India. Precamb. Res., v. 1, 55–73.

———, and ——— (1977a). Structural pattern and strain history of a superposed fold system in the Precambrian of central Rajasthan, India; I, Structural pattern in the "Main Raialo Syncline," central Rajasthan. Precamb. Res., v. 4, 39–84.

———, and ——— (1977b). Structural pattern and strain history of a superposed fold system in the Precambrian of central Rajasthan, India; II, Strain history. Precamb. Res., v. 4, 85–111.

———, and Majumder, A. (1971). Reinterpretation of the Aravalli basal conglomerate at Morchana, Udaipur District, Rajasthan, western India. Geol. Mag., v. 108, 111–114.

———, and Roy, A. B. (1983). The geology of the Precambrian basement in Rajasthan, western India. Precamb. Res., v. 19, 217–223.

———, Chaudhuri, A. K., and Bhattacharyya, A. C. (1966). Superposed folding in the older Precambrian rocks around Sangat, central Rajasthan, India. Neues Jahrb. Geol. Paleontol. Abh., v. 126, 205–230.

———, Venkatasubramanyam, C. S., and Singh, R. P. (1969). Upright folding of varying intensity on isoclinal folds of diverse orientation; A study from the early Precambrian of western India. Geol. Rundschau, v. 58, 929–950.

———, Mukhopadhyay, D. K., Mohanty, R., Mitra, S. K., and Biswal, T. K. (1984). Signifi-cance of contrast in the early stages of of the structural history of the Delhi and pre-Delhi groups in the Proterozoic of Rajasthan, western India. Tectonophysics, v. 105, 193–206.

Nair, N. G. K., and Agarwal, N. K. (1976). Primary and secondary structures in the polymetallic ores of Rajpura-Dariba, Rajasthan, India. Mineralium Deposita, v. 11, 352–356.

Narain, H., and Kaila, K. L. (1982). Inferences about the Vindhyan basin from geophysical data. In Geology of Vindhyanchal (ed. K. S. Valdiya, S. B. Bhatia, and V. K. Gaur), Hindustan Publ. Co., New Delhi, 179–191.

Narayan Das, G. R., Sharma, D. K., Singh, G., and Singh, R. (1980). Uranium mineralisation in Sikar District, Rajasthan. Geol. Soc. India J., v. 21, 432–439.

Narayana, B. L. (1974). The mode of occurrence, petrography, metamorphism and origin of amphibolites of Devgad, Baria, Panchmahals District, Gujarat State. Geol. Soc. India J., v. 15, 246–255.

Negi, R. S., and Ravindra, R. (1980). On the occurrence of stromatolites in the Kushalgarh Formation of Delhi Supergroup from Baraud, Alwar District, Rajasthan. Geol. Surv. India Misc. Publ. 44, 90–95.

Niyogi, D. (1966). Petrology of the alkalic rocks of Kishangarh, Rajasthan, India. Geol. Soc. Amer. Bull., v. 77, 65–82.

Pandya, M. K. (1981). Petromineralogy and petrochemistry of the rocks of Hammerhead syncline of Kankroli, Udaipur District, Rajasthan—A case of high grade regional metamorphism in Aravalli (Abstract). In Symposium on Three Decades of Development in Petrology, Mineralogy and Petrology in India. Geol. Surv. India, Jaipur, p. 90.

———, Solanki, S. L., and Pandya, T. K. (1980). Diagenetic features in sulphides of Dariba-Rajpura deposit, Udaipur District, Rajasthan. Geol. Soc. India J., v. 21, 425–431.

Pant, A. (1980). Resource status of rock phosphate deposits in India and areas of future potential. In Fertilizer Mineral Potential in Asia and Pacific (ed. R. P. Sheldon and W. C. Burnett), East-West Center Publ., Honolulu, 331–358.

Pareek, H. S. (1981a). Petrochemistry and petrogenesis of the Malani igneous suite, India. Geol. Soc. Amer. Bull., v. 92; Part I, p. 67–70; Part II, p. 206–273.

——— (1981b). Basin configuration and sedimentary stratigraphy of western Rajasthan. Geol. Soc. India J., v. 22, 517–527.

——— (1983). Stratigraphy of northwestern India and its correlation with that of Indian Basin, Pakistan, Malagasy and South Africa. Geophytology, v. 13, no. 1, 1–21.

Paul, D. K. (1979). Isotopic composition of strontium in Indian kimberlites. Geochim. Cosmochim. Acta, v. 43, 389–394.

———, Potts, P. J., and Gibson, I. L. (1975a). Rare earth abundances in Indian kimberlites. Earth Planet. Sci. Lett., v. 25, 151–158.

———, Rasa, D. C., and Harris, P. G. (1975b). Chemical characteristics and K/Ar ages of Indian kimberlites. Geol. Soc. Amer. Bull., v. 86, 364–366.

———, Gale, N. H., and Harris, P. G. (1977). Uranium and thorium abundances in Indian kimberlites. Geochim. Cosmochim. Acta, v. 41, 335–339.

Phadke, A. K., and Jhingran, A. G. (1968). On the carbonatite at Newania, Udaipur District, Rajasthan. Geol. Soc. India J., v. 9, 165–170.

Poddar, B. C. (1966). An example of contrasted tectonic regimes from Precambrians of Udaipur District, Rajasthan. Ind. Minerals, v. 20, 192–194.

Prakash, R., and Dalela, I. K. (1982). Stratigraphy of the Vindhyans in Uttar Pradesh: A brief review. In Geology of Vindhyanchal (ed. K. S. Valdiya, S. B. Bhatia, and V. K. Gaur), Hindustan Publ. Co., New Delhi, 54–79.

———, Swarup, P., and Srivastava, R. N. (1975). Geology and mineralisation in southern parts of Bundelkhand in Lalitpur District, Uttar Pradesh. Geol. Soc. India J., v. 16, 43–56.

Prasad, B. (1975). Lower Vindhyan formations of Rajasthan. Geol. Surv. India Records, v. 106, Part 2, 31–53.

Raja Rao, C. S. (1976). Precambrian sequences of Rajasthan. Geol. Surv. India Misc. Publ. 23, Part 2, 497–516.

———, and Mahajan, U. D. (1965). Note on the stromatolites and probable correlation of the Bhagwanpur Limestone, Chittorgarh District, Rajasthan. Curr. Sci., v. 34, 82–83.

———, Poddar, B. C., Basu, K. K., and Dutta, A. K. (1971). Precambrian stratigraphy of Rajasthan—A review. Geol. Surv. India Records, v. 101, Part 2, 60–79.

Rajarajan, K. (1971). Studies on Shahgarh Granite. Geol. Surv. India Records, v. 96, 199–204.

Rao, K. S., Lal, C., and Ghosh, D. B. (1977). Algal stromatolites in the Bhander Group of formations, Vindhyan Supergroup, Satna District, Madhya Pradesh. Geol. Surv. India Records, v. 109, Part 2, 38–47.

Rao, N. K., and Rao, G. V. U. (1968). Ore microscopic study of copper ore from Kolihan, Rajasthan, India. Econ. Geol., v. 63, 277–287.

Ray, S. K. (1974). Structural history of the Saladipura pyrite-pyrrhotite deposit and associated rocks, Khetri copper belt, Rajasthan. Geol. Soc. India J., v. 15, 227–238.

Roday, P. P. (1979). Structural pattern in the Ajabgarh rocks around Anakhar, District Ajmer. Geol. Soc. India J., v. 20, 441–449.

———, and Bhatt, A. K. (1980). Tectonic history

of the Bijawar rocks at the Barmhanghat section of the Narmada valley. Geol. Soc. India J., v. 21, 546–557.

Roy, A. B., and Jain, A. K. (1974). Polyphase deformation in the lead-zinc bearing Precambrian strata of Zawarmala, District Udaipur, southern Rajasthan. Geol., Min. Metall. Soc. India Quart. J., v. 46, 81–86.

———, and Paliwal, B. S. (1981). Evolution of Lower Proterozoic epicontinental sediments: Stromatolite bearing Aravalli rocks of Udaipur, Rajasthan, India. Precamb. Res., v. 14, 49–74.

———, ———, and Goel, O. P. (1971). Superposed folding in the Aravalli rocks of the type area around Udaipur, Rajasthan. Geol. Soc. India J., v. 12, 342–348.

———, Nagori, D. K., Golani, P. R., Dhakar, S. P., and Choudhri, R. (1980). Structural geometry of the rock phosphate bearing Aravalli rocks around Jhamarkotra mines area, Udaipur District, Rajasthan. Ind. J. Earth Sci., v. 7, 191–202.

———, Somani, A. K., and Sharma, N. K. (1981). Aravalli-pre Aravalli relationship: A study from the Bhindar region, southern Rajasthan. Ind. J. Earth Sci., v. 8, 119–130.

Roy Chowdhary, M. K., and Das Gupta, S. P. (1965). Ore localization in the Khetri copper belt, Rajasthan, India. Econ. Geol., v. 60, 69–88.

Sahni, M. R. (1975). Vindhyan paleobiology, stratigraphy and depositional environments: A critical review. Paleontol. Soc. India J., v. 20, 289–304.

———, and Srivastava, R. N. (1954). New organic remains from the Vindhyan system and the probable systematic position of Fermoria Chapman. Curr. Sci., v. 23, 39–41.

Salujha, S. K. (1982). Petrology of the surface and subsurface Vindhyan sediments. In Geology of Vindhyanchal (ed. K. S. Valdiya, S. B. Bhatia, and V. K. Gaur), Hindustan Publ. Co., New Delhi, 113–124.

———, Rehman, K., and Arora, C. M. (1971). Plant microfossils from the Vindhyans of Son Valley, India. Geol. Soc. India J., v. 12, 24–33.

Sant, V. N., Srikantan, B., Sharma, S. B., Ravindra, R., Siddiqui, M. A., Bakliwal, P. C., Joshi, S. M., Basu, S. K., Saha, A. K., Bhat, M. L., Datta, A. K., Sinha, P. N., and Chattopadhyay, N. (1980). Geology and mineral resources of Alwar District, Rajasthan. Geol. Surv. India Mem. 110, 118 p.

Sarkar, B. (1983). Complex reworked oolites in the Bhander Limestone Formation (Proterozoic), central India. Geol., Min. Metall. Soc. India Quart. J., v. 55, 38–45.

Sarkar, S. C., Bhattacharyya, P. K., and Mukherjee, A. D. (1980). Evolution of the sulfide ores of Saladipura, Rajasthan, India. Econ. Geol., v. 75, 1152–1167.

Sastri, B. B. S., and Krishna Rao, J. S. R. (1982). Organic matter in quartzites of Ambamata-Deri. Ind. J. Earth Sci., v. 9, 76–78.

Sastry, C. A., Gopalan, K., and Choudhary, A. K. (1984). Proposal for the classification of the Precambrian rocks and events from Rajasthan and NE Gujarat (Abstract). Proc. Seminar on Crustal Evolution of the Indian Shield and Its Bearing on Metallogeny, 12–14 January, 1984, Dept. of Geology, Presidency College, Calcutta, 11–12.

Sastry, V. V., Bhandari, L. L., Raju, A. T. R., and Datta, A. K. (1971). Tectonic framework and subsurface stratigraphy of the Ganga basin. Geol. Soc. India J., v. 12, 223–233.

Sen, S. (1983). Stratigraphy of the crystalline Precambrians of central and northern Rajasthan—A review. Recent Res. Geol., v. 10, 26–39.

Sen Gupta, S. (1983). Superposed folding in the Aravalli rocks around Deola, southeastern Rajasthan. Recent Res. Geol., v. 10, 120–126.

Sharma, A. K. (1980). Petrological study of the lower Vindhyan rocks of western Son Valley, District Sidhi, M. P. Geol., Min. Metall. Soc. India Quart. J., v. 52, 57–62.

Sharma, K. K. (1978). Structural and geochronological data from Gangapur area, central Rajasthan, and their bearing on the Precambrian structural history of Rajasthan, India. Recent Res. Geol., v. 7, 118–132.

——— (1980). Structure, petromineralogy, and petrochemistry of Ojagar Granite, central Rajasthan, India., Recent Res. Geol., v. 6, 385–418.

———, Bal, D. C., Prashad, R., Nand Lal, and Nagpaul, K. K. (1984). Paleo-uplift and cooling rates from various orogenic belts of India, as revealed by radiometric ages—Reply (2). Tectonophysics, v. 107, 167–172.

Sharma, R. P. (1979a). Coexisting phases in the Al_2O_3-SiO_2-H_2O system with special reference to pyrophyllite-diaspore deposits of Bundelkhand complex. Ind. Natl. Sci. Acad. Proc., Part A, v. 45, 119–128.

——— (1979b). Origin of the pyrophyllite-diaspore deposits of the Bundelkhand complex, central India. Mineralium Deposita, v. 14, 343–352.

——— (1982). Lithostratigraphy, structure and petrology of the Bundelkhand Group. In Geology of Vindhyanchal (ed. K. S. Valdiya, S. B. Bhatia and V. K. Gaur), Hindustan Publ. Co., New Delhi, 30–46.

——— (1983). Structure and tectonics of the Bundelkhand complex, central India. Recent Res. Geol., v. 10, 198–210.

Sharma, R. S. (1977). Deformational and crystallization history of the Precambrian rocks in north-central Aravalli, Rajasthan, India. Precamb. Res., v. 4, 133–162.

——— (1982). Mineralogy of scapolite bearing rock from Rajasthan, northwest peninsular India. Lithos, v. 14, 165–171.

——— (1983). Basement-cover relation in north-central Aravalli range: A tectonic and metamorphic synthesis. Recent Res. Geol., v. 10, 53–71.

———, and Joshi, M. (1984). Sand Mata granulites—A case of deep level exposure of Precambrian crust in northwest Indian shield (Abstract). Proc. Seminar on Crustal Evolution of Indian Shield and Its Bearing on Metallogeny, 12–14 January, 1984, Dept. Geology, Presidency College, Calcutta, 35–36.

———, and Macrae, N. D. (1981). Paragenetic relations in gedrite-cordierite-staurolite-biotite-sillimanite-kyanite gneisses at Ajitpura, Rajasthan, India. Contrib. Mineral. Petrol., v. 78, 48–60.

———, and Narayana, V. (1975a). Distribution of elements between coexisting garnet-biotite and muscovite-biotite pairs from poly-metamorphic schists of south-east Beawar, Rajasthan, India. Schweiz. Mineral. Petrog. Mitteilungen, v. 55, 61–77.

———, and ——— (1975b). Petrology of polymetamorphic schists from an Archaean complex terrain, southeast of Beawar, Rajasthan, India. Neues Jahrb. Mineral. Abh., v. 124, 190–222.

———, and ——— (1975c). Wollastonite paragenesis in a regional metamorphic terrain southeast of Beawar, Rajasthan, India. Neues Jahrb. Mineral. Monatsh., no. 12, 561–569.

———, and Ray, A. K. (1979). On the polymineralic paragneisses from Karera, District Bhilwara, Rajasthan—I. Petrography and mineralogy. Ind. J. Earth Sci., v. 6, 67–81.

———, and ——— (1980). On the polymineralic paragneisses from Karera, District Bhilwara, Rajasthan—II. Phase relations and parageneses. Ind. J. Earth Sci., v. 7, 131–145.

———, and Upadhyaya, T. P. (1975). Multiple deformation in the Precambrian rocks to the southeast of Ajmer, Rajasthan, India. Geol. Soc. India J., v. 16, 428–440.

Shukla, R. T., and Sharma, B. L. (1970). A sedimentary petrographic study of Alwar rocks near Debari, Udaipur, Rajasthan. Geol. Soc. India J., v. 11, 142–154.

Singh, I. B. (1973). Depositional environment of the Vindhyan sediments in Son Valley area. Recent. Res. Geol., v. 1, 146–152.

Singh, S. P. (1983). Pillow lavas from Delhi Supergroup, near Bambholai, Pali District, Rajasthan. Geol. Soc. India J., v. 24, 208–211.

——— (1985). Synsedimentary tectonics of the Proterozoic Delhi Supergroup around Dausa uplift, northeastern Rajasthan, India (Abstract). 6th Internat. Conf. on Basement Tectonics, Inst. Geophys. Planet. Phys., Los Alamos, New Mexico, p. 35.

———, and Ravindra, R. (1975). Sedimentary

structures and paleocurrent analyses in Delhi Group of rocks from northeastern parts of Rajasthan. Proc. Symp. on Recent Advances in Geology of Rajasthan, Geol. Surv. India, Udaipur.

Sivaraman, T. V., and Odom, A. L. (1982). Zircon geochronology of Berach Granite of Chittorgarh, Rajasthan. Geol. Soc. India J., v. 23, 575–577.

Somayajulu, P. V. (1975). On the origin of Deogarh amphibolites, Udaipur District, Rajasthan, India. Geol. Soc. India J., v. 16, 287–293.

——— (1979). Petrological studies in Sand Mata basic granulites, central Mewar, Rajasthan. Ind. Acad. Geosci. J., v. 22, 44–56.

Srinivasa Rao, K., and Neelakantam, S. (1978). Stratigraphy and sedimentation of Vindhyans in parts of Son Valley area, Madhya Pradesh. Geol. Surv. India Records, v. 110, Part 2, 180–193.

Srivastava, A. K. (1977). On the anatectic origin of the granitic rocks of Jhirgadandi pluton, Mirzapur District, Uttar Pradesh. Ind. Geol. Assoc. Bull., v. 10, no. 2, 37–44.

Srivastava, B. J., Singh, B. P., and Lilley, F. E. M. (1984). Magnetometer array studies in India and the lithosphere. Tectonophysics, v. 105, 355–371.

Srivastava, B. N., Rana, M. S., and Verma, N. K. (1983). Geology and hydrocarbon prospects of the Vindhyan basin. Petroleum Asia J. (Dehra Dun), v. 1, 179–189.

Tugarinov, A. I., Shanin, L. L., Kazakov, G. A., and Arakalyants, M. M. (1965). On the glauconite ages of the Vindhyan system (India). Geokhimiya, no. 6, 652–660.

Venkataraman, K. (1960). Petrology of the Majhagawan agglomeratic tuff and associated rocks. Geol., Min. Metall. Soc. India Quart. J., v. 32, 1–10.

Verma, K. K., and Barman, G. (1975). On the origin of phosphorite in the Aravalli rocks of Rajasthan, India (Abstract). Proc. Seminar on Recent Advances in Precambrian Geology and Mineral Deposits, with Special Reference to Rajasthan, Geol. Surv. India, Udaipur, 43–45.

———, and ——— (1980). On the discovery of algal stromatolites from Delhi Supergroup, Rajasthan, India. Geol. Surv. India Misc. Publ. 44, 86–89.

10. RIFT VALLEYS

The Precambrian rocks of the Indian subcontinent are traversed by three major lineaments whose most apparent features are topographic valleys (grabens) that contain Proterozoic and/or Phanerozoic sediments. The lineaments are (1) Godavari (or Pranhita-Godavari), which forms the border between the Bhandara and Dharwar cratons; (2) Mahanadi, between the Singhbhum craton and the Eastern Ghats and Bhandara craton; and (3) the combined Narmada and Son Valleys, along the southeastern margin of the Aravalli craton (Figs. 1.1 and 1.5).

Three other rift features are present in addition to the lineaments mentioned above (Fig. 10.1; Krishna Brahmam and Negi, 1973; Valdiya, 1984). These other rifts include the Cambay graben (Sabermati rift) and its northern extension along the southwestern side of the Aravalli craton and two rifts detected beneath the Deccan plateau basalts by gravity and other geophysical methods (Kurduvadi and Kayana rifts). In addition, the Damodar valley in the Chotanagpur terrain of the Singhbhum craton (Fig. 7.1) is a rift valley partly filled by Gondwana sediments.

The Narmada-Son, Godavari, and Mahanadi lineaments are clearly ancient, and some workers have proposed that they have existed since Archean time (Naqvi et al., 1974; Das and Patel, 1984). This short chapter discusses these lineaments, primarily concentrating on the Godavari area because of the accumulation of Proterozoic (Pakhal) sediments along it and their later mild deformation and metamorphism.

MAHANADI

The Gondwana (Phanerozoic) sediments of the Mahanadi lineament form a narrow band toward the eastern edge of the outcrop belt but become broader toward the west (Fig. 1.1; Mazumdar, 1978; Sarkar and Saha, 1983). The Gondwana outcrop is largely north of the present valley of the Mahanadi River, which traverses the crystalline terrain of the Eastern Ghats. The principal investigations of the Mahanadi lineament have been related to the occurrence of ultramafic intrusions containing chromite along the Sukinda trust belt north of the rift (Fig. 7.2; Ch. 7; Banerjee, 1972). Mishra (1984) correlated the Mahanadi rift with a

similar feature in the Antarctic, developed prior to the rifting of Gondwanaland.

The northern edge of the khondalite-charnockite terrain of the Eastern Ghats is in contact along the Sukinda thrust with the Iron Ore or other low-grade rocks of the Singhbhum nucleus (Ch. 7). This contact is partly covered by Gondwana sediments of the Mahanadi lineament. The general northeast-southwest trend of structures in the Eastern Ghats is abruptly truncated a few tens of kilometers southwest of the thrust (Chatterjee et al., 1964). In the block immediately southwest of the thrust, structures in the granulite-facies rocks trend northwest-southeast, at approximately a right angle to the Eastern Ghats trend. The ultramafic bodies and other blocks along the thrust zone appear to be fragmented, melange-type collections of disparate lithologies.

The preceding observations are consistent with the concept that the Sukinda thrust raised high-grade rocks of the Eastern Ghats into contact with lower-grade rocks of the Singhbhum craton. The thrust zone may be sufficiently large that it includes the entire block of NW-SE-trending granulites, possibly between two border thrusts. The melange terrain presumably represents an area disrupted by thrusting and intruded by Alpine-type ultramafic bodies. The date of thrusting is unclear, although it is clearly younger than the granulite-facies metamorphism in the Eastern Ghats, which has tentatively been regarded as Middle Proterozoic (Ch. 5). The Gondwana sediments in the Mahanadi rift may cover Proterozoic sediments at depth, a possibility consistent with the thickness of sediments required to account for the negative Bouguer gravity anomaly along the Gondwana basin (Bhatia et al., 1979).

NARMADA-SON

The valleys of the Narmada and Son rivers form a linear feature that crosses almost the entire peninsula of India in a NNE-SSW direction (Fig. 1.5; Choubey, 1971; Ghosh, 1976; Mishra, 1977; Crawford, 1978; Roday, 1983). The Narmada and Son valleys do not contain a suite of rocks unique to the valleys. Rather, they contain outcrops of rocks broadly similar to those in cratons on either side. Much of the Narmada Valley is covered by

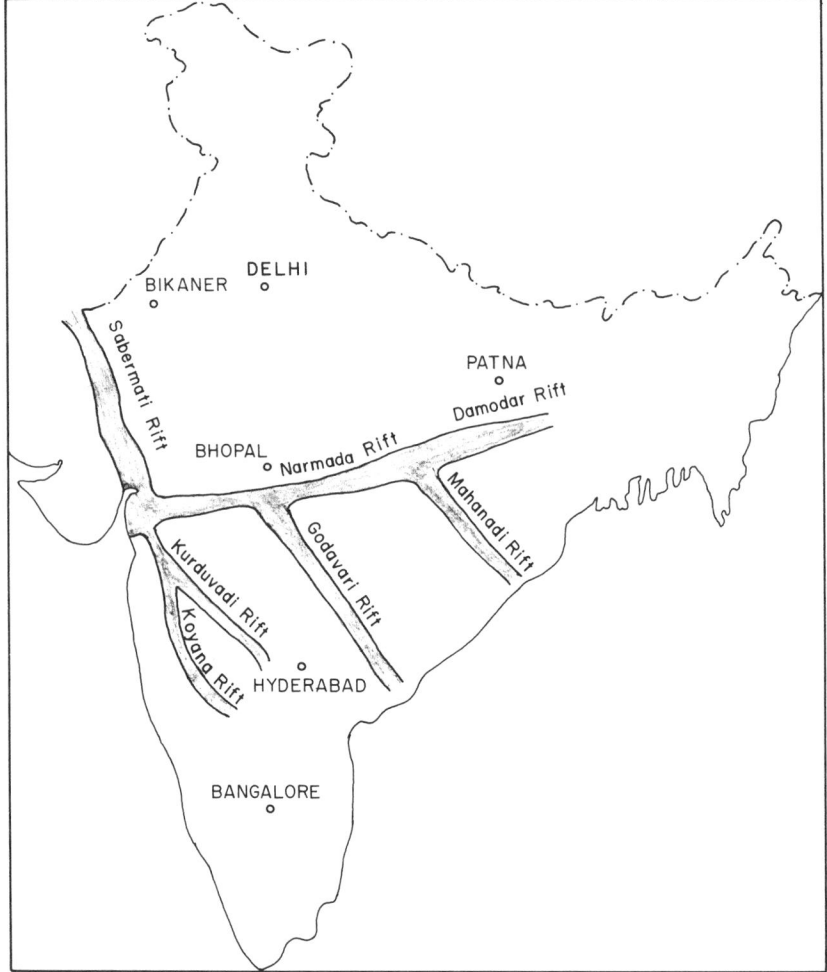

Fig. 10.1. Distribution of rifts (Krishna Brahmam and Negi, 1973).

Deccan basalt. The eastern part of the lineament, along the Son Valley, is occupied by Vindhyan sediments (Ch. 9). Vindhyan sediments do not extend south of a major southward-dipping thrust fault mapped by Tewari (1968), slightly south of the Son Valley (Fig. 1.5). Other Precambrian rocks are also exposed at various places along the lineament, including scattered outcrops of pre-Vindhyan rocks commonly regarded as the Bijawar Group (Chapter 9). Thus, the rock suites of the lineament have been described in appropriate chapters (6, 7, and 9).

Geologic relationships north and south of the Narmada-Son lineament provide two important pieces of information regarding its evolution. One is that the Satpura orogenic belt, south of the Narmada lineament, was characterized by northwest-directed thrusting and overfolding, accompanied by metamorphism up to granulite facies. Thrusts verge southward in the northern part of the Satpura orogen, implying that the present rift valley

occupies a formerly thick part of the crust underlain by stacked thrust sheets. Rifting of orogenically thickened crust is apparently a common phenomenon because of weakening of the crust by radioactive heating (Glazner and Bartley, 1985).

The second observation is that the Vindhyan sediments north of the lineament are mildly deformed and metamorphosed adjacent to the lineament but are undisturbed elsewhere in the Vindhyan outcrop farther north. The implication is that the Satpura orogeny, which is only poorly dated, was younger than the age of deposition of some of the Vindhyan Supergroup and that the Narmada valley was sufficiently coherent that it could transmit compressional stress from an orogenic event on its south side across to its northern side.

Movement directions in the Narmada-Son lineament are uncertain, but considerable strike-slip motion has apparently occurred. Das and Patel (1984) found mostly right-lateral strike-slip

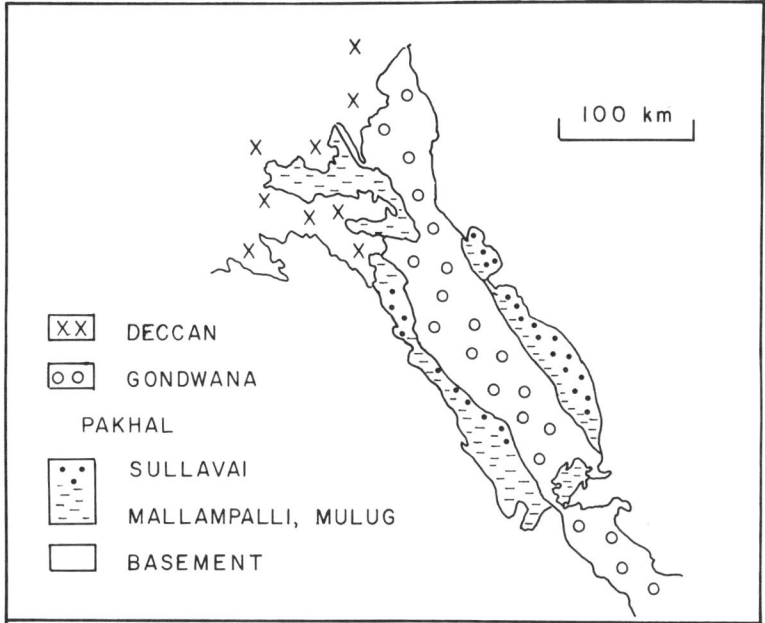

Fig. 10.2. Generalized may of Godavari rift zone (based on Basumallick, 1967b; Chaudhuri, 1970; and Srinivasa Rao et al., 1979a, 1979b).

movements on faults oriented ENE-WSW, parallel to the valley. Left-lateral motion was proposed for conjugate shears oriented NW-SE. These findings are consistent with the right-lateral movement proposed by Crawford (1978) for the period 750 to 450 m.y. ago. Crawford correlated the lineament into Madagascar at a time when India and Madagascar were both joined within Gondwanaland. Many of the faults in the lineament have dip-slip components, and the present valley is clearly a rift.

Activity along the lineament has continued until the late Phanerozoic. The lineament affected volcanism during Deccan time, contains Mesozoic or younger carbonatite and other intrusions, and has been a site of recent mineralization.

GODAVARI (PRANHITA-GODAVARI)

The Godavari and Pranhita valleys form a major rift zone between the Dharwar and Bhandara cratons (Figs. 1.5 and 10.1). The lineament thus formed extends for more than 440 km in a northwest-southeast direction and probably joins the Narmada-Son lineament under the Deccan basalts (Qureshy et al., 1968). The lineament has been active at least since the Middle Proterozoic (Naqvi et al., 1974; Srinivasa Rao et al., 1979a, 1979b; Subba Raju et al., 1978). Precambrian sediments in the Godavari Valley (Pakhal Supergroup) occur in two parallel belts along the margins of the valley separated by about 50 km of coal-bearing Gondwana sediments (Fig. 10.2). A rough age was provided by a K/Ar date of more

than 1,400 m.y. on glauconite (Vinogradov et al., 1964). Probably the best age criteria are sparse stromatolites, which merely indicate deposition in the Middle to Late Proterozoic (Raha and Sastry, 1982).

The outcrops of the Pakhal Supergroup and the major boundary faults of the graben both extend parallel to the general strike of foliation in the surrounding crystalline terrains (Fig. 10.2). This relationship suggests that the rift valley developed along the early Precambrian structural grain, where the Dharwar and Singhbhum (combined Singhbhum and Bhandara) cratons had joined (Naqvi et al., 1974). The rift zone is completely devoid of volcanism and was regarded as a "crevice type" of platform rift by Srinivasa Rao et al. (1979a).

Structural data and lithologic distribution shown in Figures 10.2 and 10.3 indicate periodic rejuvenation of the rift valley since the Middle Proterozoic. The younger rift zones inherited the

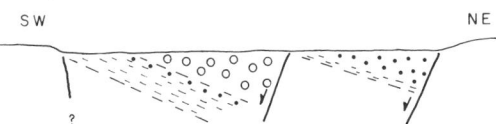

Fig. 10.3. Schematic cross-section of Godavari rift (from Fig. 10.2 and data in Basumallick, 1967b; Chaudhuri, 1970; and Srinivasa Rao et al., 1979a, 1979b). Symbols are as in Figure 10.2. Major faults are snown only on the northeastern side because of indicators that unconfortable relationships are at least partly preserved on southwestern side of rift.

TABLE 10.1. *Stratigraphy of Pakhal Group (Subba Raju et al., 1978)*

Mulug Area		Albaka Area	
Gondwana — Unconformity, partly faulted contact		Gondwana — Faulted contact	
Sullavai Group (1403 m)	9. Sullavai Sandstone (1400 m)	Red brown and mottled sandstone with arkose and shale and white sandstone	Chalamala Sandstone	Green, brown or gray sandstone and quartzite
	8. Sullavai Conglomerate (3 m)	Boulder conglomerate (Local)	Chalamala Conglomerate	Conglomerate (Local)
— Partly angular unconformity —		—Disconformity—	
Mulug Group (3110 m)	7. Mulug Shale (2500 m)	Quartz veins / Shale phyllite with numerous interbeds of quartzite & dolomite	Tippapuram Shale	Quartz veins / Shale-phyllite with numerous interbeds of quartzite and dolomite
	6. Mulug Orthoquartzite (400 m)	Quartzite with interbeds of dolomitic limestone		
	5. Mulug Arkose (200 m)	Arkose with glauconitic sandstone and siltstone		
	4. Jakaram Conglomerate (10 m)	Chert conglomerate		
—Disconformity—			
Mallampalli Group (2500 m)	3. Pandikunta Shale (200 m)	Shale with dolomite bands		
	2. Mallampalli Dolomite (2,200 m)	Dolomite and dolomitic limestone with interbeds of grit, arkose, glauconitic sandstone, chert, and shale		
	1. Mallampalli Conglomerate (100 m)	Boulder conglomerate, arkose, and quartzite		
—Nonconformity—		—Nonconformity—	
Archean		Archean	

primary rift structures or adjusted to them by producing en echelon faults and new grabens within the original graben (Subba Raju et al., 1978). Bouguer gravity values suggest a stepwise rift structure with 2 to 4 km total thickness of sediments (Qureshy et al., 1968). High heat flow also occurs in the valley (Gupta and Venkateswara Rao, 1970; Gupta and Gaur, 1984), suggesting crustal thinning and mantle upwarping.

Stratigraphy of Pakhal Supergroup

Considerable work has been done on the stratigraphy of the Pakhal Supergroup, a term that refers to all Precambrian sediments along the Godavari rift system. The group lies unconformably on crystalline basement. Conflicting views on the stratigraphy have principally concerned the southeastern part of the belt, where it is deformed and metamorphosed up to amphibolite facies. Farther northwest, the sequence is unmetamorphosed and only mildly deformed.

Subba Raju et al. (1978) designated the Mulug-Albaka section in the central part of the Godavari valley as the type area of the Pakhal Supergroup and divided it into the Mallampalli, Mulug, and Sullavai Groups from base to top (Table 10.1). The Mallampalli Group contains a basal conglomerate with pebbles and boulders of granite, vein quartz, quartzite, and banded magnetite quartzite. A disconformity occurs between the Mallampalli and Mulug Groups, and the basal member of the Mulug Group (the Jakaram conglomerate) contains pebbles of chert and vein quartz. The contact between the Mulug and Sullavai Groups is marked by a 3-m-thick conglomerate, and both disconformable and angularly unconformable relationships occur. The total thickness of the Pakhal Supergroup is about 7,000 m.

Some common stratigraphic terms are not used in the classification shown in Table 10.1. The Panganga sequence of King (1881) is equivalent to the Mallampalli or Mulug Groups. The sedi-

TABLE 10.2. *Stratigraphy and depositional environments of Pakhal sediments (after Chaudhuri and Howard, 1985)*

Rock Stratigraphic Succession, Ramgundam Area			
Time	Time Rock	Informal Stratigraphic Units	Lithologies
Cretaceous Carboniferous	Gondwana Supergroup		
Late Proterozoic	Sullavai Group (Pakhal Supergroup / Pakhal Group)	Venkatpur Sandstone	Fine-grained, clay-rich subarkosic sandstone; K-Ar date 871 ± 14 m.y.
		Encharani Quartzite/ Ramgiri Fm.	Ferruginous quartz sandstone/ pebbly arkose, arkosic sandstone
	Mulug Subgroup	---- *Unconformity* ---- Rajaram Limestone (735 m)	Calcareous sandstone with lenses of sandy limestone, stromatolitic limestone, intraclastic limestone, limestone and dolomite interbedded with green calcareous shale
Middle Proterozoic	Mallampalli Subgroup	Ramgundam Sandstone (120 m)	Arkosic and subarkosic sandstone with interbedded shale
		Damala Gutta Conglomerate (90 m)	Coarse sandstone and pebbly arkose, chert-pebble conglomerate
		---- *Unconformity* ---- Pandikunta Limestone (340 m)	Flat-bedded limestone and dolomite with lenses of calcareous and ferruginous shale
			Glauconitic sandstone; K-Ar date 1330 ± 53 m.y.
		Jonalarasi Bodu Formation (50 m)	Interbedded quartz sandstone, limestone and dolomite, arkosic limestone
Early Proterozoic & Archean(?)		Crystalline basement	

ments of the Albaka belt are also equivalent to the Mallampalli or Mulug sequences (Srinivasa Rao et al., 1979b). The Pakhal suite has been widely correlated with other middle to late Precambrian, generally undeformed (Purana) sequences, such as those in the Cuddapah and Vindhyan basins, but the validity of the correlations is unknown.

The Pakhal sediments are mostly clastic, although minor dolomite is present. There is a general tendency for increased development of shales and carbonate rocks toward the southeast. A summary of the lithology and inferred depositional environments is shown in Table 10.2 (Chaudhuri and Howard, 1985). Evidence of shallow marine deposition consists of glauconite, algal stromatolites (Chaudhuri, 1970), mud cracks, flat-topped ripple marks, interference ripple marks, ripple drift lamination, penecontemporaneous folds, and limestone-pebble conglomerates. Parakh and Parthasarathy (1970) inferred a fluvial environment for the Sullavai Group, which is higher in the section than the Mallampalli Group. Chaudhuri (1977) proposed that both eolian and aqueous transport mechanisms accounted for the mixing of rounded grains of quartz and feldspar (wind transport) with angular to subangular grains of feldspar (water transport). This conclusion implies comparatively arid conditions in the source areas. The abundance of coarse debris in parts of the sequence indicates nearby source areas for at least part of the depositional period. All observations are consistent with deposition during rifting.

Deformation and metamorphism

The Pakhal Supergroup is mostly undeformed and unmetamorphosed except toward the southeast. In the northwestern part of the outcrop area, the lower formations show gentle folds, with axes oriented north-south or NNW-SSE and variable plunge directions (Basumallick, 1967a, 1967b; Srinivasa Rao et al. 1979a; Subba Raju et al., 1978). In some areas, these folds are cross-folded along east-west axes. The upper, Sullavai, suite is generally unaffected by the folding, as indicated by the presence of angular unconformity along part of its contact.

Both the grade of metamorphism and intensity of deformation increase towaard the southeast along the Pakhal outcrop belt. The sedimentary suites have been converted to dolomitic marbles, quartzites, slates and slaty phyllites. Metamorphic index minerals include garnet, staurolite and andalusite (Ramamohana Rao, 1971; Srinivasa Rao et al., 1979a). Metamorphic temperatures were probably in the range of 500° to 600°. Ramamohana Rao and Borreswara Rao (1972) found evidence of thrusting of cratonic rocks over part of the Pakhal sequence near the extreme southeastern part of the Pakhal outcrop.

REFERENCES

Banerjee, P. K. (1972). Geology and geochemistry of the Sukinda ultramafic field, Cuttack District, Orissa. Geol. Surv. India Mem. 103, 171 p.

Basumallick, S. (1967a). Problems of the Purana stratigraphy of the Godavari valley with special reference to the type area in Warangal District, Andhra Pradesh, India. Geol., Min. Metall. Soc. India Quart. J., v. 39, 115–127.

—— (1967b). Purana sedimentation in parts of the Godavari valley. Geol. Soc. India J., v. 8, 130–141.

Bhatia, S. C., Subrahmanyam, C., Aravamadhu, P. S., and Subba Rao, D. V. (1979). Regional gravity surveys in Bihar and adjoining regions of India. Geophys. Res. Bull. (Hyderabad), v. 17, 139–146.

Chatterjee, P. K., Perraju, P., Banerjee, P. K., Nag, P., Banerjee, R. N., and Ghose, D. P. (1964). The Archaean stratigraphy in Orissa—A review. 22nd Internat. Geol. Cong. Rept., Delhi, Sect. X, 22–35.

Chaudhuri, A. (1970). Precambrian stromatolites in the Pranhita-Godavari valley, South India. Paleogeogr., Paleoclimatol. and Paleoecol., v. 7, 309–340.

—— (1977). Influence of eolian processes on Precambrian sandstones of the Godavari valley, South India. Precamb. Res., v. 4, 339–360.

——, and Howard, J. D. (1985). Ramgundam Sandstone: a middle Proterozoic shoal-bar sequence. J. Sed. Petrol., v. 55, 392–397.

Choubey. V. D. (1971). Narmada-Son lineament, India. Nature Phys. Sci., v. 232, 38–40.

Crawford, A. R. (1978). Narmada-Son lineament of India traced into Madagascar. Geol. Soc. India J., v. 19, 144–153.

Das, B., and Patel, N. P. (1984). Nature of the Narmada-Son lineament. Geol. Soc. India J., v. 25, 267–276.

Ghosh, D. B. (1976). The nature of the Narmada-Son lineament. Geol. Surv. India Misc. Publ. 34, Part 1, 119–133.

Glazner, A., and Bartley, J. M. (1985). Evolution of lithospheric strength after thrusting. Geology, v. 13, 42–45.

Gupta, M. L., and Gaur, V. K. (1984). Surface heat flow and probable evolution of Deccan volcanism. Tectonophysics, v. 105, 309–318.

——, and Venkateswara Rao, C. (1970). Heat flow studies under Upper Mantle Project. National Geophys. Res. Inst. Bull., Hyderabad, v. 8, 87–172.

King, W. (1881). The geology of the Pranhita-Godavari valley. Geol. Surv. India Mem. 18, 151–311.

Krishna Brahmam, N., and Negi, J. G. (1973). Rift valleys beneath the Deccan Traps (India). Geophys. Res. Bull. (Hyderabad), v. 11, 207–237.

Mazumdar, S. K. (1978). Precambrian geology of eastern India between the Ganga and the Mahanadi: A review. Geol. Surv. India Records, v. 110, Part 2, 60–116.

Mishra, D. C. (1977). Possible extensions of the Narmada-Son lineament towards Murray Ridge (Arabian Sea) and the eastern syntaxial bend of the Himalayas. Earth Planet. Sci. Lett., v. 36, 301–308.

——— (1984). Magnetic anomalies—India and Antarctica. Earth Planet. Sci. Lett. v.71, 173–180.

Naqvi, S. M., Divakara Rao, V., and Narain, H. (1974). The protocontinental growth of the Indian shield and the antiquity of its rift valleys. Precamb. Res., v. 1, 345–398.

Parakh, S. D., and Parthasarathy, A. (1970). Some sedimentological aspects of the Sullavai rocks around Nawar Gaon Burzug, Chandrapuri District, Maharashtra. Proc. Symp. on Purana Formations of Peninsular India, Sagar Univ., Sagar.

Qureshy, M. N., Krishna Brahmam, N., Garde, S. C., and Mathur, B. K. (1968). Gravity anomalies and the Godavari rift, India. Geol. Soc. Amer. Bull., v. 79, 1221–1229.

Raha, P. K., and Sastry, M. V. A. (1982). Stromatolites and Precambrian stratigraphy in India. Precamb. Res., v. 18, 293–318.

Ramamohana Rao, T. (1971). Metamorphism of the Pakhals of Yellandlapad area, Andhra Pradesh. In Studies in Earth Science (ed. T. V. V. G. R. K. Murthy and S. S. Rao), Today and Tomorrow Printers, New Delhi, 225–235.

———, and Borreswara Rao, C. (1972). Trace elements in minerals and rocks of the Precambrian group of the Yellandlapad area, Andhra Pradesh. Geol. Soc. India J., v. 13, 165–171.

Roday, P. P. (1983). Post-crystallization defor-

mation in syntectonic and neotectonic veins and its relationship with tectonic movement in a part of the Narmada valley, central India. Recent Res. Geol., v. 10, 158–177.

Sarkar, S. N., and Saha, A. K. (1983). Structure and tectonics of the Singhbhum-Orissa Iron Ore craton, eastern India. Recent Res. Geol., v. 10, 1–25.

Srinivasa Rao, K., Subba Raju, M., Sreenivasa Rao, T., Imam Ali Khan, Md., and Silekar, V. S. (1979a). Tectonic evolution of the Godavari graben. In Fourth International Gondwana Symposium (ed. B. Laskar and C. S. Raja Rao), Hindustan Publ. Co., New Delhi, vol. II, 889–900.

———, ———, ———, ———, ——— (1979b). Stratigraphy of the upper Precambrian Albaka belt east of the Godavari River in Andhra Pradesh and Madhya Pradesh. Geol. Soc. India J., v. 20, 205–213.

Subba Raju, M., Srinivasa Rao, T., Setti, D. N., and Reddi, B. S. R. (1978). Recent advances in our knowledge of the Pakhal Supergroup, with a special reference to the central part of the Godavari valley. Geol. Surv. India Records, v. 110, Part 2, 39–59.

Tewari, A. P. (1968). A new concept of the paleotectonic set up of a part of northern peninsular India with special reference to the great boundary faults. Geol. Mijnbouw, v. 47, 21–27.

Valdiya, K. S. (1984). Aspects of Tectonics—Focus on South Central Asia. Tata McGraw-Hill Publ. Co., New Delhi, 219 p.

Vinogradov, A., Tugarinov, A., Zhykov, C., Stapnikova, N., Bibikova, E., and Khorre, K. (1964). Geochronology of Indian Precambrian. 22nd Internat. Geol. Cong. Rept., Sect. X, 553–567.

11. HIMALAYAS

The Himalayas (Fig. 11.1) show geological activity intermittent from the Early Proterozoic to the present. Most of the range is composed of Precambrian rocks that have undergone thrusting and reactivation during Cenozoic subduction and collision. Gansser (1981) stated that "the Himalaya is not strictly a mountain range born from the Tethys since the greater part consists of the reworked northern edge of the Indian shield (Indian plate), with Gondwanic affinities." Similar conclusions have been drawn by Valdiya (1980, 1981, 1984), and others. The importance of Precambrian rocks in the Himalayas is also shown by geophysical data (Helene and Molnar, 1983; Klootwijk, 1984; Seeber and Armbruster, 1984).

The amount of information on the Himalayas is beyond the scope of this book and has been summarized in several recent volumes (Saklani, 1978; Farah and DeJong, 1979; Valdiya, 1980, 1984; Gupta and Delaney, 1981; Tahirkeli, 1982; the continuing series of volumes on Himalayan Geology published by the Wadia Institute of Himalayan Geology, in Dehra Dun, India). In this chapter, we merely review those aspects of the Precambrian geology of the Himalayas that will help to understand the evolution of the Indian shield and its fate during continental collision. All of the Phanerozoic rocks in the Himalayas have been deposited on this Precambrian basement, and thus the importance of the Indian shield in the orogenic process cannot be overemphasized.

The broad elements of the Himalayas consist of a subducting slab (the Indian shield), an overriding slab (the Tethyan suite), and a sedimentary wedge contained between, and decoupled from, the two converging slabs. The Himalayan ranges can be subdivided into the following regions: the Tibetan (Tethyan) Himalayas, to the north; the Great (central) Himalayas, consisting of the central crystalline axis (Vaikrita Group); the Lesser, or lower, Himalayas; and the Subhimalayas (Outer Himalayas, foothills belt) to the south (Fig. 11-1). The central crystalline axis of the Great Himalayas has overridden suites of the Lesser Himalayas along the Main Central Thrust (probably the same as the Main Mantle Thrust to the west). The Lesser Himalayas are bounded on the south by the Main Boundary Thrust, which is largely covered by Tertiary sediments washed southward from the mountains. These sediments now constitute the Subhimalayas. The crystalline rocks of the Great Himalayas grade northward into overlying supracrustal assemblages of the Tibetan Himalayas.

The central crystalline axis of the Great Himalayas consists of high-grade metamorphic rocks, migmatites, and Cenozoic granites. It contains Precambrian rocks overthrust by other Precambrian and Phanerozoic rocks. The Lesser Himalayas consists largely of nappes that override autochthonous Precambrian sedimentary sequences.

The transverse faults and fractures of the Lesser Himalayas and Subhimalayas are parallel to structures in the basement of the Ganges basin and the Precambrian rocks of the Aravalli craton (Fig. 9.1). Faults that frame major mountain ranges of the craton extend into the Himalayas, and magnetotelluric and deep magnetic sounding studies indicate a conductive ridge structure projecting from the northern Aravalli ranges into the Lesser Himalayas (Srivastava et al., 1984). Basement ridges that extend into the Himalayas include the Bundelkhand-Faizabad ridge, to the southeast, and the Delhi-Haridwar ridge, continuing the trend of the Delhi suite in the Aravalli Mountains (Ch. 9).

The Indus-Tsangpo suture zone (Fig. 11.1) is considered by most workers as the northern limit of the Indian peninsula (Gansser, 1981; Klootwijk, 1981). Gansser stated that the northern edge of peninsular India could never have been far from the southern margin of a complex Eurasian land mass, thus arguing against the existence of a major intervening Tethyan Ocean. A contrary model for the interaction of India and Eurasia considers a much larger "Greater India" in Gondwana time (Veevers et al., 1975). In this view, the northern edge of India is underthrust beneath the Tibetan plateau, with the trace of the major thrust being the Main Central Thrust of the Great Himalayas (Figs. 11.1 and 11.2). This enlarged Gondwanic India fills the gap left between India and Australia in reconstructions of Gondwanaland.

The question of the extent of underthrusting by India is obviously important in reconstructing Precambrian India. Various estimates of the amount of subduction have been proposed by different workers as follows:

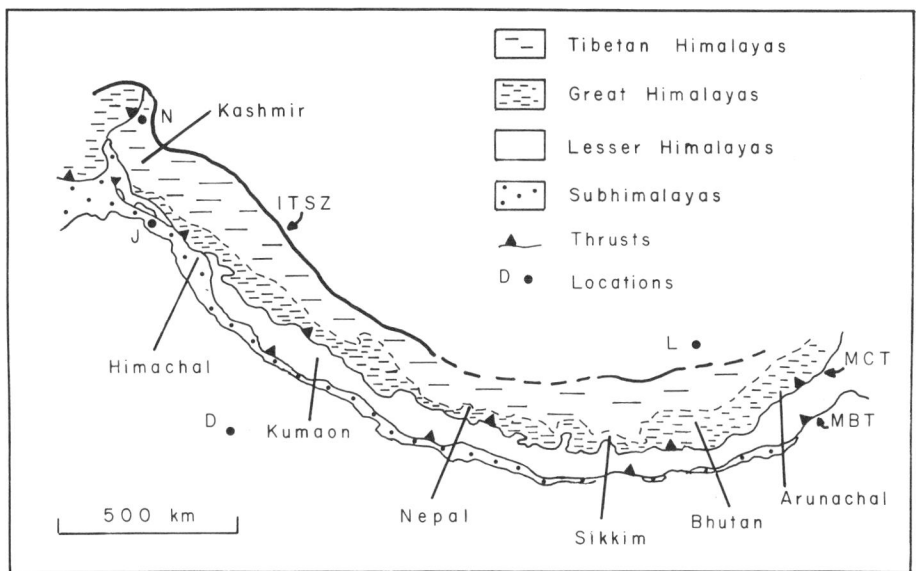

Fig. 11.1. Himalayas—subdivisions. ITSZ, Indus-Tsangpo suture zone; MCT, Main Central Thrust; MBT, Main Boundary thrust; N, Nanga Parbat; J, Jammu; L, Lhasa; D, Delhi.

Klootwijk (1981, 1984) used paleomagnetic and other data to indicate 200 km of crustal shortening within the Himalayas and 50 km of subduction. These data are consistent with extent of India as far north as the Indus-Tsangpo suture zone.

Based primarily on gravity data, Helene and Molnar (1983) suggested that the northern edge of India is approximately 150 km north of the southern edge of the Tibetan plateau.

Powell and Conaghan (1975) proposed that the major underthrusting did not occur at the

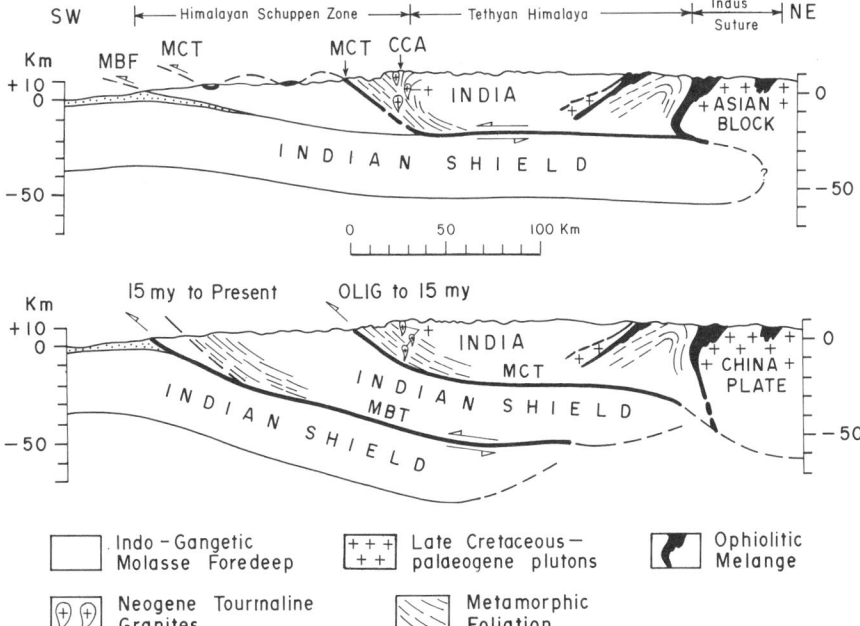

Fig. 11.2. Two different models of underthrusting of the Indian plate beneath the Himalayas (Powell, 1979). The upper model was preferred by Powell, and the lower one was suggested by LeFort (referenced by Powell).

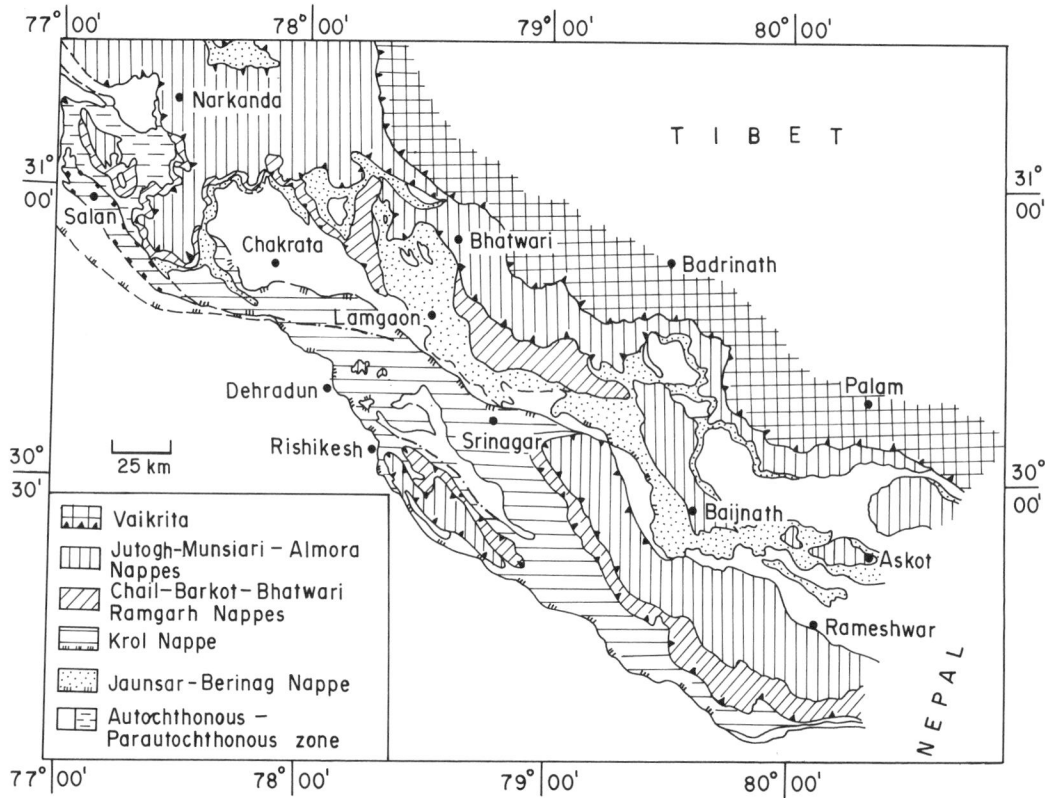

Fig. 11.3. Structural map of Kumaun, Lesser Himalayas (Valdiya, 1980). Reproduced with permission from the Geology of Kumaon Lesser Himalayas, published by the Wadia Institute of Himalayan Geology, p. 12.

northern edge of the Indian plate but along a deep crustal dislocation within it. Thus, the entire region between the central crystalline axis and the Main Boundary Fault is a wedge-shaped schuppen structure containing the sedimentary cover and slivers of basement scraped off the Indian shield.

PRECAMBRIAN ROCKS

The principal criterion used by early workers for recognition of Precambrian rocks in the Himalayas was their unfossiliferous nature and extent of metamorphism. Geochronologic work, however, has now demonstrated that many rocks formerly thought to be Precambrian are Paleozoic or even younger (Sinha Roy, 1982, 1983). A large number of stratigraphic names have been used for the various Precambrian suites. In the Great Himalayas, the principal one is the Vaikrita Group, and outliers occur in the Lesser Himalayas as klippen above the autochthonous rocks. The Vaikrita Group is the basement on which the Tethyan rocks of the northern edge of the Indian craton were deposited. The Tethyan rocks are now the primary suites in the Tibetan Himalayas. The

Lesser Himalayas contain a series of thrust suites, many of which contain Precambrian rocks, and an autochthonous zone, also containing Precambrian suites.

Radioactive age measurements in the crystalline axis of the Great Himalayas include Rb/Sr determinations on two gneissic suites: 1,980 ± 135 m.y. and 2,236 ± 196 m.y. (Bhanot et al., 1980), with initial $^{87}Sr/^{86}Sr$ ratios that are so variable that the significance of the age values is unclear; and ages in the range of 500 to 600 m.y. (Frank et al., 1977; Mehta, 1977), which almost certainly indicate Late Proterozoic/early Paleozoic metamorphic resetting.

Several Rb/Sr measurements have shown Proterozoic ages within the nappe sheets of the Kumaon Lesser Himalayas (Fig. 11.3; Valdiya, 1980). They include about 1,500 m.y. for rocks in the Chail thrust (Bhanot et al., 1978); 1,860 ± 65 m.y. for quartz porphyries in the Rampur nappe and 1,200 m.y. for associated granites (Trivedi et al., 1982); 1,840 ± 70 m.y. for granitic gneisses interbedded with Berinag quartzites in the Kulu-Rampur window (Frank et al., 1977); 1,800 m.y. with initial $^{87}Sr/^{86}Sr$ ratio of 0.7235 ± 0.0046 for granitic rocks of the Ramgarh nappe (Trivedi et al., 1984); 1,620 ± 90 m.y. (Powell et al., 1979) for gneisses in the Kumaon-Almora-Askot region regarded by Crawford (1981) as having the oldest reliable date in the Himalayas; and 550 m.y. for the Champawat granodiorite/Almora granite in the Almora nappe (Trivedi et al., 1984), which presumably represents resetting of older material.

Specific correlations between rocks in the Himalayas and suites in the Aravalli craton are generally impossible. The exception is the proposed relationship between sediments of the Vindhyan basin (Ch. 9) and various suites in the Lesser Himalayas. One of the correlative suites is the Jammu limestone and other carbonates in the predominantly autochthonous assemblages, although some similar rocks occur within the thrust belts. These carbonates contain stromatolites and lithologies similar to those in the Vindhyan Supergroup (Valdiya, 1984).

REGIONAL RELATIONSHIPS OF PRECAMBRIAN ROCKS

Lesser Himalayas of Kashmir and Punjab

The Precambrian rocks in the Kashmir Himalayas have been designated as the Salkhala and Dogra suites (Wadia, 1934; Fig. 11.1). The Salkhala Group consists of schistose and phyllitic quartzites, carbonaceous slates, sericite schists, and intercalated marbles. They are thrust over Paleozoic and Mesozoic sediments and at Nanga Parbat are highly metamorphosed and granitized (Ghosh et al., 1966). At some places they are overlain by the Dogra Slates. The Salkhala Group shows evidence of metamorphism only in the Cenozoic. The Dogra Slates may be equated with the Attock and Hanzara slates west of Nanga Parbat and with the Simla slates in the Himachal region to the east. They are generally comparable to the Vindhyan sediments of the Aravalli craton (Ch. 9).

The Jammu Limestone occurs to the east of the Dogra Slates in apparent autochthonous or parautochthonous inliers north of the Great Boundary Fault. It contains a stromatolitic fauna similar to that of the Bhander Limestone of the Vindhyan Supergroup (Ch. 9) and almost certainly represents a northern extension of the Vindhyan basin (Raha, 1980; Shah, 1980).

The Salkhala Group abuts against the ophiolites of the Indus-Tsangpo suture zone. Tahirkeli et al. (1979) suggested that this area contains an island arc (Kohistan) obducted onto the Paleozoic cover of the Indian shield during the collision of India and Eurasia. The extension of the Salkhala and Dogra suites to the suture zone indicates that the Precambrian rocks of Gondwanic India that have been thrust southward over the subducting Indian plate extended at least as far north as the suture.

Lesser Himalayas of Kumaon and Himachal Pradesh

The best development of Precambrian rocks in the Himalayas showing affinities with rocks in peninsular India is in the Kumaon region (Fig. 11.1; Valdiya, 1980). Gansser (1964) distinguished three lithotectonic zones that have Precambrian constituents: the Krol nappe, which contains the Berinag quartzite; the Almora-Dudatoli thrust sheet, which contains the Garhwal suite and also the Munsiari suite (attributed by some workers to the Vaikrita Group); and the Deoban-Tejan autochthonous zone. More recent publications use some of this terminology but have redefined many terms, and in this book we do not attempt to resolve the various conflicts.

In Himachal Pradesh, to the west of the Kumaon area, similar stacking of thrust slices has presented a complex pattern of Precambrian and younger rocks (Gansser, 1964; Sinha, 1980). Gansser believed that the Simla Slates were the oldest rocks and correlated them with the Dogra Slates in the Kashmir region; he regarded the suite as occurring both in thrust slices and in autochthonous blocks. The Simla Slates are dark bluish gray, argillaceous and micaceous, pencil slates and graywackes intercalated with microcrystalline limestones, some of which contain stromatolites similar to those of the Vindhyan Supergroup (Valdiya, 1980).

Valdiya (1980) divided the Lesser Himalayas of the Kumaon region into outer and inner belts. The outer zone contains Precambrian rocks in the Damtha Group, whereas the inner zone contains Precambrian suites in both the Damtha and Tejan groups. These groups are autochthonous and conformable. The Damtha Group contains the Chakrata and Rautgara formations, and the Tejan Group contains the Deoban and Mandhali formations. These suites have been correlated tentatively throughout the entire Lesser Himalayas (Table 11.1). The inner zone also contains exposures of the Vaikrita Group. The Chakrata Formation consists of a thick succession of turbiditic graywackes and siltstones rhythmically alternating with slates, and Valdiya proposed that it may be considerably older than 1,000 m.y. The Chakrata Formation is overlain by the Rautgara Formation, which consists of graywackes with minor conglomerates and abundant mafic sills, dikes, and flows.

The Rautgara slates are abruptly overlain by the limestones and dolomites of the Deoban Formation of the Tejan Group. The Deoban Formation contains stromatolites such as *Bariculia baicallica* and *Kussiella kussiensis,* which also occur

TABLE 11.1. *Correlation of suites in Lesser Himalayan autochthonous zone (Valdiya, 1980)*

Probable Age	Kashmir	Himachal	Kumaun	Darjeeling	Bhutan	Arunachal
Lr. Eocene	Subathu	Subathu	Subathu			
Lr. & Mid. Permian	"Agglomeratic Slate" S of			Damuda	Damuda	Khelong (Bhareli)
	Panjal Thrust			Rangit	Diuri	Rangit Saleri
Up. Riphean-Vendian		Basantpur	Mandhali (Sor)			
Mid. Riphean		Shali	Deoban (Gangolihat)	Buxa	Buxa	Dedza
Lr. Riphean		Sundernagar	Rautgara Chakrata	Jainti Sinchular	Phuntsholing	Bichom (Miri)

The Damtha suite includes the Rautgara and Chakrata Formations. The Tejan suite includes the Deoban and Gangolihat Formations.

Prepared with permission from the Geology of Kumaon Lesser Himalaya, published by the Wadia Institute of Himalayan Geology, p. 43.

in the Vindhyan Supergroup (Valdiya, 1980). The Deoban Formation is succeeded upward by the Mandhali Formation, which shows only local discordance at its base. The Mandhali Formation consists of grayish green and black, carbonaceous, pyritic phyllites and slates interbedded with limestones and conglomerates. The limestones have yielded stromatolites with ages estimated as 950 to 680 m.y.

SEDIMENTATION IN NORTHERN GONDWANIC INDIA

Prior to Cretaceous rifting, India was attached to present southern continents as part of Gondwanaland (e.g., Veevers et al., 1975; Fig. 1.1). During this time, the only active margin of "Gondwanic India" was its present northern edge, which was bordered by the Tethyan Ocean. There are two major contrasting points of view regarding the sedimentation patterns in Gondwanic India. One is that the northern margin of India was never far south of the southern margin of Eurasia, with only a narrow intervening Tethyan Ocean. This closeness of a northern landmass permitted clastic debris to be shed southward onto the Indian margin during late Precambrian, Paleozoic and early Mesozoic time (e.g., Rupke, 1974; Crawford, 1979). The contrary view is that sediments along the northern margin of India were derived from the south and transported by currents flowing away from the Indian shield (Valdiya, 1980). Some workers have proposed a crystalline divide between basins in the south, which received debris from the Indian craton, and basins in the north, which had sediment sources outside the craton (Fuchs, 1967).

A major difference in depositional facies of lower Paleozoic (to Upper Proterozoic) rocks in northern India is shown in the distribution of salt. Western basins contain abundant salt (e.g., in the Salt Range of Pakistan), whereas eastern basins are comparatively free of evaporites (Burbank, 1983). This difference in lithologies permitted much greater decollement thrusting of Himalayan rocks southard in Pakistan than in most of India.

The equivalents of Vindhyan sediments have been proposed at various places in the Lesser Himalayas. Carbonate sequences, with stromatolites, are particularly well correlated (e.g., the Jammu Limestone, discussed previously). It is possible that the Vindhyan basin extended to the northern edge of the Indian shield (Krishnan and Swaminath, 1959; Valdiya, 1980, 1984). Several workers have proposed that the Vindhyan lithologies changed progressively northward, generally becoming richer in clay and poorer in sand to the north (Valdiya, 1980; Srikantia, 1981). For example, in the Chakrata Formation of the autochthonous zone of the Kumaon Himalayas, flute casts and scour marks demonstrate that the turbidity currents that deposited this suite of Vindhyan rocks flowed in a generally northward direction throughout the outcrop area. Clearly this part of the Vindhyan basin was filled by currents flowing down slopes normal to the coast, depositing sediments derived from the Aravalli craton.

Following Vindhyan time, the major sites of sedimentation appear to have shifted northward across the Indian platform. Evaporitic sediments occur between the Vindhyan outcrops of the Aravalli craton and the Lesser Himalayas, including the Saline Group of the Salt Range of Pakistan, deposits in the Marwar Group of the western Aravalli craton, and the Mandi Formation of Himachal Pradesh. No stratigraphic break has been discovered between apparent Late Proterozoic and apparent early Paleozoic sedimentary suites in northern India.

REFERENCES

Bhanot, V. B., Kwatra, S. K., Kansal, A. K., and Pandey, B. K. (1978). Rb-Sr whole-rock age for

Chail series of northwestern Himalaya. Geol. Soc. India J., v. 19, 224–227.

——, Pandey, B. K., Singh, V. P., and Kansal, A. K. (1980). Rb-Sr ages for some granitic and gneissic rocks of Kumaun and Himachal Himalaya. In Stratigraphy and Correlation of Lesser Himalaya Formations (ed. K. S. Valdiya and S. B. Bhatia), Hindustan Publ. Co., New Delhi, 139–142.

Burbank, D. W. (1983). The chronology of intermontane basin development in the northwestern Himalaya and the evolution of the Northwest Syntaxis. Earth Planet. Sci. Lett., v. 64, 77–92.

Crawford, A. R. (1979). The Indus suture line, the Himalaya, Tibet and Gondwanaland. Geol. Mag., v. 111, 369–383.

—— (1981). Isotopic age data for the eastern half of the Alpine Himalayan belt. In Zagros, Hindukush, Himalaya and Geodynamic Evolution (ed. H. K. Gupta and F. M. Delaney), Geodynamics Series, v. 3, Amer. Geophys. Union, Washington, 189–205.

Farah, A., and DeJong, K. A. (eds.) (1979). Geodynamics of Pakistan. Geol. Surv. Pakistan, Quetta, 361 p.

Frank, W., Thoni, M., and Purtschaller, F. (1977). Geology and petrography of Kulu-Lahul area. In Colloq. Internat. Ecologie et Geologie de Himalaya, Comm. Nat. Rech. Sci., Paris, v. 268, 147–172.

Fuchs, G. (1967). Zur Bau des Himalaya. Osterr. Akad. Wiss., Math. Nat. Kl. Vienna, v. 113, 211 p.

Gansser, A. (1964). Geology of the Himalayas. Wiley Interscience, New York, 289 p.

—— (1981). The geodynamic history of the Himalayas. In Zagros, Hindukush, Himalaya and Geodynamic Evolution (ed. H. K. Gupta and F. M. Delaney). Geodynamics Series, v. 3, Amer. Geophys. Union, Washington, 111–121.

Ghosh, S. K., Ray, S. K., and Sen, S. (1966). Age and metamorphism of the rocks of the Simla area—A critical study. Geol. Soc. India Bull., v. 6, 57–60.

Gupta, H. K., and Delaney, F. M. (eds.) (1981). Zagros, Hindukush, Himalaya and Geodynamic Evolution. Geodynamics Series, v. 3, Amer. Geophys. Union, Washington, 323 p.

Helene, L. C., and Molnar, P. (1983). Constraints on the structure of the Himalaya from an analysis of gravity anomalies and a flexural model of the lithosphere. J. Geophys. Res., v. 88, 8171–8191.

Klootwijk, C. T. (1981). Greater India's northern extent and its underthrust of the Tibetan plateau: Paleomagnetic constraints and implications. In Zagros, Hindukush, Himalaya and Geodynamic Evolution (ed. H. K. Gupta and F. M. Delaney), Geodynamics Series, v. 3, Amer. Geophys. Union, Washington, 313–323.

—— (1984). A review of Indian Phanerozoic palaeomagnetism: Implications for the India-Asia collision. Tectonophysics, v. 105, 331–353.

Krishnan, M. S., and Swaminath, J. (1959). The great Vindhyan basin of northern India. Geol. Soc. India J., v. 1, 10–30.

Mehta, P. K. (1977). Rb-Sr geochronology of the Kulu-Mandi belt: Its implications for the Himalayan tectonogenesis. Geol. Rundschau, v. 66, 156–175.

Powell, C. McA. (1979). A speculative tectonic history of Pakistan and surroundings: Some constraints from the Indian Ocean. In Geodynamics of Pakistan (ed. A. Farah and K. A. DeJong), Geol. Surv. Pakistan, Quetta, 5–24.

——, and Conaghan, P. J. (1975). Tectonic models of the Tibetan plateau. Geology, v. 3, 727–731.

——, Crawford, A. R., Armstrong, R. L., Prakash, R., and Wynne-Edwards, H. R. (1979). Reconnaissance Rb-Sr dates for the Himalayan central gneiss, northwest India. Ind. J. Earth Sci., v. 6, 139–151.

Raha, P. K. (1980). Stratigraphy and depositional environment of the Jammu Limestone, Udhampur District, Jammu. In Stratigraphy and Correlations of Lesser Himalayan Formations (ed. K. S. Valdiya and S. B. Bhatia), Hindustan Publ. Co., New Delhi, 145–151.

Rupke, J. (1974). Stratigraphic and structural evolution of the Kumaon Lesser Himalaya. Sediment. Geol., v. 11, 81–265.

Saklani, P. S. (ed.) (1978). Tectonic Geology of the Himalaya, v. 1. Today and Tomorrow Printers, New Delhi, 350 p.

Seeber, L., and Armbruster, J. G. (1984). Some elements of continental subduction along the Himalayan front. Tectonophysics, v. 105, 263–278.

Shah, S. K. (1980). Stratigraphy and tectonic setting of the Lesser Himalayan belt of Jammu. In Stratigraphy and Correlations of Lesser Himalayan Formations (ed. K. S. Valdiya and S. B. Bhatia), Hindustan Publ. Co., New Delhi, 152–162.

Sinha, A. K. (1980). Tectono-stratigraphic problem in the Lesser Himalayas zone of Simla region, Himachal Pradesh. In Stratigraphy and Correlations of Lesser Himalayan Formations (ed. K. S. Valdiya and S. B. Bhatia), Hindustan Publ. Co., New Delhi, 99–116.

Sinha Roy, S. (1982). Himalayan main central thrust and its implications for Himalayan inverted metamorphism. Tectonophysics, v. 84, 225–246.

—— (1983). Some problems of Precambrian-Proterozoic stratigraphy of the Himalayas. Geol. Surv. India Records, v. 112, Part 2, 33–39.

Srikantia, S. V. (1981). The lithostratigraphy, sedimentation and structure of the Proterozoic-

Phanerozoic formations of Spiti basin in the higher Himalaya of Himachal Pradesh. Himal. Geol., v. 4, 396–413.

Srivastava, B. J., Singh, B. P., and Lilley, F. E. M. (1984). Magnetometer array studies in India and the lithosphere. Tectonophysics, v. 105, 355–371.

Tahirkeli, R. A. K. (1982). Geology of the Himalaya, Karakoram, and Hindukush in Pakistan. Peshawar Univ. Bull. Geol., Special Issue, v. 15, 51 p.

———, Mattauer, M., Proust, F., and Tapponier, P. (1979). The India-Eurasia suture zone in northern Pakistan: Synthesis and interpretation of recent data at plate scale. In Geodynamics of Pakistan (ed. A. Farah and K. A. DeJong), Geol. Surv. Pakistan, Quetta, 125–130.

Trivedi, J. J., Gopalan, K., Sharma, K. K., Gupta, K. R., and Choubey, V. M. (1982). Rb-Sr ages of Gaik Granite, Ladakh batholith, northwest Himalaya. Ind. Acad. Sci. Proc., Earth Planet. Sci. Sect., v. 91, 65–73.

———, Gopalan, K., and Valdiya, K. S. (1984). Rb-Sr ages of granitic rocks within the Lesser Himalayan nappes, Kumaun, India. Geol. Soc. India J., v. 25, 641–653.

Valdiya, K. S. (1980). Geology of Kumaun Lesser Himalaya. Wadia Inst. Himalayan Geology, Dehra Dun, 291 p.

——— (1981). Tectonics of the central sector of the Himalaya. In Zagros, Hindukush, Himalaya and Geodynamic Evolution (ed. H. K. Gupta and F. M. Delaney), Geodynamics Series, v. 3, Amer. Geophys. Union, Washington, 87–111.

——— (1984). Aspects of Tectonics—Focus on South Central Asia. Tata McGraw-Hill Publ. Co., New Delhi, 319 p.

Veevers, J. J., Powell, C. McA., and Johnson, B. D. (1975). Greater India's place in Gondwanaland and in Asia. Earth Planet. Sci. Lett., v. 27, 383–387.

Wadia, D. N. (1934). The Cambrian-Trias sequence of northwestern Kashmir (parts of Muzaffarabad and Baramula Districts). Geol. Surv. India Records, v. 68, Part 2, 121–176.

AUTHOR INDEX

SUBJECT INDEX

Note: The symbol (*R*) after a term indicates that the term refers to the name of a rock suite, including super-groups, groups, formations, members, plutonic bodies, etc.